Lecture Notes in Mathematics

Edited by A. Dold and B. Eckmann

1345

J. Seade

s.)

c Dynamics

ond International Colloquium
held in Mexico, July 1986

Springer-Verlag

Berlin Heidelberg New York London Paris Tokyo

Editors

Xavier Gomez-Mont
José A. Seade
Instituto de Matemáticas, Area de la Investigación Científica
Circuito Exterior, Ciudad Universitaria
México 04510, D.F., México

Alberto Verjovski
International Centre for Theoretical Physics
P.O. Box 586, 34100 Trieste, Italy

Mathematics Subject Classification (1980): 34C35, 58F18, 53C12, 57R30

ISBN 3-540-50226-2 Springer-Verlag Berlin Heidelberg New York
ISBN 0-387-50226-2 Springer-Verlag New York Berlin Heidelberg

Printed in Germany

Printing and binding: Druckhaus Beltz, Hemsbach/Bergstr.
2146/3140-543210

Dedicated to Professors

Abdus Salam

Jacob Palis

and

Christopher Zeeman

in Recognition of the Influence that the International Centre for Theoretical Physics (ICTP), Trieste, Italy has on the Development of Mathematics in Mexico.

PREFACE

The Semester of Activities on Dynamical Systems was held at the Instituto de Matemáticas of the National University of Mexico from January to June 1986, and it concluded with the Second Colloquium of the Mexican Mathematical Society held in the same Institute and in the University of Guadalajara, in Chapala, Mexico, in July 1986. We would like to thank the participants for their mathematical interest and the following institutions for their financial support:

CONACYT, MEXICO

SECRETARIA DE EDUCACION PUBLICA, MEXICO

UNIVERSIDAD DE GUADALAJARA, MEXICO

UNIVERSIDAD NACIONAL AUTONOMA DE MEXICO

CNP_q, BRASIL

CNRS, FRANCE

MINISTERE DES AFFAIRES ETRANGERES, FRANCE

THIRD WORLD ACADEMY OF SCIENCES, ITALY

We would like to thank specially Professor Robert Moussu for his support to the Colloquium.

The Editors

TABLE OF CONTENTS

First Integrals for Singular Holomorphic Foliations With Leaves of Bounded Volume

J. C. Alexander* and Alberto Verjovsky†

Department of Mathematics and Institute for Physical Science and Technology
University of Maryland
College Park, MD 20742 USA

Departamento de Matemáticas
Centro de Investigación del IPN
Apartado Postal 14-740
México 14, D. F., México

We consider the germ of k-dimensional holomorphic foliation in $\mathbb{C}^n$ with an isolated singularity at the origin. Under the assumption that the germs of the leaves have bounded k-volume, it is proved that all leaves are closed and that at least one separatrix exists. If the k-volume (or k-dimensional Hausdorff measure) of the separatrix set is also finite, the germ has a very regular structure. In particular, the leaf space is a complex analytic space. The problem is motivated by the study of singularities of complex differential equations. Illustrative examples and a partial converse are presented.

1. Introduction

The subjects of complex dynamics, and more generally, of holomorphic foliations, have characters different from their real counterparts, due to the rich structure of complex analysis. Many of the results of complex analytic geometry have important implications for holomorphic foliations. In this report we consider one such implication. Bishop [1] has shown that a bound on the volumes and Hausdorff measure of analytic sets has geometric consequences. We study the consequences for the structure of a holomorphic foliation in the neighborhood of an isolated singularity. The foliation has a very regular structure. It contains separatrices. Leaves which are not separatrices are closed. The leaf space has a complex analytic structure, so that the foliation has the maximal number of first integrals. In this report we develop such consequences of a bound on the volume of leaves.

A nonsingular foliation of a manifold is a decomposition of the manifold into disjoint immersed submanifolds, called *leaves*. Foliations with singularities correspond to integrable systems of forms. It is convenient to begin with the following [22, def. 3.1, ch. III, pp. 106–107]. Let $\Omega \subset \mathbb{C}^n$ be an open subset and let $0 < k < n$. An $(n-k)$*-dimensional holomorphic Frobenius structure* F on Ω is a collection of $n-k$ holomorphic one-forms $F = \{\omega_1, \dots, \omega_{n-k}\}$ on Ω such that for each $i = 1, \dots, n-k$, the integrability condition

$$d\omega_i \wedge \omega_1 \wedge \cdots \wedge \omega_{n-k} = 0$$

* Partially supported by the National Science Foundation. Este autor agradace al Centro de Investigación del IPN y CONACYT (México) cuyo apoyo durante la visita al Centro hizo posible el presente trabajo.

† Present address: International Centre for Theoretical Physics, Strada Costiera 11, Miramare, 34100 Trieste, Italy

is satisfied. For each $z \in \Omega$, let

$$K_z = \bigcap_{i=1}^{n-k} \ker \omega_i(z),$$

a subspace of the tangent space at z. The *singular locus* of F is the set

$$S(F) = \{z \in \Omega : \dim_{\mathbb{C}}(K_z) > k\}.$$

This is an analytic subset of Ω. The Frobenius system F is *regular* if $\dim_{\mathbb{C}}(S(F)) < k$. On the complement of $S(F)$ in Ω, the forms $\omega_i \in F$ are linearly independent and thus determine a nonsingular k-dimensional foliation $\mathcal{F}(F)$ of $\Omega - S(F)$.

More generally, a *holomorphic foliation of codimension* q with singularities in the complex manifold M is a nonsingular foliation of codimension q in $M - A$, where A is an analytic set of codimension bigger than 1. If A has codimension bigger than q, we say the foliation is *regular*. The forms that define the foliation in A may be taken to be those local 1-forms which are tangent to the foliation in $M - A$.

In particular, a foliation of codimension $n-1$ in a manifold of dimension n may be given by the solutions of an ordinary complex differential equation

$$\frac{dz}{dT} = f(z), \quad T \in \mathbb{C}. \tag{1.1}$$

The orbits of (1.1) are the leaves and the stationary points constitute the singular set. However note that if $n > 2$, the resulting foliation is not in general an $(n-1)$-dimensional holomorphic Frobenius structure.

A (holomorphic) *first integral* of a foliation defined on Ω is a (nontrivial) holomorphic function $\rho: \Omega \to \mathbb{C}$ which is constant on leaves. There are a number of adjectives ('strong', 'weak', 'formal') that can be put in front of the term, depending on the particular context, and a number of results concerning the existence and number of such integrals can be found in [7, 13, 16, 17, 18, 20, 21, 27, 28]. In the context of Frobenius structures, first integrals are related to the *integrability problem* [22]. A first integral is a function defined on the leaf space of the foliation, that is a map to a one-dimensional variety. If a foliation admits r first integrals, they form a map from the leaf space to an r dimensional variety. If the map does not factor through an $(r-1)$-dimensional variety, the first integrals are *independent*. In this paper, we introduce a condition of a differential-geometric nature, essentially that the k-volumes of the leaves of a k-dimensional foliation are bounded, and under this condition, prove the existence of the maximal number $(n-k)$ of independent first integrals. Indeed we determine the structure of the leaf space of the germ of the foliation. Our results are somewhat analogous to those of Epstein [9] and Edwards-Millett-Sullivan [8].

We recall some terminology and results. A leaf of a non-singular foliation is (locally) an analytic variety if and only if it is (locally) closed [13]. A variety V may be the union of a finite number of irreducible components. The *dimension* of V is the maximum of the dimensions of its components. It is *purely k-dimensional* if all of its components are exactly k-dimensional. For k-dimensional V, let $\mathrm{Vol}_{2k}(V)$ denote the Euclidean $2k$-dimensional volume of V as a (possibly singular) submanifold of $\mathbb{C}^n$. Given any subset S of Ω, let $\mathcal{L}(S) = \mathcal{L}_\Omega(S)$, called the *saturation* of S in Ω, be the union of the leaves which intersect S. A subset of Ω is *saturated* if it is it own saturation. If F is a holomorphic foliation defined on $\Omega \subset \mathbb{C}^n$, nonsingular in $\Omega - A$, a *separatrix* of F is an analytic set $W \subset \Omega$ such that $A \cap W \neq \emptyset$ and $W - A$ is a leaf of $\mathcal{F}(F)$. Let $\Sigma(F)$ denote the union of all separatrices; $\Sigma(F)$ is called the *separatrix set* of F. An *orbifold* (or V-manifold) is the quotient of a finite group action on a complex manifold [6, 14, 24]. An orbifold is a normal space [6]. We also use Hausdorff measure for subsets of $\mathbb{C}^n$ and the Hausdorff metric on the set of closed subsets of $\mathbb{C}^n$, see e.g. [23]. We collect our results in an omnibus theorem.

Theorem. *Let $\mathcal{F}$ be a holomorphic foliation of codimension $n-k$ defined on a neighborhood $\mathcal{U}$ of the origin in $\mathbb{C}^n$, $0 < k < n$, nonsingular in $\mathcal{U} - \{0\}$. Suppose there exists a positive constant K such that for any leaf L of $\mathcal{F}$,*

$$\mathrm{Vol}_{2k}(L) \leq K. \tag{1.2}$$

Then every leaf is closed in $\mathcal{U} - \{0\}$ and is thus a k-dimensional variety. Those leaves which are not closed in $\mathcal{U}$ are precisely the separatrices. There exists at least one separatrix. Let $\Sigma = \Sigma(\mathcal{F})$ denote the union of the separatrices. If the $2k$-dimensional Hausdorff measure $\mathcal{H}_{2k}(\Sigma)$ is finite, then there exists a subneighborhood $\mathcal{V}$ of the origin such that in $\mathcal{V}$:

1. *Σ is a purely k-dimensional subvariety of $\mathcal{V}$ and in particular has a finite number of irreducible components;*
2. *if F_i is a sequence of closed leaves converging to any subset of Σ, it converges to all of Σ.*
3. *there are an $(n-k)$-dimensional singular space S, a point $p \in S$, and a holomorphic map $\pi: \mathcal{V} \to S$ such that $\pi^{-1}(p) = \Sigma$ and $\pi^{-1}(q)$ is a leaf of the foliation distinct from Σ, for $q \neq p$ in S.*

Thus we have $n-k$ first integrals in the map π. The proof occupies section 4.

The question of the existence of separatrices is very old. It was proposed by Briot-Bouquet [3] in 1856 for the case of holomorphic differential equations in $\mathbb{C}^2$ with an isolated singularity at the origin. The existence of a separatrix in this case was settled affirmatively in [5].

A partial converse of the theorem is valid. If there exists a map $\pi: \mathcal{U} \to V$, by Fubini's theorem the integral of the k volumes of the leaves (= fibers) is integrable over V. Several questions can be raised. Is the main theorem valid if the k-volumes are only integrable in some sense instead of uniformly bounded? Or does the existence of π ensure that the k-volumes are uniformly bounded? In particular, if all the leaves that are not separatrices are closed, are the volumes uniformly bounded? Also suppose $\mathcal{F} = \{\omega_1, \ldots, \omega_k\}$ is a regular holomorphic Frobenius system near the origin in $\mathbb{C}^n$, $0 < k < n$, and suppose that in some neighborhood of the origin,

$$\int_{\mathcal{U}} \log \|\omega_1(z) \wedge \omega_2(z) \wedge \cdots \wedge \omega_k(z)\| \, dz < \infty.$$

Does $\mathcal{F}$ have k independent holomorphic first integrals? The volume of the leaves is related to the integral.

The theorem implies that the foliation is transversally Riemannian off of the origin [22].

The authors would like to thank Xavier Gómez-Mont for helpful discussions and a careful reading of the paper.

2. Examples

We consider several examples, all differential equations, which illustrate some aspects of the theorem. These are derived from [4, 10]. Consider the complex differential system

$$\frac{dz}{dT} = Az, \qquad z \in \mathbb{C}^n, \quad A \in GL(n, \mathbb{C}). \tag{1.3}$$

For simplicity, suppose A is diagonal, with entries $\lambda_1, \ldots, \lambda_n$. The solution (leaf) through a point $(z_1, \ldots, z_n)$ is given by

$$\phi(z_1, \ldots, z_n, T) = (e^{\lambda_1 T}, \ldots, e^{\lambda_n T}). \tag{1.4}$$

1. If all the λ_i are equal, the leaves are all the punctured complex lines. Each leaf is a separatrix. Any continuous function constant on the leaves must be constant; there are no first integrals. Although the volumes of the leaves are bounded, the 2-dimensional Hausdorff measure of the separatrix set Σ is infinite.
2. Suppose A is hyperbolic and in the Poincaré domain (i.e., the convex hull of the eigenvalues of A does not contain the origin and the eigenvalues are independent over the reals). In this case, there is a nonzero λ_0 such that $\arg(\lambda - i/\lambda_0) < \pi/2$ for all $i = 1, \ldots, n$. For each $i = 1, \ldots, n$, let $T_N = -\lambda_0/N\lambda_i$ for $N = 1, \ldots, \infty$; we see that every nonsingular leaf contains at least one eigenspace in its closure. The separatrices are the eigenspaces. The closures of the other orbits are not analytic. It can be verified explicitly that the volumes of the leaves are not uniformly bounded near the origin.
3. Let $n = 3$ and suppose the convex hull of the eigenvalues contains the origin in its interior. The solution through a point (z_1, z_2, z_3) with all $z_i \neq 0$ is closed in $\mathbb{C}^n$. For suppose $a > 0$ and let $P(a) = \{(z_1, z_2, z_3) : |z_i| \leq a, i = 1, 2, 3\}$ be a polydisk, then $\{T \in \mathbb{C} : \phi(z_1, z_2, z_3, T) \in P(a)\}$ is a compact convex subset of $\mathbb{C}$. There are leaves $\phi(z_1, z_2, z_3, T)$ with some of the $z_i = 0$ which contain eigenspaces in their closures. The leaves are generically closed, but the volumes are not uniformly bounded.
4. Suppose $n = 2$. Consider the hyperbolic resonant case with $\lambda_1/\lambda_2 = -p/q$ for real positive integers p, q. Then the flow has the first integral $f\colon \mathbb{C}^2 \to \mathbb{C}$ given by $f(z_1, z_2) = z_1^q z_2^p$. It is easy to verify directly from the uniformization of the leaves given by the flow that the leaves have uniformly bounded volume. Note however that a transversal to the separatrix $\{z_2 = 0\}$ intersects each closed leaf p times, whereas a transversal to the separatrix $\{z_1 = 0\}$ intersects each closed leaf q times. The group Γ is cyclic of order q or p, depending on the separatrix. In this case, because the group is cyclic, the orbifold V is not singular, although the projection π is. It would be interesting to have an example with non-cyclic group.
5. Consider the elliptic resonant case for $n = 2$ with $\lambda_1/\lambda_2 = +p/q$ for real positive integers p, q (this case is related to example 1). Then there is a first integral on the complement of the origin, to wit $(z_1, z_2) \mapsto z_1^q z_2^{-p}$. However this integral does not extend across the origin. In this case all leaves are separatrices.

From these examples it is evident that having leaves with uniformly bounded volume is a highly non-generic situation. However having first integrals is also non-generic. We mention for example, the result of Mattei-Moussou [18] (which subsumes part of ours in the case of codimension one). Their result states that a codimension-one foliation admits a first integral if and only if the leaves are closed in $\mathcal{U} - \{0\}$ and if the set of leaves containing 0 in their closures in countable. There may be some kind of 'sliced' or 'fibered' version of our theorem: if there is some kind of r-codimensional 'slice' of $\mathbb{C}^n$ such that the leaves have finite $(k-r)$-volume, then are there $n-k-r$ first integrals? Making sense of the words is part of the question.

3. Bishop's results

We will need some results of Bishop relating k-volumes of subsets and analyticity. For details see [1, 25]. For convenience we collect them here. A sequence of subsets (in particular varieties) $\{V_i\}$, $i = 1, 2, \ldots$ in Ω has a (set) limit V_∞ if for each compact $C \subset \Omega$, the Hausdorff metric $d(V_i \cap C, V_\infty \cap C) \to 0$ as $i \to \infty$.

Bishop 1. *Let $\{V_i\}$ be a sequence of purely k-dimensional varieties in an open subset $\Omega \subset \mathbb{C}^n$ with* *uniformly bounded 2k-volumes; that is* $\mathrm{Vol}_{2k}(V_i) \leq K$ *for all i. Suppose* $\lim_i V_i = V_\infty$. *Then* V_∞ *is also a purely k-dimensional variety in* Ω *and* $\mathrm{Vol}_{2k}(V_\infty) \leq K$.

Bishop 2. *Let* V_1 *be a subvariety of an open set* $\Omega \subset \mathbb{C}^n$. *If* V *is a purely k-dimensional subvariety of* $\Omega - V_1$ *such that* $\overline{V} \cap V_1$ *has zero 2k-dimensional Hausdorff measure (*$\overline{V}$ *denotes closure in* Ω*), then* $\overline{V}$ *is a k-dimensional variety in* Ω.

Bishop 3. *Let* V_1 *be a subvariety of an open set* $\Omega \subset \mathbb{C}^n$. *If* V *is a purely k-dimensional subvariety of* $\Omega - V_1$ *with* $\mathrm{Vol}_{2k}(V) < \infty$, *then* $\overline{V}$ *is a purely k-dimensional variety in* Ω.

4. The proof

In this section we assume that $\mathcal{F}$ is a regular foliation of codimension $n-k$ defined in a neighborhood $\mathcal{U}$ of 0 in $\mathbb{C}^n$. Moreover we assume that (1.2) holds. For the first proposition, we do not need to assume that 0 is an isolated singularity. Let $S(\mathcal{F})$ be the singularity set in $\mathcal{U}$.

Proposition. *Under the above hypotheses, every leaf is closed in* $\mathcal{U} - S(\mathcal{F})$ *and hence is an analytic subvariety of* $\mathcal{U} - S(\mathcal{F})$. *For any* $z \in S(\mathcal{F})$, *there is at least one separatrix containing z in its closure. The separatrices are precisely the leaves* $L \subset \mathcal{U} - S(\mathcal{F})$ *such that the closure* $\overline{L}$ *of L in* $\mathcal{U}$ *intersect* $S(\mathcal{F})$.

Proof. Suppose there is a leaf L in $\mathcal{U} - S(\mathcal{F})$ which is not closed. Then there is a sequence $\{z_i\} \subset L$ which converges to $z \notin L$. Let $\mathcal{W}$ be a foliated chart of z in $\mathcal{U} - S(\mathcal{F})$. That is, $\mathcal{W}$ is holomorphically equivalent, say by f to a product $\mathcal{W}_1^k \times \mathcal{W}_2^{n-k}$, where $\mathcal{W}_i^r$ is open in $\mathbb{C}^r$ and the leaves of $\mathcal{F}|_\mathcal{W}$ are $f^{-1}(\mathcal{W}_1^k \times \{z_2\})$, called *plaques*. The k-volumes of subsets of $\mathcal{W}$ with the metric of $\mathbb{C}^n$ and with the metric of $\mathcal{W}_1^k \times \mathcal{W}_2^{n-k} \subset \mathbb{C}^k \times \mathbb{C}^{n-k}$ are not the same. However because f is Lipschitz, each is bounded by some constant multiple of the other. In particular, the volumes of a sequence of sets is unbounded in one metric if and only if it is unbounded in the other. The set $L \cap \mathcal{W}$ consists of an infinite number of plaques converging to the plaque containing z. Hence the $2k$-volume of $L \cap \mathcal{W}$ is infinite, contradicting the assumption (1.2). Thus every leaf is closed in $\mathcal{U} - S(\mathcal{F})$. By the regularity of $\mathcal{F}$, the Hausdorff measure $\mathcal{H}_{2k}(S(\mathcal{F})) = 0$. Thus by Bishop 2, the closure $\overline{L}$ of L in $\mathcal{U}$ is a purely k-dimensional analytic subvariety of $\mathcal{U}$, so if $\overline{L} \cap S(\mathcal{F}) \neq \emptyset$, then L is a separatrix.

Let $z \in S(\mathcal{F})$. We show there is a separatrix containing z in its closure. Let $\{z_i\}$ be a sequence in $\mathcal{U} - S(\mathcal{F})$ which converges to z. Let $L(z_i)$ denote the leaf through z_i. Let $\mathcal{U} = \cup C_j$, $C_j \subset C_{j+1}$ be a description of $\mathcal{U}$ as an increasing sequence of compact sets containing all the z_i. Then for each j, $\overline{L}(z_i) \cap C_j$ is a sequence of compact subsets of C_j. The set of closed subsets of a compact set endowed with the Hausdorff metric is compact (Blaschke's selection lemma [2], see [15,§42.II, 23]). Hence there is a convergent subsequence of $\overline{L}(z_i) \cap C_j$. By Cantor's diagonal process, there is a convergent subsequence of $\overline{L}(z_i)$. Let $W(z)$ denote the limit. By Bishop 1, $W(z)$ is a purely k-dimensional variety containing z. Hence $W(z) - S(\mathcal{F})$ must be a finite union of leaves of $\mathcal{U} - S(\mathcal{F})$. At least one of them has to contain z in its closure. The result is proved.

Now suppose in addition that $S(\mathcal{F}) = \{0\}$ and that $\mathcal{H}_{2k}(\Sigma(\mathcal{F})) < \infty$, where Σ is the separatrix set. By Bishop 3, $\Sigma(\mathcal{F})$ is a purely k-dimensional variety and hence is the finite union of irreducible components $\Sigma_1 \cup \cdots \cup \Sigma_r$. Each Σ_i is an irreducible variety which is possibly singular only at the origin.

We recall the cone theorem of Milnor [19, thm. 2.10], which is also valid for analytic varieties [11]: Let $\Sigma^l \subset \mathbb{C}^n$ be an l-dimensional variety which is singular (possibly) only at the origin. Then there exists $\epsilon > 0$ such that every sphere $S_\eta^{2n-1} = \{z \in \mathbb{C}^n : |z| = \eta\}$ with $\eta \leq \epsilon$ intersects Σ transversally in a real nonsingular analytic variety $\mathrm{Lk}_\eta(\Sigma)$, called the *link* of Σ^l. Furthermore, if D^{2n}_ϵ denotes the closed disk of radius ϵ, the pair $(D^{2n}_\epsilon, \Sigma^l \cap D^{2n}_\epsilon)$ is homeomorphic to the pair $(D^{2n}_\epsilon, \mathrm{Cone}\,\mathrm{Lk}_\epsilon(\Sigma))$. Actually more is proved. The homeomorphism is a real analytic equivalence on $D^{2n}_\epsilon - \{0\}$, so that for any $0 < \eta < \epsilon$, the intersection of Σ^l with the set $S^{2n}_{\eta,\epsilon} = \{z \in \mathbb{C}^n : \eta \leq |z| \leq \epsilon\}$ is real analytically the product $\mathrm{Lk}_\epsilon(\Sigma) \times [\eta, \epsilon]$.

Let ϵ be so small that $D^{2n}_\epsilon \subset \mathcal{U}$ and so that the conclusions of the cone theorem are valid for this ϵ for all the components Σ_i of the separatrix, $i = 1, \ldots, r$. Let $M_i = \mathrm{Lk}_\epsilon(\Sigma_i)$. Consider $\mathcal{F}|_{S^{2n-1}_\epsilon}$. It defines a foliation of S^{2n-1}_ϵ which is possibly singular. The leaves are the components of the intersections of the leaves of $\mathcal{F}$ with S^{2n-1}_ϵ. To distinguish them from the leaves in $\mathcal{U}$, we denote the leaf in S^{2n-1}_ϵ containing $x \in S^{2n-1}_\epsilon$ by $L_\epsilon(x)$. By transversality, the foliation of S^{2n-1}_ϵ is nonsingular in a closed tubular neighborhood $T_{\delta_i}(M_i)$ of radius δ_i of each M_i, $i = 1, \ldots, r$. In $T_{\delta_i}(M_i)$, each leaf is an irreducible real analytic variety and they are closed.

Lemma 1. *For each $i = 1, \ldots, r$, there exists $\delta_i > 0$ and $c_i > 0$ such that $\mathrm{Vol}_{2k-1}(L_\epsilon) < c_i$ for all leaves L_ϵ which intersect $T_{\delta_i}(M_i)$.*

Proof. Let $x_j \in T_{\delta_i}(M_i)$, $j = 1, 2, \ldots$ be a sequence converging to $x \in M_i$ such that $\mathrm{Vol}_{2k-1}(L_\epsilon(x_j)) \to \infty$. We claim eventually all the $L_\epsilon(x_j) \subset T_{\delta_i}(M_i)$. If not, choose a subsequence such that $L_\epsilon(x_i) \cap \partial T_{\delta_i}(M_i) \neq \emptyset$ for all j. By Blaschke's selection lemma, there is a further subsequence which converges in the Hausdorff-metric topology, say to $M \subset S^{2n-1}_\epsilon$. Cover the compact $T_{\delta_i}(M_i)$ with a finite number of foliated charts in $\mathbb{C}^n$. Denote the union of these by Y and let $L_Y(x_j) = \mathcal{L}_Y(\{x_j\})$ be the leaf in Y containing x_j. There is a further subsequence such that the $L_Y(x_j)$ converge in Y. By Bishop 1, the limit of the $L_Y(x_j)$ is a purely k-dimensional complex analytic variety which is thus a finite union of leaves in Y. By transversality $M \cap T_{\delta_i}(M_i)$ is a finite union of closed nonsingular leaves in $T_{\delta_i}(M_i)$. On the other hand, M is a connected subset which contains both $x \in M_i$ and some point of $\partial T_{\delta_i}(M_i)$ (in a compact space, the limit of closed connected subsets is connected). However the previous two sentences state incompatible facts. Thus eventually the $L_\epsilon(x_j) \subset T_{\delta_i}(M_i)$. Consider again the covering of $T_{\delta_i}(M_i)$ by a finite number of foliated charts. The intersections of these charts with S^{2n-1}_ϵ are foliated charts of $T_{\delta_i}(M_i)$. The volumes (respectively $2k$-dimensional and $(2k-1)$-dimensional) of the plaques are bounded above and below. Thus since $\mathrm{Vol}_{2k-1}(L_\epsilon(x_j)) \to \infty$, there exists some chart that the number of intersections of the $L_\epsilon(x_j)$, $j = 1, \ldots \infty$, with the chart is unbounded. Thus the $L(x_j)$ have unbounded $2k$-volumes. This contradicts the assumption (1.2). The proof is complete.

This lemma states that the phenomenon of [26] cannot occur in the present context. On the contrary, the structure of the foliation on S^{2n-1}_ϵ is regular. In particular, the results of [8,9] are valid, and we obtain the following corollary.

Corollary. *Each M_i, $i = 1, \ldots, r$, has an arbitrarily small open tubular neighborhood $\tau_{\delta_i}(M_i)$ in S^{2n-1}_ϵ such that $\tau_{\delta_i}(M_i)$ and its closure $\overline{\tau}_{\delta_i}(M_i)$ are saturated and in $\tau_{\delta_i}(M_i)$ all holonomy groups are finite.*

(The subscript δ_i is not necessarily a distance, but is only an index for the neighborhood.) In particular the holonomy group of M_i in $\tau_{\delta_i}(M_i)$ is finite. By Cartan's theorem [6],, we may find coordinates of a transversal to M_i in $\tau_{\delta_i}(M_i)$ such that the holonomy group Γ_i is a subgroup of $U(n-k)$, and by a result of Haefliger thesis (see [22]), the foliation in $\tau_{\delta_i}(M_i)$ is obtained locally by suspending this representation. Note that those leaves corresponding to fixed points of the holonomy group have nontrivial holonomy, so there is an open dense set of leaves that have trivial holonomy. The leaf space of the foliation in $\tau_{\delta_i}(M_i)$ is the germ of the complex analytic space $(\mathbb{C}^{n-k}/\Gamma, 0) = S_i$ [6]. Fixing the model in $\tau_{\delta_i}(M_i)$ of the foliation given by the suspension of Γ_i, the fact that the foliation in $\mathcal{U}$ has leaves with finite volume implies, by an argument similar to the one of Lemma 1, that the number of leaves of $\tau_{\delta_i}(M_i)$ which belong to the same leaf in $\mathcal{U}$ is bounded by some number N.

Given any neighborhood $\mathcal{U}$ of the origin, we construct a subneighborhood. Choose ϵ so that the $D_\epsilon^{2n} \subset \mathcal{U}$ and so that all the components Σ_i of Σ are the cones of their links M_i in D_ϵ^{2n}.

Lemma 2. *Suppose $\{z_j\}$, $j = 1, 2, \ldots$, is a sequence of points in D_ϵ^{2n} converging to $z \in \Sigma$. Let $L(z_j)$ be the leaf containing z_j in D_ϵ^{2n}, and let L be the limit of any subsequence of the $L(z_j)$. Then $L \subset \Sigma$.*

Proof. If not there is $y \in L - \Sigma$. Let $y_j \in L(z_j)$ converge to y. Since each $L(z_j)$ is connected, L is connected (limit of connected closed sets is connected), and contains both y and the origin. Consider $D_{\epsilon'}^{2n}$ for ϵ' slightly larger than ϵ (close enough to ϵ that $D_{\epsilon'}^{2n} \subset \mathcal{U}$ and the cone structure for the separatrix Σ' in $D_{\epsilon'}^{2n}$ still holds). Let $L'(z_j)$ and L denote the leaves and the limit (possibly with respect to a subsequence), respectively, in the interior of $D_{\epsilon'}^{2n}$. By Bishop 1, L' is a purely k-dimensional variety and $L \subset L'$. Thus y is connected to the origin in L'. Thus there is an irreducible component $L'_0 \subset L' - \Sigma'$. Note that $0 \notin L'_0$. Thus L'_0 intersects Σ' somewhere in $D_{\epsilon'}^{2n}$. However this is impossible since the foliation is nonsingular off of the origin. The lemma is proved.

For each $i = 1, \ldots, r$, consider the saturation $\mathcal{L}_{D_\epsilon^{2n}}(\tau_{\delta_i}(M_i))$. A leaf $L \in \mathcal{L}_{D_\epsilon^{2n}}(\tau_{\delta_i}(M_i))$ can intersect S_ϵ^{2n} at points not in $\tau_{\delta_i}(M_i)$. By Lemma 2, δ_i can be made small enough that all $L \in \mathcal{L}_{D_\epsilon^{2n}}(\tau_{\delta_i}(M_i))$ intersect S_ϵ^{2n} transversally. Suppose each δ_i has been so chosen. Let $\mathcal{V}$ be the the interior of the union of the $\mathcal{L}_{D_\epsilon^{2n}}(\tau_{\delta_i}(M_i))$, $i = 1, \ldots, r$. Relabel $\Sigma \cap \mathcal{V}$ to Σ.

Lemma 3. *$\mathcal{V}$ is a connected neighborhood of Σ.*

Proof. Lemma 2 implies $\mathcal{V}$ is a neighborhood (consider a sequence $\{z_j\}$ converging to the origin). Consider the connected component of $\mathcal{V}$ containing the origin. This component contains all of Σ. By transversality, the closure of the component contains all of $\overline{\Sigma}$, hence each $\tau_{\delta_i}(M_i)$, hence each $\mathcal{L}_{D_\epsilon^{2n}}(\tau_{\delta_i}(M_i))$. By transversality again, the component contains all of $\mathcal{V}$.

Lemma 4. *Suppose $\tau_{\delta_j}(M_j) \cap \mathcal{L}_{D_\epsilon^{2n}}(\tau_{\delta_i}(M_i)) \neq \emptyset$ for all sufficiently small $\tau_{\delta_i}(M_i)$. Then $\tau_{\delta_j}(M_j) \cap \mathcal{L}_{D_\epsilon^{2n}}(\tau_{\delta_i}(M_i))$ contains a neighborhood of M_j.*

Proof. Consider the relation $R \subset \overline{\tau}_{\delta_i}(M_i) \times \overline{\tau}_{\delta_j}(M_j)$. Namely $(p, q) \in R$ if either p and q are on the same leaf in D_ϵ^{2n} or if $p \in M_i$ and $q \in M_j$. We claim this is a closed relation. For suppose $\{(p_i, q_i)\}$, $i = 1, 2, \ldots$, is a sequence of points in R converging to (p, q). We can suppose the sequence of leaves $\{L(p_i) = L(q_i)\}$ converges in D_ϵ^{2n}. If $p \in M_i$ or $q \in M_j$, then $(p, q) \in R$ by Lemma 2. Otherwise, extending to $D_{\epsilon'}^{2n}$ and using, as above, connectivity, Bishop 1 and the nonsingularity off of Σ, we see that p and q lie on the same leaf and hence $(p, q) \in R$. We have shown that R is closed and hence compact in $\overline{\tau}_{\delta_i}(M_i) \times \overline{\tau}_{\delta_j}(M_j)$. By projection to $\overline{\tau}_{\delta_i}(M_i)$, we see the sets

$$R_i = \{p \in \overline{\tau}_{\delta_i}(M_i) \colon (p, q) \in R \text{ for some } q\} \subset \overline{\tau}_{\delta_i}(M_i),$$
$$R_j = \{q \in \overline{\tau}_{\delta_i}(M_i) \colon (p, q) \in R \text{ for some } p\} \subset \overline{\tau}_{\delta_j}(M_j),$$

are closed. We study the points $(p, q) \in R$ with $q \in \partial R_j$ (where boundaries are with respect to S_ϵ^{2n}). One possibility for such a (p, q) is that $q \in \partial \overline{\tau}_{\delta_j}(M_j)$. A second possibility is that $p \in \partial \overline{\tau}_{\delta_i}(M_i)$. A third possibility that is $p \in \tau_{\delta_i}(M_i) - M_i$ *and* $q \in \tau_{\delta_j}(M_j) - M_j$. We claim the third is in fact not possible. For suppose there is such a (p, q) on a common leaf L. Then L has a saturated tubular neighborhood Y with finite holonomy [8, 9]. Each leaf in Y intersects both $\tau_{\delta_i}(M_i)$ and $\tau_{\delta_j}(M_j)$. Thus R_i contains a neighborhood of p and R_j contains a neighborhood of q. That is, $p \notin \partial R_i$ and $q \notin \partial R_j$ and the claim is proved. Now suppose the lemma is false. Then there exist points $q \in \partial R_j - M_j$

arbitrarily close to M_j. However, by the claim just proved, the leaf $L(q)$ for any such q must satisfy $L(q) \cap \tau_{\delta_i}(M_i) \subset \partial\tau_{\delta_i}(M_i)$. However this contradicts Lemma 2. The result is proved.

Let $\mathcal{L}^{\circ}_{D^{2n}_\epsilon}(\tau_{\delta_i}(M_i))$ denote the intersection of $\mathcal{L}_{D^{2n}_\epsilon}(\tau_{\delta_i}(M_i))$ with the interior of D^{2n}_ϵ.

Lemma 5. *For any $i = 1, 2, \ldots, r$, $\mathcal{L}^{\circ}_{D^{2n}_\epsilon}(\tau_{\delta_i}(M_i))$ is a connected neighborhood of Σ.*

Proof. Note that $\cup_{i=1}^{r} \mathcal{L}^{\circ}_{D^{2n}_\epsilon}(\tau_{\delta_i}(M_i)) = \mathcal{V}$. We claim that if $\mathcal{L}_{D^{2n}_\epsilon}(\tau_{\delta_i}(M_i)) \cap \mathcal{L}_{D^{2n}_\epsilon}(\tau_{\delta_j}(M_j)) \neq \emptyset$ and $\mathcal{L}_{D^{2n}_\epsilon}(\tau_{\delta_i}(M_i)) \cap \mathcal{L}_{D^{2n}_\epsilon}(\tau_{\delta_k}(M_k)) \neq \emptyset$, then also $\mathcal{L}_{D^{2n}_\epsilon}(\tau_{\delta_j}(M_j)) \cap \mathcal{L}_{D^{2n}_\epsilon}(\tau_{\delta_k}(M_k)) \neq \emptyset$. For $\mathcal{L}_{D^{2n}_\epsilon}(\tau_{\delta_i}(M_i)) \cap \mathcal{L}_{D^{2n}_\epsilon}(\tau_{\delta_j}(M_j)) \neq \emptyset$ is equivalent to $\mathcal{L}_{D^{2n}_\epsilon}(\tau_{\delta_i}(M_i)) \cap \mathcal{L}_{D^{2n}_\epsilon}(\tau_{\delta_j}(M_j)) \neq \emptyset$, which is equivalent to $\mathcal{L}_{D^{2n}_\epsilon}(\tau_{\delta_i}(M_i)) \cap \tau_{\delta_j}(M_j) \neq \emptyset$. By Lemma 4, $\mathcal{L}_{D^{2n}_\epsilon}(\tau_{\delta_i}(M_i)) \cap \tau_{\delta_j}(M_j)$ contains a neighborhood of M_j, as does $\mathcal{L}_{D^{2n}_\epsilon}(\tau_{\delta_k}(M_k)) \cap \tau_{\delta_j}(M_j)$. Hence $\mathcal{L}_{D^{2n}_\epsilon}(\tau_{\delta_i}(M_i)) \cap \mathcal{L}_{D^{2n}_\epsilon}(\tau_{\delta_k}(M_k)) \neq \emptyset$ and thus $\mathcal{L}_{D^{2n}_\epsilon}(\tau_{\delta_i}(M_i)) \cap \mathcal{L}_{D^{2n}_\epsilon}(\tau_{\delta_k}(M_k)) \neq \emptyset$, as claimed. Accordingly we may define an equivalence relation among the indices $i = 1, 2, \ldots, r$; namely, two indices i and j are equivalent if $\mathcal{L}_{D^{2n}_\epsilon}(\tau_{\delta_i}(M_i)) \cap \mathcal{L}_{D^{2n}_\epsilon}(\tau_{\delta_j}(M_j)) \neq \emptyset$. If there is more than one equivalence class, $\mathcal{V}$ is decomposed into the disjoint union of two open subsets. Since $\mathcal{V}$ is connected, this cannot be and there is only one equivalence class. Thus any $\mathcal{L}^{\circ}_{D^{2n}_\epsilon}(\tau_{\delta_i}(M_i))$ is an open set containing Σ and the lemma is proved.

At this point we can improve Lemma 2.

Corollary. *Suppose $\{z_j\}$, $j = 1, 2, \ldots$, is a sequence of points in $\mathcal{V}$ converging to $z \in \Sigma$. Then any convergent subsequence of the leaves $L(z_j)$ converges to all of Σ.*

Proof. If not, there is more than one equivalence class in the proof of Lemma 5.

Lemma 6. *For $i = 1, 2, \ldots$, let z_i and w_i be points in $\mathcal{V}$ on the same leaf. Suppose the sequences $\{z_i\}$ and $\{w_i\}$ converge to z and w in $\mathcal{V}$ respectively, and $z \notin \Sigma$. Then w and z lie on the same leaf.*

Proof. A subsequence of the leaves $L(z_i)$ converges to an analytic space L of dimension k by Bishop 1. Since each $\overline{L}(z_i)$ intersects S^{2n}_ϵ transversally, the limit of the $\overline{L}(z_i)$ is $\overline{L}$. As a limit of closed connected subsets, $\overline{L}$ is connected, and since the foliation is nonsingular off of 0, $\overline{L}$ is a leaf in D^{2n}_ϵ. By transversality, L is a leaf in $\mathcal{V}$. However $w \in L$, and the lemma is proved.

Recall that S_i is the leaf space of $\tau_{\delta_i}(M_i)$. We introduce an equivalence relation $\sim$ in each S_i; namely, $p \sim q$ if the leaves in $\tau_{\delta_i}(M_i)$ represented by p and q are contained in the same leaf in $\mathcal{V}$. By Lemma 6, this equivalence relation is closed and Hausdorff. Since it is holomorphic, the orbit space $T_i = S_i/\sim$ has the structure of a complex analytic space [12]. Recall that the germ of the leaf space $S_i = (\mathbb{C}^{n-k}/\Gamma, 0)$ as germs of analytic varieties. Thus the germ of T_i is a quotient of $(\mathbb{C}^{n-k}/\Gamma, 0)$. Denote these germs by $g(S_i)$ and $g(T_i)$. There is a natural analytic map $\tilde{G}_{ji}: g(S_i) \to g(T_j)$; namely a class of a leaf L in S_i is mapped to the class of L in T_j. Lemma 5 implies $\tilde{G}_{ji}$ is defined on the germ. Moreover $\tilde{G}_{ji}$ factors to a natural analytic map $G_{ji}: g(T_i) \to g(T_j)$. Clearly G_{ij} is the inverse of G_{ji}. Thus all the $g(T_i)$ are naturally isomorphic. Finally let $g(\Sigma)$ be the germ of the leaf space of a neighborhood of Σ in $\mathcal{V}$ and hence in the interior of D^{2n}_ϵ, with the added equivalence that all of Σ is identified to a point. From Lemma 5, we see that $g(\Sigma)$ is naturally isomorphic to any $g(T_i)$.

Thus we have detailed the structure of $g(\Sigma)$ and also proved our theorem.

5. A final remark

Let $D_i = D_i^{2(n-k)}$ be a disk transversal to the foliation in $\tau_{\delta_i}(M_i)$, with center $0_i = D_i \cap M_i$. The projection maps $D_i \to S_i$ are the quotients of the holonomy group Γ_i of M_i (which is the same as the holonomy of Σ_i, since Σ_i is a cone over M_i). The projection maps $D_i \to S_i \to T_i$ are surjective finite holomorphic mappings. Thus in the complement of nowhere dense closed analytic subsets, they are coverings. Assume that they are Galois coverings, and let $\mathcal{G}$ be the group of deck transformations. The elements of $\mathcal{G}$ are bounded holomorphic functions on the complement of a nowhere dense analytic subset, so by Riemann's extension theorem, they extend to biholomorphisms of the D_i. These extended elements preserve 0_i, but only the elements of $\Gamma_i \to \mathcal{G}$ correspond to the holonomy of M_i. For example, in example 2.4, the separatrices Σ_1 and Σ_2 are the axes with $\Gamma_1 = Z/pZ$, $\Gamma_2 = Z/qZ$, and $\mathcal{G} = Z/pqZ$. It would be interesting to know how the foliations in $\tau_{\delta_i}(M_i)$ amalgamate to form the foliation in $\mathcal{V}$.

References

[1] E. Bishop, "Conditions for the analyticity of certain sets," *Michigan Math. J.*, **11** (1964), 289–304.

[2] W. Blaschke, *Kreis und Kugel*, B. G. Teubner, Leipzig, 1916.

[3] C. A. Briot & J. C. Bouquet, "Propiétés des fonctions définies par équations différentielles," *Jour. de l'Ecole Polytechnique*, **36** (1856), 133–198.

[4] C. Camacho, N. H. Kuiper & J. Palis, "The topology of holomorphic flows with singularity," *Inst. Hautes Études Sci. Publ. Math.*, **48** (1978), 5–38.

[5] C. Camacho & P. Sad, "Invariant varieties through singularities of holomorphic vector fields," *Ann. of Math. (2)*, **115** (1982), 579–595.

[6] H. Cartan, "Quotients of complex analytic spaces," in *Contributions to Function Theory*, K. Chandrasekharan, ed., TATA Institute of Fundamental Research, Bombay, 1960, 1–15.

[7] C. Cerveau, Université de Dijon, Thèse d'Etat.

[8] R. Edwards, K. Millett & D. Sullivan, "Foliations with all leaves compact," *Topology*, **16** (1977), 13–32.

[9] D. B. A. Epstein, "Foliations with all leaves compact," *Ann. Inst. Fourier (Grenoble)*, **26:1** (1976), 265–282.

[10] J. Guckenheimer, "Hartman's theorem for complex forms in the Poincaré domain," *Compositio Math.*, **24** (1972), 75–82.

[11] H. Hamm, "Lokale Topologische Eigenschaften Komplexe Räume," *Math. Ann.*, **191** (1971), 235–252.

[12] H. Holman, "Komplex Räume mit Komplexen Transformationsgruppen," *Math. Ann.*, **150** (1963), 327–360.

[13] H. Holman, "Holomorphe Blätterungen Komplexer Räume," *Comment. Math. Helv.*, **47** (1972), 185–204.

[14] S. Kobayashi, *Hyperbolic Manifolds and Holomorphic Mappings*, Pure and Applied Mathematics **#2**, Marcel Dekker, New York, NY, 1970.

[15] K. Kuratowski, *Topology* (vol. II), Academic Press, New York, NY, 1968.

[16] B. Malgrange, "Frobenius avec singularité I: codimension 1," *Inst. Hautes Études Sci. Publ. Math.*, **46** (1976), 163–173.

[17] B. Malgrange, "Frobenius avec singularité II: le cas general," *Invent. Math.*, **39** (1977), 67–89.

[18] J. F. Mattei & R. Moussou, "Holonomie et intégrales premières," *Ann. Sci. École Norm. Sup. (4)*, **13** (1980), 469–523.

[19] J. Milnor, *Singular Points of Complex Hypersurfaces*, Ann. of Math. Studies **#61**, Princeton Univ. Press, Princeton, NJ, 1968.

[20] H. J. Reiffen, "Leafspace and integrability," 1986, preprint.

[21] H. J. Reiffen & U. Vetter, "Pfaffsche Formen auf Komplexen Räumen," *Math. Ann.*, **167** (1966), 338–350.

[22] B. Reinhart, *Differential Geometry of Foliations*, Ergebnisse der Mathematik und Ihre Grenzgebiete #**99**, Springer-Verlag, New York–Heidelberg–Berlin, 1983.

[23] C. A. Roger, *Hausdorff Measures*, Cambridge Univ. Press, Cambridge, 1970.

[24] I. Satake, "On a generalization of the notion of manifolds," *Proc. Nat. Acad. Sci. U.S.A.*, **42** (1956), 359–363.

[25] G. Stolzenberg, *Volumes, Limits and Extensions of Analytic Varieties*, Lect. Notes in Math. #**19**, Springer-Verlag, New York–Heidelberg–Berlin, 1966.

[26] D. Sullivan, "A counterexample to the periodic orbit conjecture," *Inst. Hautes Études Sci. Publ. Math.*, **46** (1976), 5–14.

[27] R. Thom, "On singularities of foliations," in *Manifolds–Tokyo 1973*, A. Hattori, ed., Math. Soc. of Japan, Tokyo, 1973, 171–174.

[28] H. E. Winkelkemper, "The graph of a foliation," *Ann. Global Anal. Geom.*, **1** (1983), 51–75.

SURGERY ON COMPLEX POLYNOMIALS

BODIL BRANNER

Mathematical Institute
The Technical University of Denmark
Building 303
DK-2800 Lyngby
Denmark

ADRIEN DOUADY

Université de Paris-Sud
Département de Mathématiques
Bâtiment 425
F-91405 Orsay

École Normale Supérieure
Centre de Mathématiques
45 rue d'Ulm
F-75230 Paris Cedex 05
France

TABLE OF CONTENTS

Typeset by $\mathcal{A}_{\mathcal{M}}\mathcal{S}$-TEX

1. Introduction.

We shall present two results concerning the parameter spaces for quadratic and cubic polynomials, considered as dynamical systems. The results are obtained by *surgery* .

From any quadratic polynomial

$$P_c(z) = z^2 + c$$

with c in the *limb* $M_{1/2}$ of the Mandelbrot set M we shall construct

(A) a quadratic polynomial $P_{c'}$ with

$$c' = \Phi_A(c)$$

in the limb $M_{1/3}$ of the Mandelbrot set M

and

(B) a cubic polynomial Q_a with

$$a = \Phi_B(c)$$

in the limb F_+ of the set F (to be described in section 2), where

$$Q_a(z) = z^3 - 3a^2 z + 2a^3 - 2a \ .$$

The two constructions take place in several steps, different in (A) and (B) but parallel :

$$P_c \curvearrowright f \curvearrowright g \curvearrowright h \curvearrowright \begin{cases} P_{c'} \\ Q_a \end{cases}$$

(1) From part of the dynamical plane for P_c with $c \in M_{1/2}$ we obtain through cutting and sewing a new plane region and a new map f which has a line of discontinuities.

(2) By smoothing f we obtain a quasi-regular mapping g .

We say that g is obtained from P_c by topological surgery.

(3) Furthermore we introduce a g-invariant almost complex structure σ. The almost complex structure σ can be integrated by a quasi-conformal homeomorphism φ. The mapping

$$h = \varphi \circ g \circ \varphi^{-1}$$

is polynomial-like of degree 2 in (A) and of degree 3 in (B).

(4) Hence h is hybrid equivalent to a polynomial $P_{c'}$ or Q_a.

We say that the polynomial is obtained from P_c by holomorphic surgery.

This ends the construction in the dynamical plane.

There are choices to be made in intermediate steps, some of which do not depend continuously on the parameter c, so it is not clear that we can get a result for the parameter spaces. Nevertheless we prove

(5) the mapping Φ is continuous

and

(6) injective.

We have the following results in the parameter spaces :

THEOREM A. *The map*

$$\Phi_A \colon M_{1/2} \to M_{1/3}$$

is a homeomorphism of $M_{1/2}$ onto its image.

The image will be described precisely in section 15.

As a consequence we have

COROLLARY A.
The image $\Phi_A([-2, -\frac{3}{4}])$ is a topological arc in $M_{1/3}$.

This arc is the *principal vein* of $M_{1/3}$.

We hope that this result can be extended to all veins of M.

It is conjectured that M is locally connected. Since any connected compact metric space which is locally connected is arcwise connected, the result can be viewed as a step towards that conjecture.

THEOREM B. *The map*

$$\Phi_B : M_{1/2} \to F_+$$

is a homeomorphism.

The set F_+ is a subset of the boundary of the *connectedness locus* $\mathcal{C}(3)$. The boundary can be approached from outside the connectedness locus. As a consequence we have

COROLLARY B. *The limit as* $r \to 0$ *of* $E_{r,0}$ *is homeomorphic to* $M_{1/2}$.

The set $E_{r,0}$ is defined in section 3 .

Our results rely in an essential way on [DH 2], that is the theory of polynomial-like mappings and the theory of quasi-conformal mappings. A combinatorial remark of J.-C. Yoccoz was the starting point of theorem B , and the proof of theorem A makes use of an inequality of Yoccoz (see [Y] and [L]) estimating the size of the limbs of M . Our work has also been influenced by M. Shishikura [S] . Shishikura pointed out that, while smoothing maps in the process of surgery, it is essential to concentrate the non-holomorphy to regions where orbits pass at most once.

The paper is divided into 3 parts.

In the first part, I Context, we fix the terminology, summarize the tools and describe the dynamical behavior which is characteristic for polynomials P_c with $c \in M_{1/2}$.

In the second part, II Cubic and quadratic polynomials, we prove theorem B . We give that proof first, since it is simpler than that of theorem A .

Finally in the third part, III A vein of M , we give the proof of Theorem A .

Part II and III follow the steps (1) - (6) . Step (1) is contained in II.6 and III.10 . Step (2) in II.6 and III.10-13 ; in this step the proof of theorem A becomes much more delicate than that of theorem B . Step (3) and (4) are in II.7 and III.14 . Step (5) in II.8 and III.14 . Step (6) in II.9 and III.15 .

We thank J.-C. Yoccoz, E.Ghys and M. Shishikura for many helpful conversations and Yuval Fisher for the computer drawings in this paper. Most of all we thank John H. Hubbard. This paper is clearly a continuation of both [DH 2] and [BH] .

The central part of this work was done during the Symposium on Dynamical Systems in Mexico.

2. Notation, terminology and results.

For a polynomial $f : \mathbf{C} \to \mathbf{C}$ we denote by

$$K(f) = \{ z \mid f^n(z) \not\to \infty \}$$

the *filled in Julia set* . It is a compact set which may be connected or not, and if connected, then locally connected or not.

For a quadratic polynomial

$$P_c(z) = z^2 + c$$

we set $K_c = K(P_c)$.

The *Mandelbrot set* M is the set in the parameter plane $\mathbf{C}$

$$M = \{ c \mid K_c \textit{ connected} \} = \{ c \mid 0 \in K_c \} .$$

Denote by Ω a *hyperbolic component* of M, that is a connected component of $\overset{\circ}{M}$ such that P_c has an attractive cycle for every $c \in \Omega$.

We denote by φ_Ω the conformal representation

$$\varphi_\Omega : \Omega \to D$$

defined by mapping $c \in \Omega$ to the multiplier of the attractive cycle and we denote by c_Ω the *center* of Ω defined by

$$c_\Omega = \varphi_\Omega^{-1}(0) .$$

It is proved in [D 1] that the multiplier gives the conformal representation. The mapping φ_Ω extends to a homeomorphism of the closures. For each hyperbolic component Ω we define

$$\gamma_\Omega(t) = \lim_{r \to 1} \varphi_\Omega^{-1}(re^{2\pi it}) ,$$

giving a parametrization

$$\gamma_\Omega \colon \mathbf{T} = \mathbf{R}/\mathbf{Z} \ \to \ \partial\Omega$$

of the boundary. For $t \in \mathbf{T}$ and $c \in \partial\Omega$ we say that

$$t \text{ is the } \textit{internal argument} \text{ of } c \iff \gamma_\Omega(t) = c .$$

The unit for arguments is one turn.

In particular we denote by Ω_0 the set of values of c such that P_c has an attractive fixed point α_c , the conformal representation

$$\varphi_{\Omega_0} : \Omega_0 \to D$$

is defined by $\varphi_{\Omega_0}(c) = P'_c(\alpha_c)$ and

$$\gamma_{\Omega_0}(t) = \frac{1}{2}e^{2\pi it} - \frac{1}{4}e^{4\pi it}$$

parametrizes the big cardioid of M .

Furthermore for $t \in \mathbf{T}-\{0\} = \mathbf{Q}/\mathbf{Z}-\{0\}$ let Ω_t denote the hyperbolic component satisfying

$$\overline{\Omega}_t \cap \overline{\Omega}_0 = \gamma_{\Omega_0}(t) = \gamma_{\Omega_t}(0) ,$$

see figure 2.1.

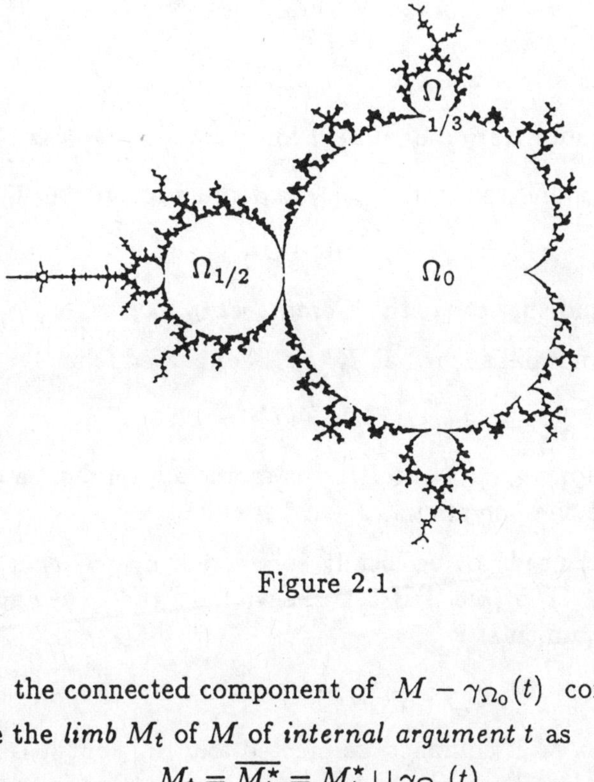

Figure 2.1.

Set

$M_t^* =$ the connected component of $M - \gamma_{\Omega_0}(t)$ containing Ω_t

and define the *limb* M_t of M of *internal argument* t as

$$M_t = \overline{M_t^*} = M_t^* \cup \gamma_{\Omega_0}(t) .$$

For $K \subset \mathbf{C}$ any compact set, connected and *full* (i.e. $\mathbf{C} - K$ connected) we denote by φ_K the conformal representation

$$\varphi_K : \mathbf{C} - K \quad \to \quad \mathbf{C} - \overline{D}_{r(K)}$$

satisfying

$$\frac{\varphi_K(z)}{z} \to 1 \qquad \text{when } |z| \to \infty$$

($r(K) =$ the *radius of capacity = the transfinite diameter).*

The potential G_K created by K satisfies

$$G_K = \log \frac{|\varphi_K|}{r_K} .$$

We set $G_K(z) = 0$ for $z \in K$.

The *external ray* of K of *argument* t is defined as

$$\mathcal{R}_K(t) = \varphi_K^{-1}(\{ re^{2\pi it} \mid r \in]r(K), +\infty[\}) .$$

Set

$$\arg_K(z) = t \qquad \text{if } z \in \mathcal{R}_K(t)$$

and

$$\gamma_K(t) = \lim_{r \to r(K)} \varphi_K^{-1}(re^{2\pi it})$$

if this limit exists. For $z \in \partial K$,

$$t \text{ is an } \textit{external argument} \text{ for } z \iff \gamma_K(t) = z .$$

If K is locally connected, then $\gamma_K(t)$ is defined for all $t \in \mathbf{T}$ and

$$\gamma_K : \mathbf{T} \to \partial K$$

is continuous and surjective, the *Carathéodory loop* of K .

If f is a polynomial such that $K(f)$ is connected, then

$$\varphi_f = \varphi_{K(f)} : \mathbf{C} - K(f) \ \to \ \mathbf{C} - D ,$$

we write $\gamma_f(t)$ for $\gamma_{K(f)}(t)$, etc. If f is monic and of degree d , then φ_f conjugates f to the polynomial $z \mapsto z^d$.

The set $K(f)$ need not be locally connected, but $\gamma_f(t)$ is always defined for all $t \in \mathbf{Q}/\mathbf{Z}$ (see [DH 3]) and we have the following action on the external arguments

$$f(\gamma_f(t)) = \gamma_f(d\,t) .$$

We know that M is compact, connected and full, but it is not known whether M is locally connected. However $\gamma_M(t)$ is defined for all $t \in \mathbf{Q}/\mathbf{Z}$. In particular

$$\gamma_M(\frac{1}{3}) = \gamma_M(\frac{2}{3}) = \gamma_{\Omega_0}(\frac{1}{2}) = -\frac{3}{4} ,$$

and

$$\gamma_M(\frac{1}{7}) = \gamma_M(\frac{2}{7}) = \gamma_{\Omega_0}(\frac{1}{3}) ,$$

see figure 2.2.

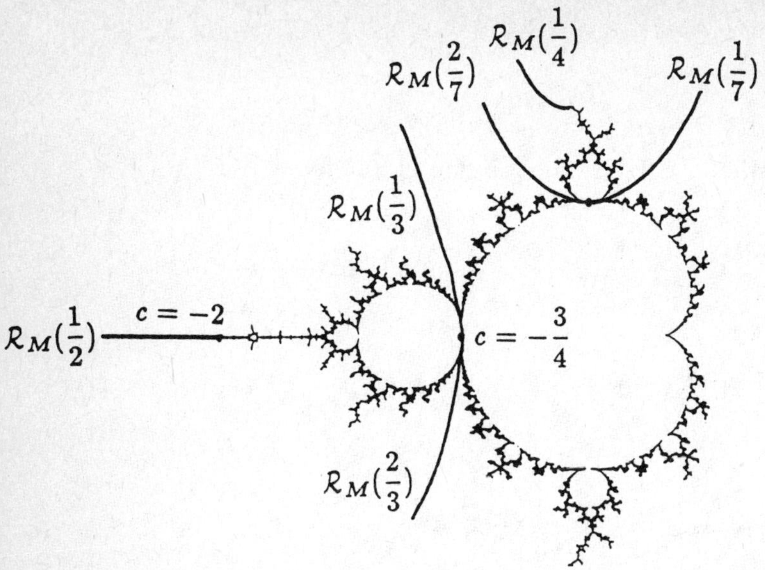

Figure 2.2.

THEOREM A. *There exists a continuous, injective mapping*

$$\Phi_A \colon M_{1/2} \to M_{1/3}$$

such that

$$-\frac{3}{4} = \gamma_{\Omega_0}(\frac{1}{2}) \quad \mapsto \quad \gamma_{\Omega_0}(\frac{1}{3})$$

and

$$-2 = \gamma_M(\frac{1}{2}) \quad \mapsto \quad \gamma_M(\frac{1}{4}) \ .$$

COROLLARY A. *There exists in M a topological arc which connects* 0 *to* $\gamma_M(\frac{1}{4})$.

Let $Q_{a,b}$ denote the cubic polynomial

$$Q_{a,b}(z) = z^3 - 3a^2 z + b \ ,$$

we set $K_{a,b} = K(Q_{a,b})$. The critical points are : a and $-a$.

The *connectedness locus* $\mathcal{C}(3)$ is the set in the parameter space $\mathbf{C}^2$ defined by

$$\mathcal{C}(3) = \{ \ (a,b) \mid K_{a,b} \text{ connected } \} = \{ \ (a,b) \mid \{a,-a\} \subset K_{a,b} \ \} \ .$$

For $(a,b) \in \mathcal{C}(3)$,

$$\beta_+ = \beta_+(a,b) = \gamma_{a,b}(0) \ , \ \beta_- = \beta_-(a,b) = \gamma_{a,b}(\frac{1}{2})$$

are fixed points of $Q_{a,b}$. If $\beta_- \neq \beta_+$, then there is another fixed point $\alpha(a,b)$; if not then there can be 2 other fixed points $\alpha_+ = \alpha_+(a,b)$ and $\alpha_- = \alpha_-(a,b)$.

Let $\mathcal{F}$ denote the one-parameter subfamily of cubic polynomials defined by

$$\begin{aligned}\mathcal{F} &= \{ (a,b) \mid Q_{a,b}(a) = -2a \} = \{ (a,b) \mid b = 2a^3 - 2a \} \\ &= \{ (a,b) \mid Q_{a,b}(a) \text{ is a fixed point, but } a \text{ is not fixed} \} \cup \{ (0,0) \} .\end{aligned}$$

Set

$$Q_a(z) = Q_{a,2a^3-2a}(z) = z^3 - 3a^2 z + 2a^3 - 2a ,$$

$K_a = K(Q_a)$ and

$$F = \{ a \mid K_a \text{ connected } \} = \{ a \mid -a \in K_a \} = \mathcal{F} \cap \mathcal{C}(3) .$$

The set F is symmetric with respect to 0 . We have

$$r(F) = 2^{-\frac{2}{3}} , \ \gamma_F(\frac{1}{3}) = \gamma_F(\frac{2}{3}) = -\frac{1}{3} ,$$

see figure 2.3.

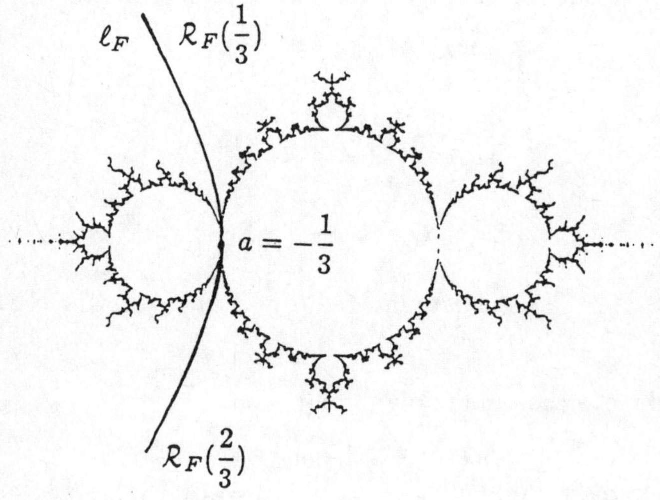

Figure 2.3.

Let F_+ denote the limb of F

$$\begin{aligned}F_+ &= \{ a \in F \mid Q_a(a) = -2a = \beta_+ \} \\ &= \text{the part of } F \text{ placed to the left of } \ell_F\end{aligned}$$

where

$$\ell_F = \mathcal{R}_F(\frac{1}{3}) \cup \{ -\frac{1}{3} \} \cup \mathcal{R}_F(\frac{2}{3}) .$$

THEOREM B. *There exists a homeomorphism*

$$\Phi_B : M_{1/2} \to F_+$$

such that

$$\Phi_B(-\frac{3}{4}) = -\frac{1}{3} .$$

Figure 2.4(a) shows the limb $M_{1/2}$ and (b) the limb F_+ .

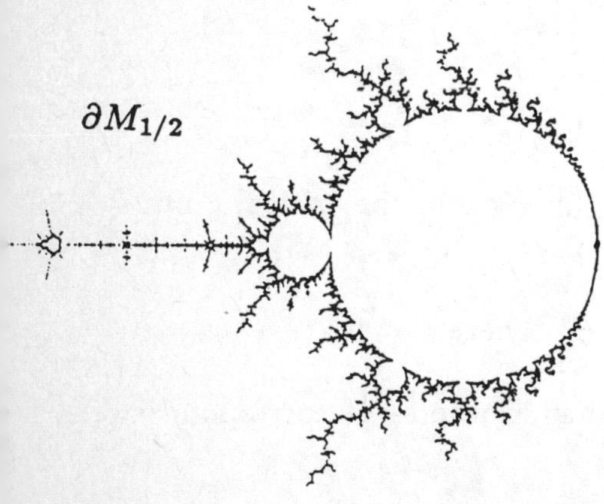

Figure 2.4(a).

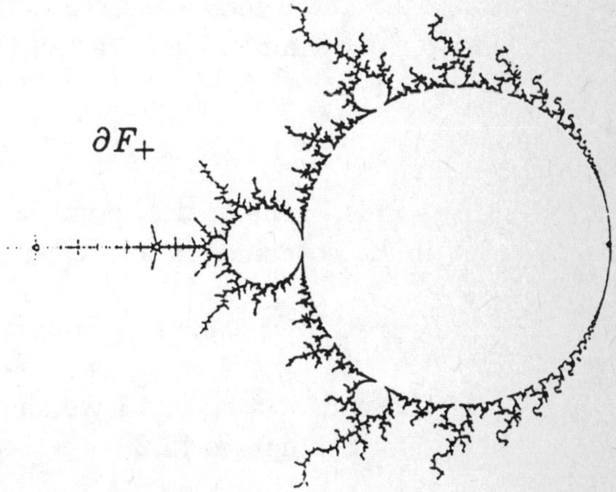

Figure 2.4(b).

In figure 2.5(a) we show the Julia set of $P_{-3/4}$ and in 2.5(b) the Julia set of $Q_{-1/3}$. We have marked some important external rays.

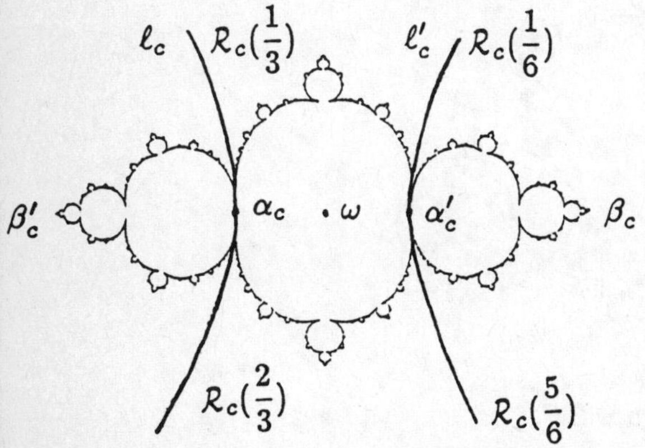

Figure 2.5(a).

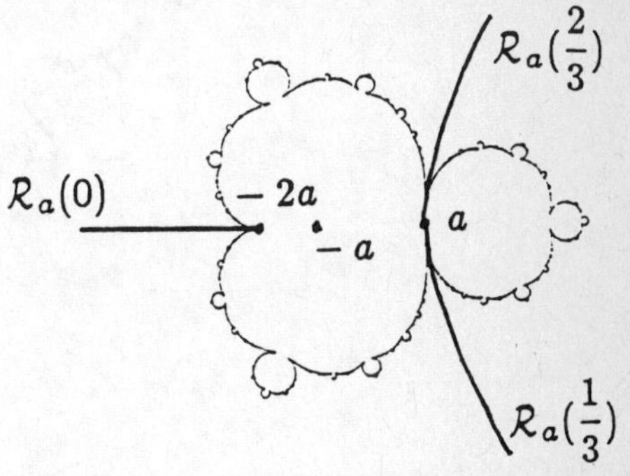

Figure 2.5(b).

As a consequence of the proof of theorem B we shall see that the dynamical behavior of P_c on the filled in Julia set K_c is related to the dynamical behavior of Q_a with $a = \Phi_B(c)$ on the filled in Julia set K_a in the following way :
If we remove the part of K_c to the left of

$$\mathcal{R}_c(\frac{1}{3}) \cup \{\alpha_c\} \cup \mathcal{R}_c(\frac{2}{3})$$

and change the map P_c to the first return map f_c on that part of K_c , then the dynamical behavior of f_c is homeomorphically equivalent to the dynamical behavior of Q_a on K_a .
The point

$$\gamma_c(\frac{1}{6}) = \gamma_c(\frac{5}{6}) \text{ in } K_c$$

corresponds to the critical point a of Q_a in K_a and the critical point 0 of P_c in K_c corresponds to the critical point $-a$ of Q_a in K_a .

In figure 2.6 we show the Julia set of $P_{c'}$ where $c' = \Phi_A(-\frac{3}{4})$.

At the end of section 15 we show other examples of corresponding Julia sets. See figures 15.2 – 4 .

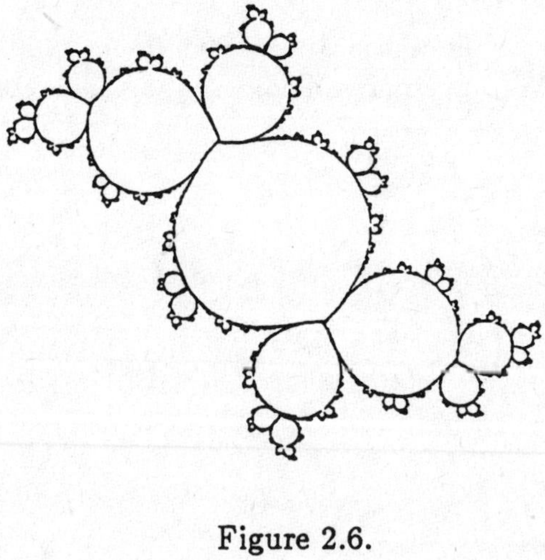

Figure 2.6.

3. Motivation for theorem B.

In [BH] we have studied cubic polynomials $Q_{a,b}$ with (a,b) in the complement Ξ of the connectedness locus $\mathcal{C}(3)$ in $\mathbf{C}^2$. In particular we have defined a homeomorphism from Ξ onto $\mathbf{R}_+ \times S^3$, hence a mapping

$$G\colon \Xi \to \mathbf{R}_+$$

and a one parameter group $(S_t)_{t\in\mathbf{R}_+}$ (the stretching operator) operating on Ξ . The function G is given by

$$G(a,b) = sup(G_{a,b}(a), G_{a,b}(-a))$$

where

$$G_{a,b}(z) = \lim_{r\to\infty} \frac{1}{3^n} \log \mid Q^n_{a,b}(z) \mid .$$

Therefore the function G measures the escape rate towards infinity of the fastest escaping of the two critical points . For $r > 0$ the topological sphere $S_r = G^{-1}(r)$ is decomposed into two symmetrical parts S_r^+ and S_r^- corresponding to $G(a,b) = G_{a,b}(a)$ and $G(a,b) = G_{a,b}(-a)$ respectively.

For $(a,b) \in S_r$ let $H_{a,b}$ denote the set

$$H_{a,b} = \{\ z \mid G_{a,b}(z) \le r\ \} .$$

The set is compact, connected and full. The map

$$\varphi_{a,b}\colon \mathbf{C} - H_{a,b} \ \to\ \mathbf{C} - \overline{D}_{\exp r}$$

conjugates $Q_{a,b}$ to the polynomial $z \mapsto z^3$.

The mappings

$$(a,b) \ \mapsto\ \varphi_{a,b}(\mp a) \qquad \text{where } (a,b) \in S_r^{\pm}$$

are fibrations. Both S_r^+ and S_r^- are therefore fibrations over $\mathbf{T}$ and with the same fiber : a trefoil clover leaf, the monodromy acts by $\frac{1}{3}$-turn . The central circle Γ_r , the same for S_r^+ and S_r^- , consists of the polynomials of S_r of the form $z \mapsto z^3 + b$.

The projection

$$\pi_+\colon S_r^+ \ \to\ \mathbf{T}$$

admits a lifting

$$\tilde{\pi}_+\colon S_r^+ - \Gamma_r \ \to\ \mathbf{T}$$

satisfying

$$\pi_+(a,b) = 3\tilde{\pi}_+(a,b)$$

and

$$\gamma_{a,b}(\tilde{\pi}_+(a,b)) = -2a \ .$$

For $\theta \in \mathbf{T}$ let $L_{r,\theta}$ denote

$$L_{r,\theta} = \{ \ (a,b) \in S_r^+ \mid \tilde{\pi}_+(a,b) = \theta \ \} \ ,$$

that is one of the leaves in the trefoil clover.

For $(a,b) \in L_{r,\theta}$ we have $Q^n_{a,b}(a) \to \infty$. Let $E_{r,\theta}$ denote the subset

$$E_{r,\theta} = \{ \ (a,b) \in L_{r,\theta} \mid Q^n_{a,b}(-a) \not\to \infty \ \} \ .$$

The set $E_{r,\theta}$ is a compact set, having a non-countable infinity of connected components. A countable infinity of them are copies of the Mandelbrot set M and presumably the others are points.

An interesting question is to understand the turning operator

$$\tau \colon L_{r,0} \ \to \ L_{r,0} \ ,$$

to classify the components of $E_{r,0}$ due to how they turn when θ makes a turn, to find out which of the components are independent of the others and which interchange with each other.

J.-C.Yoccoz has remarked, that combinatorially the copies of M in $E_{r,0}$ correspond bijectively to copies of M in the limb $M_{1/2}$.

The polynomials $Q_{a,b}$ with $(a,b) \in E_{r,0}$ are characterized by

(1) $\qquad G_{a,b}(a) = r \ , \ G_{a,b}(-a) = 0 \ ,$

(2) $\qquad Q_{a,b}(a) = Q_{a,b}(-2a) \in \mathcal{R}_{a,b}(0) \ ,$

(3) $\qquad \gamma_{a,b}(\frac{1}{3}) = \gamma_{a,b}(\frac{2}{3}) = a \ , \ \gamma_{a,b}(0) = -2a \ .$

The polynomials Q_a with $a \in F_+$ are characterized by

(1) $\qquad G(a,b) = 0 \ ,$

(2) $\qquad Q_a(a) = Q_a(-2a) = -2a \ ,$

(3) $\qquad \gamma_a(\frac{1}{3}) = \gamma_a(\frac{2}{3}) = a \ , \ \gamma_a(0) = -2a \ .$

The set F_+ is obtained as

$$F_+ = \lim_{r \to 0} E_{r,0} \ .$$

Therefore the homeomorphism between F_+ and $M_{1/2}$ that we shall construct appears as a realization of Yoccoz' correspondance.

4. The tools.

We shall use the theory of quasi-conformal mappings, in particular the theorem of integrability by Morrey-Ahlfors-Bers, and the theory of polynomial-like mappings, see [DH 2] Chapter I.

Quasi-conformal mappings.

Let U and V be two open sets in $\mathbf{C}$. A *quasi-conformal* homeomorphism

$$\varphi: U \to V$$

is a homeomorphism, which locally is in the Sobolev space H^1 . Hence it is possible for almost every $x \in U$ to define an $\mathbf{R}$-linear tangent map $T_x\varphi$ and an ellipse

$$E_x = (T_x\varphi)^{-1}(S^1) \ ;$$

furthermore we request that there exists a constant Λ , such that for almost every x the ratio of the axes of E_x is bounded by Λ. The smallest possible Λ is called the *dilatation ratio* of φ . A *quasi-regular* mapping is a mapping of the form

$$\psi = h \circ \varphi \ ,$$

where φ is quasi-conformal and h is holomorphic, but h may have critical points.

The theorem of integrability can be stated as follows :

THEOREM OF INTEGRABILITY. *Let U be an open set in $\mathbf{C}$ and let $(E_x)_{x\in U}$ be a measurable field of ellipses with the ratio of the axes bounded. Then there exists an open set V in $\mathbf{C}$ and a quasi-conformal homeomorphism*

$$\varphi: U \to V$$

such that

$$(T_x\varphi)^{-1}(S^1) = \rho(x)E_x \text{ for almost every } x \ ,$$

where $\rho(x) \in \mathbf{R}_+$.

If $U = \mathbf{C}$ then $V = \mathbf{C}$. If U is simply connected and bounded, then we can choose $V = D$.

An *almost complex structure* on U is given by a new structure of $\mathbf{R}^2$ as a $\mathbf{C}$-vector space for every $x \in U$ or — what is the same — by an ellipse E_x defined up to a real positive factor. The standard structure σ_0 is defined by the circles. An almost complex structure σ is quasi-conformally equivalent to σ_0 , if it is defined by a measurable field of ellipses with bounded dilatation ratio. To *integrate* the almost complex structure σ therefore means to find a quasi-conformal mapping φ such that $(T_x\varphi)^{-1}(S^1) = \rho(x)\ E_x$ for almost every x .

The theorem of integrability can as well be stated as follows :

THEOREM OF INTEGRABILITY. *Let U be an open set in* C *and*

$$\mu = u(z)\ \frac{d\bar{z}}{dz}$$

a measurable Beltrami form on U with

$$\|\mu\| \ := sup|u(z)| < 1 .$$

Then there exists an open set V in C *and a quasi-conformal homeomorphism $\varphi\colon U \to V$ such that*

$$\bar{\partial}\varphi = \mu\ \partial\varphi$$

where $\partial\varphi = (\partial\varphi/\ \partial z)dz$ and $\bar{\partial}\varphi = (\partial\varphi/\ \partial\bar{z})d\bar{z}$.

We shall also use the following result by Ahlfors-Bers , concerning the dependence of parameters :

Let U be an open set in C *isomorphic to D , let (μ_n) be a sequence of measurable Beltrami forms on U ,*

$$\mu_n = u_n(z)\ \frac{d\bar{z}}{dz} ,$$

and let μ be another measurable Beltrami form on U ,

$$\mu = u(z)\ \frac{d\bar{z}}{dz} .$$

Suppose $sup_n\|\mu_n\| < 1$, (i.e. the almost complex structures σ_n defined by the μ_n have a dilatation ratio bounded by a constant Λ independent of n), and that the μ_n tends to μ in the following weak sense : For every continuous function $h\colon U \to \mathbf{R}$ with compact support

$$\int_U h\ u_n \to \int_U h\ u .$$

Let $\varphi: U \to D$ be a quasi-conformal homeomorphism such that

$$\mu = \frac{\bar{\partial}\varphi}{\partial\varphi} .$$

Then there exists a sequence (φ_n) of quasi-conformal homeomorphisms.

$$\varphi_n: U \to D ,$$

tending to φ uniformly on U with

$$\mu_n = \frac{\bar{\partial}\varphi_n}{\partial\varphi_n} .$$

Polynomial-like mappings.

Furthermore we shall use the notion of a *polynomial-like mapping* , that is a proper holomorphic mapping

$$f: U' \to U ,$$

where U and U' are open sets isomorphic to D , with U' relatively compact in U . We denote by $K(f)$ the set of $z \in U'$ such that $f^n(z)$ is defined and belongs to U' for all $n \in \mathbf{N}$.

Let

$$f: U' \to U \qquad \text{and} \qquad g: V' \to V$$

be two polynomial-like mappings with $K(f)$ and $K(g)$ connected. A *holomorphic* equivalence (respectively a *quasi-conformal*) between f and g is an analytic isomorphism (respectively a quasi-conformal homeomorphism)

$$\varphi: U_1 \to V_1 ,$$

where U_1 and V_1 are neighborhoods of $K(f)$ and $K(g)$, satisfying

$$g \circ \varphi = \varphi \circ f \quad \text{on } U_1' = f^{-1}(U_1) .$$

A *hybrid* equivalence is a quasi-conformal equivalence φ with $\bar{\partial}\varphi = 0$ almost everywhere on $K(f)$. These equivalences are denoted by $\sim_h$, $\sim_{qc}$, $\sim_{hb}$ respectively.

We define the *degree* of a polynomial-like mapping by counting the inverse images of a point with their multiplicity.

THE STRAIGHTENING THEOREM. *Let f be a polynomial-like mapping of degree d with $K(f)$ connected. Then f is hybrid equivalent to a polynomial P of degree d.*

If $d = 2$, then we can choose P of the form P_c, and c is then uniquely determined. Furthermore if f depends continuously on a parameter λ, then c depends continuously on λ.

If $d = 3$, then we can choose P of the form $Q_{a,b}$. There are 4 choices for the pair (a, b). If f depends continuously on λ, then it is not in general possible to choose (a, b) to depend continuously on λ. Similarly the class formed by the 4 possible pairs (a, b) does not in general depend continuously on λ.

In the situations where we shall use the straightening theorem we also need the following result:
If P and Q are two polynomials with $K(P)$ and $K(Q)$ connected and if P and Q are hybrid equivalent, then they are conjugate by an affine map.

5. The dynamics of P_c for $c \in M_{1/2}$.

The polynomial $P_c(z) = z^2 + c$ has one critical point $\omega = 0$. For $c \in M$ we denote by β_c the fixed point satisfying

$$\beta_c = \gamma_c(0)$$

and by β'_c the other preimage of β_c

$$\beta'_c = \gamma_c(\frac{1}{2}) = -\beta_c \ .$$

For c inside the big cardioid of M the other fixed point α_c is attractive, and the Julia set, ∂K_c , is a Jordan curve. But for $c \in M_t^*$ with $t \in \mathbf{Q}/\mathbf{Z} - \{0\}$ the other fixed point α_c is repulsive and a pinching point in the Julia set. It has external arguments depending on t .

For $t = \frac{1}{2}$ we have

$$\alpha_c = \gamma_c(\frac{1}{3}) = \gamma_c(\frac{2}{3}) \ .$$

For $t = \frac{1}{3}$ we have

$$\alpha_c = \gamma_c(\frac{1}{7}) = \gamma_c(\frac{2}{7}) = \gamma_c(\frac{4}{7}) \ .$$

The two mappings, Φ_A and Φ_B , we are going to construct, will associate to a quadratic polynomial P_c with $c \in M_{1/2}$ in the first case a quadratic polynomial $P_{c'}$ with $c' \in M_{1/3}$ and in the second case a cubic polynomial Q_a with $a \in F_+$.

We start with some preliminary constructions in the dynamical plane for P_c . We shall make use of these constructions in both cases.

We denote by α'_c the other preimage of α_c

$$\alpha'_c = \gamma_c(\frac{1}{6}) = \gamma_c(\frac{5}{6}) = -\alpha_c$$

We have

$$P_c(\alpha'_c) = P_c(\alpha_c) = \alpha_c$$
$$P_c(\beta'_c) = P_c(\beta_c) = \beta_c.$$

Choose $\eta > 0$ and set

$$W_i = W_i^c = \{ \ z \in \mathbf{C} \mid G_c(z) \leq \frac{\eta}{2^i} \ \} \ ,$$

$$W = W_0$$

and

$$A_i = W_i - \overset{\circ}{W}_{i+1}.$$

The A_i are compact annuli, $P_c(A_{i+1}) = A_i$ and $\bigcup_i A_i = W - K_c$.

The lines

$$\ell_c = \mathcal{R}_c(\frac{1}{3}) \cup \{\alpha_c\} \cup \mathcal{R}_c(\frac{2}{3})$$

and

$$\ell'_c = \mathcal{R}_c(\frac{1}{6}) \cup \{\alpha'_c\} \cup \mathcal{R}_c(\frac{5}{6})$$

decompose W into 3 compact subsets V , V' and V'' with $\omega \in V$, $\beta_c \in V'$ and $\beta'_c \in V''$. See figure 5.1 .

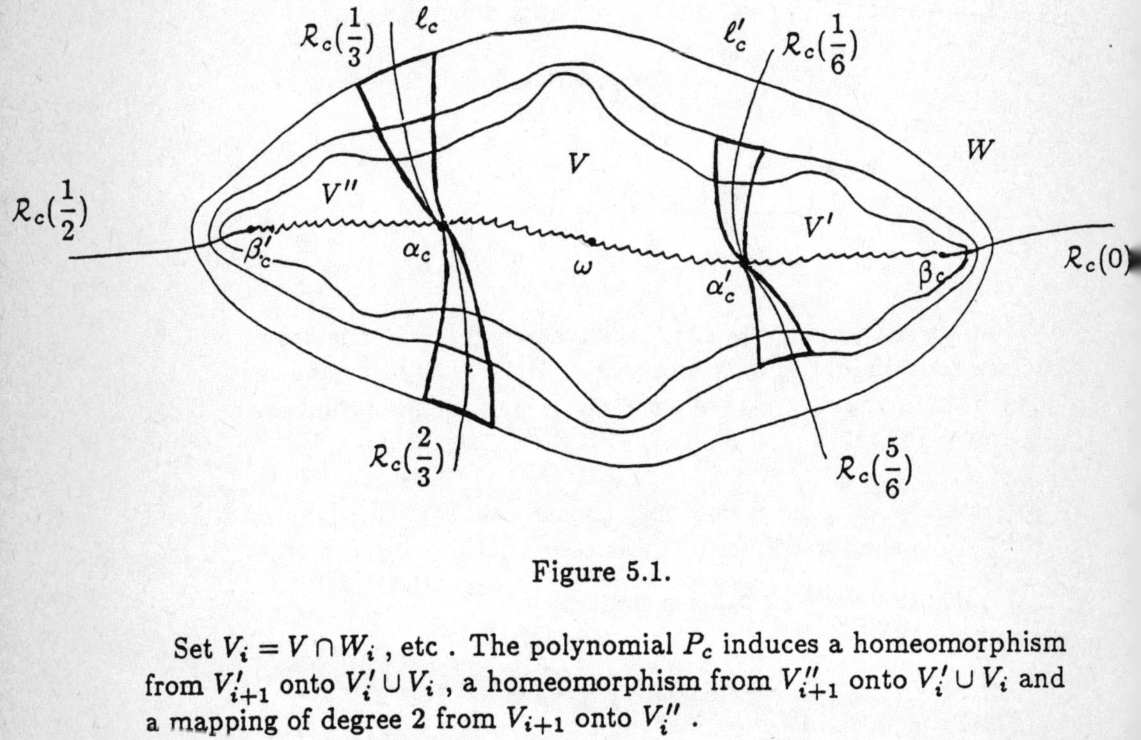

Figure 5.1.

Set $V_i = V \cap W_i$, etc . The polynomial P_c induces a homeomorphism from V'_{i+1} onto $V'_i \cup V_i$, a homeomorphism from V''_{i+1} onto $V'_i \cup V_i$ and a mapping of degree 2 from V_{i+1} onto V''_i .

Fix a $q \in]0, \frac{1}{6\eta}[$ (in section 13 we shall vary both η and q). For $\theta \in \mathbf{T}$ we define the *sector* $S(\theta) = S^c_q(\theta)$, centered at $\mathcal{R}_c(\theta)$ and with slope q as

$$S(\theta) = \{z \in W - K_c \mid \varphi_c(z) = e^{s+2\pi i t} , \mid t - \theta \mid \leq q\, s\} .$$

Set

$$S_i(\theta) = S(\theta) \cap W_i .$$

The polynomial P_c induces a homeomorphism from $S_{i+1}(\theta)$ onto $S_i(2\theta)$. In particular P_c^2 induces a homeomorphism

$$P_c^2 \colon S_{i+2}(\frac{k}{3}) \rightarrow S_i(\frac{k}{3}) \qquad \text{for } k = 1,2\ .$$

Set

$$\widetilde{P}_c = -P_c\ ,$$

then $\widetilde{P}_c$ induces a homeomorphism

$$\widetilde{P}_c \colon S_{i+1}(\frac{k}{6}) \rightarrow S_i(1 - \frac{k}{6}) \qquad \text{for } k = 1,5$$

and therefore a homeomorphism

$$\widetilde{P}_c^2 \colon S_{i+2}(\frac{k}{6}) \rightarrow S_i(\frac{k}{6}) \qquad \text{for } k = 1,5\ .$$

We set

$$S = S(\frac{1}{3}) \cup S(\frac{2}{3})$$

and

$$S' = S_1(\frac{1}{6}) \cup S_1(\frac{5}{6})\ .$$

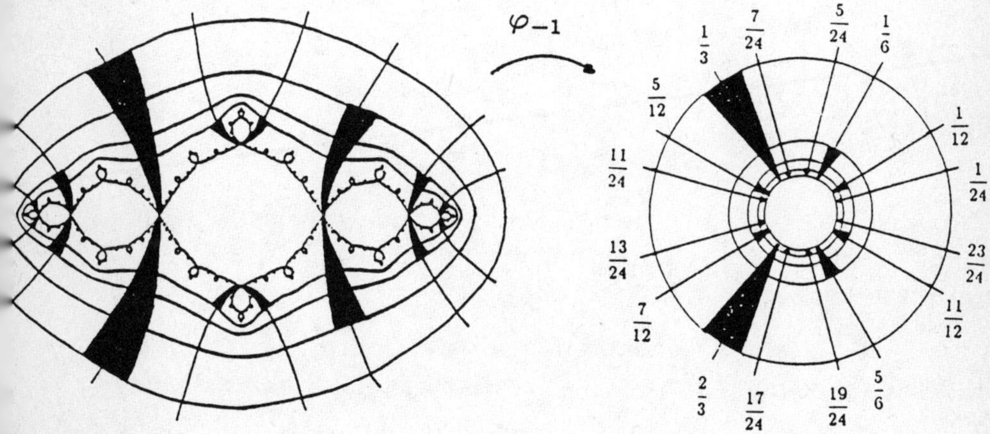

Figure 5.2.

The map P_c induces a homeomorphism

$$P_c\colon S' \to S$$

which conjugates

$$\tilde{P}_c\colon S' \cap W_2 \to S' \qquad \text{to} \qquad P_c\colon S \cap W_1 \to S\ .$$

We have $\bar{S} = S \cup \{\alpha_c\}$ and $\bar{S}' = S' \cup \{\alpha'_c\}$. The inverse images by P_c^{-n} of the fixed point α_c are

$$P_c^{-n}(\alpha_c) = \{\ \gamma_c(\frac{m}{3\ 2^n})\ |\ m < 3\ 2^n \text{ and } (3,m) = 1\}\ .$$

The inverse image by P_c^{-n} of S is

$$P_c^{-n}(S) = \bigcup_m S_n(\frac{m}{3\ 2^n}) \quad \text{where } m < 3\ 2^n \text{ and } (3,m) = 1\ ;$$

these sectors are disjoint, due to the hypothesis $q\ \eta < \frac{1}{6}$.
Figure 5.2 shows $P_c^{-n}(S)$ for $n = 0,1,2,3$, where c is chosen as the center of $\Omega_{1/2}$, i.e. $c = -1$.

6. The construction of a quasi-regular mapping of degree 3.

We start here the proof of theorem B.

From a polynomial P_c with $c \in M_{1/2}$ and with W , $S(\theta)$ etc. as in section 5 we form the space $X = X^c$ obtained from $V \cup V'$ by identifying

$$\mathcal{R}_c(\frac{1}{3}) \cap W \text{ with } \mathcal{R}_c(\frac{2}{3}) \cap W$$

by equipotential identification, that is

$$\varphi_c^{-1}(\rho\, e^{2\pi i/3}) \quad \leftrightarrow \quad \varphi_c^{-1}(\rho\, e^{4\pi i/3}) \ .$$

Let

$$\pi : V \cup V' \to X$$

denote the projection . Set

$$\begin{aligned} &\mathcal{R}_X(\theta) = \pi(\mathcal{R}_c(\theta) \cap W) \ , \\ &S_X = \pi(S \cap V) \ , \\ &\ell_X = \pi(\ell_c) \text{ and } \ell'_X = \pi(\ell'_c) \ . \end{aligned}$$

The space X can also be viewed as the quotient of $V \cup V' \cup S$ by the equivalence relation identifying $\varphi_c^{-1}(\rho\, e^{2\pi i(1/3\, +\, t)})$ to $\varphi_c^{-1}(\rho\, e^{2\pi i(2/3\, +\, t)})$. Therefore there exists on X a unique structure as a Riemann surface with $\mathbf{R}$-analytic boundary such that π is analytic, except at α_c .

We define

$$f : \pi(V_2 \cup V_1') - (\ell'_X - \alpha'_c) \quad \to \quad X$$

as the *first return map* of P_c that is

$$\pi(z) \ \mapsto \begin{cases} \pi(P_c^2(z)) \ , & \text{for } z \in V_2 \\ \pi(P_c(z)) \ , & \text{for } z \in V_1' \ . \end{cases}$$

We have

$$f(\alpha'_c) = \pi(\alpha_c) \ .$$

The mapping f is analytic on $\ell_X \cap \pi(W_2)$ and at $\pi(\alpha_c)$, but f cannot be extended to ℓ'_X , which is a line of discontinuities.

For simplicity we shall write z instead of $\pi(z)$ for $z \in V \cup V' - \ell_c$.

This ends step (1) in the construction. In order to finish the topological surgery, leaving a better ciatrix, we shall continue with step (2) .

Consider the polynomial $P_0(z) = z^2$ and let $S^0(\theta)$ denote the sector centered at $\mathcal{R}_0(\theta)$ and with slope q. Let γ be a curve which is the graph of a C^1-map

$$\gamma\colon \mathbf{T} \to [e^{\eta/4}, e^{\eta/2}]$$

satisfying

$$\gamma(\theta) = \begin{cases} e^{\eta/2}, & \text{for } \theta \in [\frac{5}{6} + q\,\frac{\eta}{2}, 1] \cup [0, \frac{1}{6} - q\,\frac{\eta}{2}] \\ e^{\eta/4}, & \text{for } \theta \in [\frac{1}{6} + q\,\frac{\eta}{4}, \frac{5}{6} - q\,\frac{\eta}{4}]. \end{cases}$$

Then γ is a curve joining the two circles of radius $e^{\eta/2}$ and $e^{\eta/4}$ in the sectors $S^0(\frac{1}{6})$ and $S^0(\frac{5}{6})$, see figure 6.1.

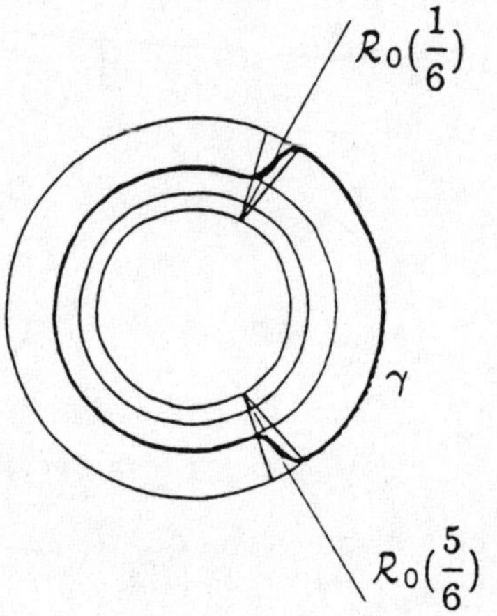

Figure 6.1.

Let $\widetilde{W}_1 \subset W$ be the compact set bounded by $\varphi_c^{-1}(\gamma)$. Then

$$\widetilde{W}_1 \cap (V' - S') = V_1' - S'$$
$$\widetilde{W}_1 \cap (V - S') = V_2 - S'$$
$$W_2 \cap S' \subset \widetilde{W}_1 \cap S' \subset W_1 \cap S'.$$

Set

$$X_1 = \pi(\widetilde{W}_1 \cap (V \cup V')),$$

$$S_X' = \pi(\widetilde{W}_1 \cap S'),$$

$$T_0 = S_X - (f^2)^{-1}(S_X),$$

$$T_i = (f^2)^{-1}(T_{i-1}) \cap S_X.$$

The restriction of f^2 to T_{i-1} is an analytic isomorphism from T_{i-1} onto T_i . Set

$$T'_0 = S'_X - (\widetilde{P}_c^2)^{-1}(S'_X) \qquad \text{where} \qquad \widetilde{P}_c = -P_c \, .$$

See figure 6.2.

The set T'_0 has two connected components $T'_0(\frac{1}{6})$ and $T'_0(\frac{5}{6})$, diffeomorphic to rectangles and contained in $S(\frac{1}{6})$ and $S(\frac{5}{6})$ respectively. For $\theta = \frac{1}{6}, \frac{5}{6}$ set

$$T'_i(\theta) = (\widetilde{P}_c^2)^{-1}(T'_{i-1}(\theta)) \cap S(\theta) \, .$$

The restriction of the mapping $\widetilde{P}_c^2$ to $T'_{i-1}(\theta)$ is an analytic isomorphism from $T'_{i-1}(\theta)$ onto $T'_i(\theta)$.

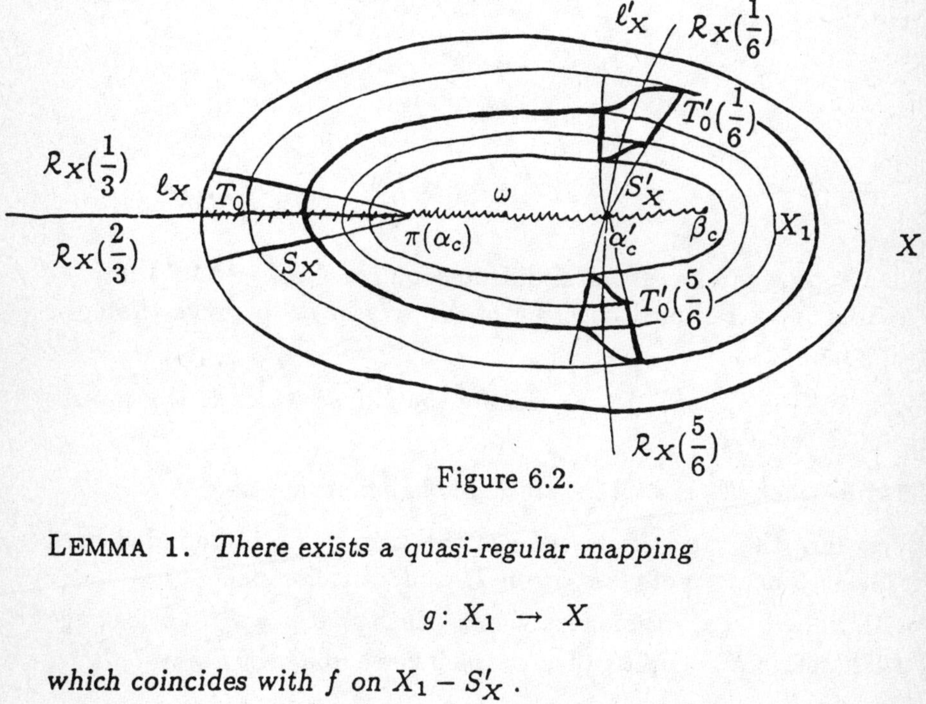

Figure 6.2.

LEMMA 1. *There exists a quasi-regular mapping*

$$g\colon X_1 \to X$$

which coincides with f *on* $X_1 - S'_X$.

PROOF:
Choose for $\theta = \frac{1}{6}, \frac{5}{6}$ a diffeomorphism

$$g_{0,\theta}\colon T'_0(\theta) \to T_0(X)$$

such that

(1) $\qquad g_{0,\theta}$ is tangent to f on $T'_0(\theta) \cap \partial S'$

(2) $f^2 \circ g_{0,\theta} = g_{0,\theta} \circ \widetilde{P}_c^2$ on $I'(\theta) = T_0'(\theta) \cap T_1'(\theta)$
and the two mappings have the same linear tangent map at every point of $I'(\theta)$.

Then define

$$g_{i,\theta} \colon T_i'(\theta) \rightarrow T_i$$

inductively by

$$f^2 \circ g_{i,\theta} = g_{i-1,\theta} \circ \widetilde{P}_c^2 \ .$$

Finally define

$$g \colon X_1 \rightarrow X$$

by

$$g = \begin{cases} f & \text{on } X_1 - S_X' \\ g_{i,\theta} & \text{on } T_i'(\theta) \end{cases}$$

$$\text{and } g(\alpha_c') = \pi(\alpha_c) \ .$$

The mapping g induces a local diffeomorphism of $X_1 - \{\omega, \alpha_c'\}$ into X, and makes X_1 a ramified covering of X. We have to prove that g is quasi-regular.

For $x \in X_1 - \{\omega, \alpha_c'\}$, let E_x denote the ellipse which is the inverse image of S^1 by $T_x g$.

If $x \notin S'$, then E_x is a circle, since g is holomorphic at x.

For $x \in T_0'$ the ratio of the axes of the ellipses E_x is bounded by a constant Λ , since g is of class C^1 on T_0' and T_0' is compact.

For all $x \in S' - \{\alpha_c'\}$ there exists an n , such that $y = \widetilde{P}_c^{2n}(x)$ belongs to T_0'. Therefore E_x is up to a factor the inverse image of E_y by $T_x \widetilde{P}_c^{2n}$, since

$$P_c^{2n} \circ g = g_0 \circ \widetilde{P}_c^{2n} \qquad \text{in a neighborhood of } x \ .$$

Furthermore $\widetilde{P}_c^{2n}$ is holomorphic, so E_x and E_y has the same ratio of the axes.

As a result the ratio of the axes for E_x is bounded by Λ for all $x \in S' - \{\alpha_c'\}$, and therefore for all $x \in X_1 - \{\omega, \alpha_c'\}$. Since g is of class C^1 except at α_c', this suffices to prove that g is quasi-regular. ∎

REMARK.

(1) The mapping g is a ramified covering of degree 3 with critical points ω and α_c' .

(2) We have $g(S_X') = S_X$ and $g(S_X \cap X_1) = S_X$, thus S_X' is a set where orbits of g pass at most once.

(3) We can smoothen f in the <u>same</u> way for all $c \in M_{1/2}$ by making the construction for the polynomial $P_0(z) = z^2$ and then obtain the mapping $g_{0,\theta}$ for any $c \in M_{1/2}$ through conjugation by φ_c .

7. The construction of a polynomial of degree 3.

After having finished the topological surgery in section 6 we shall complete it into holomorphic surgery in this section.

First we change the complex structure.

LEMMA 2. *Let g be given as in lemma 1 .*
There exsists an almost complex structure σ on X, quasi-conformally equivalent to the standard structure σ_0 , such that $g^*\sigma = \sigma$ and σ coincides with σ_0 on $\pi(K_c \cap (V \cup V'))$.

PROOF: Let Λ denote the dilatation ratio of g. For $x \in S'_X$ let E_x denote the ellipse

$$E_x = (T_x g)^{-1}(S^1) .$$

For $x \in X$, if there exists an i such that $g^i(x)$ is defined and belongs to S'_X, then such an i is unique (see remark (2) in section 6) , since $g^{i+1}(x) \in S_X$ and $g^n(x) \in S_X$ for all $n > i$ for which $g^n(x)$ is defined.

When there exists such an i , we denote by E_x the ellipse

$$E_x = (T_x g^i)^{-1}(E_{g^i(x)}) .$$

The mapping g^i is holomorphic in a neighborhood of x , since $g^j(x) \in X - S'_X$ for $j < i$; it follows, that the ratio of the axes of E_x is the same as that of $E_{g^i(x)}$, therefore bounded by Λ .

When there is no i such that $g^i(x) \in S'_X$ then we set $E_x = S^1$.

The almost complex structure σ defined by the family of ellipses $(E_x)_{x \in X}$ satisfies the requirement. ▮

Let σ be an almost complex structure as in lemma 2 , and let

$$\varphi : \mathring{X} \to U$$

be a quasi-conformal homeomorphism integrating σ . Set

$$U' = \varphi(\mathring{X}_1)$$

and

$$h = \varphi \circ g \circ \varphi^{-1} : U' \to U .$$

Since $g^*\sigma = \sigma$, the mapping h is holomorphic, and h is polynomial-like of degree 3 (see remark (1) in section 6). We have completed step (3).

By applying the straightening theorem we obtain a polynomial Q of degree 3 and a hybrid equivalence ψ from h to Q .We can choose Q of the form

$$Q_{a,b}(z) = z^3 - 3a^2z + b \ .$$

We have 4 choices for the pair (a, b) .

We started from a quadratic polynomial P_c , with critical point $\omega = 0$ and fixed points α_c , β_c .The critical points for g are ω and α'_c , and the fixed points with a single access are β_c and $\pi(\alpha_c)$. We determine one of the 4 choices by imposing that

$$\psi \circ \varphi(\alpha'_c) = a$$

and

$$\psi \circ \varphi(\pi(\alpha_c)) = \beta_+(a, b) = \gamma_{a,b}(0) \ .$$

Hence

$$\psi \circ \varphi(\omega) = -a$$

and

$$\psi \circ \varphi(\beta_c) = \beta_-(a, b) = \gamma_{a,b}(\frac{1}{2}) \ .$$

The condition $\psi \circ \varphi(\alpha'_c) = a$ force the polynomial $Q_{a,b}$ to be of the form

$$Q_a(z) = z^3 - 3a^2z + 2a^3 - 2a \ ,$$

the condition $\psi \circ \varphi(\pi(\alpha_c)) = \gamma_a(0)$ force a to be in F_+ .

Furthermore we observe that the parameter a does not depend on the choices made , i.e. the choice of the boundary defining X_1 , the choice of g on S'_X and the choice of the quasi-conformal homeomorphism. Actually if we had made other choices and obtained $a' \in F_+$, then there exists a hybrid equivalence ψ' from Q_a to $Q_{a'}$. Therefore Q_a and $Q_{a'}$ are affine conjugated, and since $\{a, \ a'\} \subset F_+$ we have $a = a'$.

This ends the construction in the dynamical plane. The uniquely determined polynomial Q_a with $a \in F_+$ is obtained from P_c through holomorphic surgery.

8. The definition and continuity of the mapping Φ_B.

The constructions given in sections 6 and 7 define a mapping

$$P_c \mapsto Q_a$$

corresponding to a mapping

$$\Phi_B \colon M_{1/2} \to F_+ \quad \text{where } \Phi_B(c) = a\,.$$

To finish the proof of theorem B we still have to prove that Φ_B is continuous and bijective. In this section we shall prove continuity.

The continuity is not obvious, since σ and φ do not depend continuously on c , and probably h even less. (We are using the notation from section 7.) Furthermore the construction uses the straightening theorem in degree 3 , a case which is in general not continuous with respect to parameters. We shall prove the continuity using a method inspired by Chapter II in [DH 2].

Since $M_{1/2}$ and F_+ are compact sets, it suffices to prove that Φ_B is a closed graph . Let (c_n) be a sequence of points in $M_{1/2}$ converging to a limit c . Let $Q_n = Q_{a_n}$ be the polynomials constructed from $P_n = P_{c_n}$, and let $Q = Q_a$ be the polynomial constructed from P_c . Suppose a_n has a limit $\tilde{a}$. We have to prove that $\tilde{a} = a$, in other words that $\tilde{Q} := Q_{\tilde{a}} = Q$.

We treat separately the case $c \in \overset{\circ}{M}$ and $c \in \partial M$. The decomposition of $\mathbf{C}$ into $\overset{\circ}{M} \cup (\mathbf{C} - M)$ and ∂M is the first Mañé-Sad-Sullivan decomposition for the family P_c . The subset $\overset{\circ}{M} \cup (\mathbf{C} - M)$ is the set of c for which P_c is structurally stable on a neighborhood of the Julia set. The case $c \in \overset{\circ}{M}$ rely on the theory of [MSS] , and the proof in this case is analogous to the one of proposition 12, Chapter II.5 in [DH 2] .

The case $c \in \partial M$ will result from the two following lemmas. The first one is analogous to proposition 7, Chapter I.6 in [DH 2] :

LEMMA 3. *Let a and a' be in $\mathbf{C}$ with $a \in \partial F$. Suppose there exists a quasi-conformal equivalence ψ from Q_a to $Q_{a'}$ such that*

$$\psi(a) = a' \quad \text{and} \quad \psi(\beta_+(a)) = \beta_+(a')\,.$$

If $\beta_+(a) = \beta_-(a)$, then we also assume that $\psi(\mathcal{R}_a(0))$ ends at $\beta_+(a') = \beta_-(a')$ at the same acces as $\mathcal{R}_{a'}(0)$. Then $a = a'$.

The proof is analogous to the proof in Chapter I.6 in [DH 2] .

LEMMA 4. *The polynomials Q and $\widetilde{Q}$ are quasi-conformally equivalent.*

PROOF: Using the notation from section 7 we denote by $g_n, \varphi_n, h_n \ldots$ the mappings $g, \varphi, h \ldots$ defined starting from P_n. We may assume (see remark (3) in section 6) that g_n is obtained by smoothing f_n in the <u>same</u> way for all n.

The mappings φ_n are quasi-conformal with a bounded dilatation ratio. Therefore they form an equicontinuous family and we may assume , by restricting to a subfamily, that the sequence φ_n tends towards a quasi-conformal homeomorphism $\widetilde{\varphi}$.

We do not necessarily have $\bar{\partial}\widetilde{\varphi} = 0$ on $K(g)$, since $K(g)$ is not in general the limit of $K(g_n)$.

But

$$h_n = \varphi_n \circ g_n \circ \varphi_n^{-1} \qquad \text{tends towards} \qquad \widetilde{h} = \widetilde{\varphi} \circ g \circ \widetilde{\varphi}^{-1},$$

and $\widetilde{h}$ is quasi-conformally equvalent to h. Using the lemma p.313, II.7 in [DH 2] , we get that $\widetilde{h}$ is quasi-conformally equvialent to $\widetilde{Q}$, hence

$$Q \sim_{hb} h \sim_{qc} \widetilde{h} \sim_{qc} \widetilde{Q} \text{ and } Q \sim_{qc} \widetilde{Q}.$$

∎

From the details of the proof of lemma 4 , we have obtained :

REMARK.
We can find a quasi-conformally equivalence satisfying the conditions in lemma 3 .

A polynomial, for which every critical point is strictly preperiodic, is called a Misiurewicz polynomial. We know that the Misiurewicz points in M , i.e. the c such that P_c is Misiurewicz, form a dense set in ∂M. For $a \in F - \{0\}$ the critical point a is always strictly preperiodic, so a is a Misiurewicz point in F, if the critical point $-a$ is also strictly preperiodic. The Misiurewicz points for F belong to ∂F.

If the c_n are Misiurewicz points in M , then the a_n are Misiurewicz points in F , therefore $\lim a_n = \widetilde{a} \in \partial F$ and it follows from lemma 4 and the remark that the assumptions in lemma 3 are fulfilled, hence $a = \widetilde{a}$.

In the general case we can find a sequence $(c_n^\star)$ of Misiurewicz points in M tending to c , and such that the $a_n^\star = \Phi_B(c_n^\star)$ has a limit $\widetilde{a}^\star$. Therefore we have $\widetilde{a}^\star \in \partial F$ and from what we have just prove that $a = \widetilde{a}^\star$, so $a \in \partial F$. Applying again lemma 3 , lemma 4 and the remark we obtain $a = \widetilde{a}$.

This ends the proof of the continuity of Φ_B.

9. Bijectivity of the mapping Φ_B.

To prove the bijectivity of Φ_B we will construct a mapping

$$\Psi_B \colon F_+ \to M_{1/2} ,$$

and then show that Ψ_B is the inverse of Φ_B .

So starting from a cubic polynomial Q_a with $a \in F_+$ we shall construct a quadratic polynomial P_c with $c \in M_{1/2}$. The construction develops in steps as before, parallel but different,

$$Q_a \curvearrowright f \curvearrowright g \curvearrowright h \curvearrowright P_c .$$

In section 6 we defined the region X^c by removing the part of W^c to the left of ℓ_c . Since we are making the opposite construction here, we shall define the region X^a by adding a copy of part of W^a similar to what we removed. In part III we shall see that the first step in the construction of the mapping Φ_A is also to define a region by adding a copy of part of W .

Start with a polynomial Q_a with $a \in F_+$. The polynomial Q_a has two critical points a and $-a$ and

$$a = \gamma_a(\frac{1}{3}) = \gamma_a(\frac{2}{3}) .$$

The polynomial has 3 fixed points of which

$$\beta_+ = \beta_+(a) = \gamma_a(0)$$

and

$$\beta_- = \beta_-(a) = \gamma_a(\frac{1}{2}) .$$

We have

$$Q_a(a) = -2a = \beta_+ .$$

Let the potential η be fixed and let

$$W_i = W_i{}^a = \{ z \in \mathbf{C} \mid G_a(z) \leq \frac{\eta}{3^i} \}$$

and $W = W_0$. The line

$$\ell'_a = \mathcal{R}_a(\frac{1}{3}) \cup \{a\} \cup \mathcal{R}_a(\frac{2}{3})$$

decomposes W into 2 compact sets $\widehat{V}$ and V' , see figure 9.1.

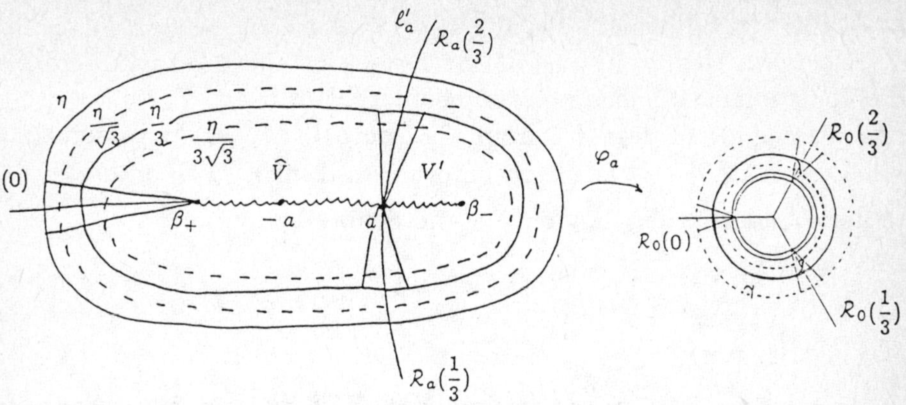

Figure 9.1.

Fix a $q \in]0, \frac{1}{2\eta}[$. For $\theta \in \mathbf{T}$ we define the sector $S(\theta) = S_q^a(\theta)$, centered at $\mathcal{R}_a(\theta)$ and with slope q as

$$S(\theta) = \{ z \in W - K_a \mid \varphi_a(z) = e^{s+2\pi i t} , \mid t - \theta \mid \leq q\, s \} .$$

Set

$$S_i(\theta) = S(\theta) \cap W_i .$$

The polynomial Q_a induces a homeomorphism from $S_{i+1}(\theta)$ onto $S_i(3\,\theta)$. In particular Q_a induces a homeomorphism

$$Q_a \colon S_{i+1}(\frac{k}{3}) \to S_i(0) \qquad \text{for } k = 0,1,2 .$$

Let $\widetilde{Q}_a$ denote the homeomorphism

$$\widetilde{Q}_a \colon S_{i+1}(\frac{k}{3}) \to S_i(1-\frac{k}{3}) \qquad \text{for } k = 1,2$$

defined by mapping $z \in S_{i+1}(\frac{k}{3})$ to the preimage of $Q_a^2(z)$ belonging to $S_i(1-\frac{k}{3})$. Then $\widetilde{Q}_a^2$ induces a homeomorphism

$$\widetilde{Q}_a^2 \colon S_{i+2}(\frac{k}{3}) \to S_i(\frac{k}{3}) \qquad \text{for } k = 1,2 .$$

Set

$$S' = S_1(\frac{1}{3}) \cup S_1(\frac{2}{3}) .$$

Define V by cutting $\widehat{V}$ along $\mathcal{R}_a(0)$ so that V contains 2 copies of $\mathcal{R}_a(0) \cap W$. Denote by ℓ_a the line formed by these 2 copies and β_+. The polynomial Q_a induces a homeomorphism from V'_{i+1} onto $V'_i \cup V_i$ and a mapping of degree 2 from V_{i+1} onto $V'_i \cup V_i$. In particular we shall use that $Q|_{V'_1}{}^{-1}$ is a homeomorphism from $V'_0 \cup V_0$ onto V'_1.

Let V'' be another copy of V', and denote by

$$\tau: V' \to V''$$

the identity mapping.

We construct a Riemann surface $X = X^a$ with $\mathbf{R}$-analytic boundary by gluing ℓ_a of $V_{1/2} \cup V'_{1/2}$ to $\tau(\ell'_a)$ of V''_1. For a point $x \in \ell'_a$ of potential s, the point $\tau(x)$ is identified with one of the two points in ℓ_a of potential $\sqrt{3}s$. In figure 9.2 is shown which one.

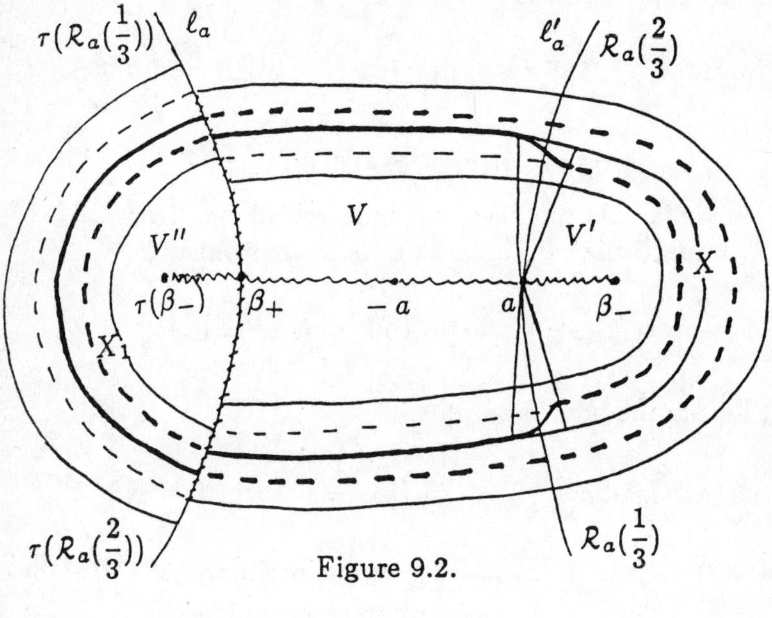

Figure 9.2.

Set

$$\widehat{X}_1 = V_1 \cup V'_{3/2} \cup V''_{3/2} ,$$

and define

$$f: \widehat{X}_1 - (\ell'_a - \{a\}) \to X$$

by

$$f = \begin{cases} \tau \circ Q_a|_{V'_1}{}^{-1} \circ Q_a & \text{on } V_1 \\ Q_a & \text{on } V'_{3/2} \\ Q_a \circ \tau^{-1} & \text{on } V''_{3/2} . \end{cases}$$

The mapping f is holomorphic along ℓ_a, but discontinuous along ℓ'_a. Notice that if we remove $V''_{3/2}$ then the first return map of f is Q_a. This ends step (1) in the construction. As in section 6 we continue with step (2).

Let γ be a curve which is the graph of a C^1-map

$$\gamma\colon \mathbf{T} \to [e^{\eta/(3\sqrt{3})}, e^{\eta/3}],$$

see figure 9.1, joining the two circles of radius $e^{\eta/(3\sqrt{3})}$ and $e^{\eta/3}$ in the sectors $S_1(\frac{1}{3})$ and $S_1(\frac{2}{3})$. Let $\widetilde{W}_1 \subset W$ be the compact set bounded by $\varphi_a^{-1}(\gamma)$, and let $X_1 \subset X$ be the set which coincides with $\widehat{X}_1$ outside S' and with $S'_X = S' \cap \widetilde{W}_1$ in S'. From $\tau(S' \cap V')$ and $S(0)$ we define the sectors $S_X(0)$ and $S_X(1)$ in the obvious way. For $\theta = 0, 1$ the mapping f induces a homeomorphism

$$f\colon S_X(\theta) \cap (V_{3/2} \cup V''_2) \to S_X(1-\theta).$$

For $\theta = 0, 1$ set

$$T_0(\theta) = S_X(\theta) - (f^2)^{-1}(S_X(\theta))$$

and

$$T_i(\theta) = (f^2)^{-1}(T_{i-1}(\theta)) \cap S_X(\theta).$$

For $\theta = \frac{1}{3}, \frac{2}{3}$ set

$$T'_0(\theta) = S'_X(\theta) - (\widetilde{Q}_a^2)^{-1}(S'_X(\theta))$$

and

$$T'_i(\theta) = (\widetilde{Q}_a^2)^{-1}(T'_{i-1}(\theta)) \cap S'_X(\theta).$$

Choose diffeomorphisms

$$g_{0,1/3}\colon T'_0(\frac{1}{3}) \to T_0(0)$$
$$g_{0,2/3}\colon T'_0(\frac{2}{3}) \to T_0(1)$$

satisfying (1) and (2) as in section 6 with $\widetilde{P}_c$ replaced by $\widetilde{Q}_a$. Then define

$$g_{i,1/3}\colon T'_i(\frac{1}{3}) \to T_i(0)$$
$$g_{i,2/3}\colon T'_i(\frac{2}{3}) \to T_i(1)$$

inductively by

$$f^2 \circ g_{i,\theta} = g_{i-1,\theta} \circ \widetilde{Q}_a^2 .$$

Finally define $g\colon X_1 \to X$ similarly to g in section 6. This mapping g is a quasi-regular mapping. This ends step (2).

The complex stucture is changed as in section 7 to an almost complex structure σ on X , quasi-conformally equivalent to the standard structure σ_0 , such that $g^*\sigma = \sigma$ and σ coincides with σ_0 on $X - S'_X$. The essential point in the proof is that g is holomorphic except at S'_X and that orbits pass S'_X at most once : if $x \in S'_X$ then $g^n(x) \in S_X$ for all n where $g^n(x)$ is defined. Let

$$\varphi\colon \overset{\circ}{X} \to U$$

be a quasi-conformal homeomorphism which integrates σ , and set

$$h = \varphi \circ g \circ \varphi^{-1}\colon U' \to U .$$

The mapping h is a polynomial-like mapping of degree 2 , and it follows from the straightening theorem, that h is hybrid equivalent to a uniquely determined polynomial P_c . We define the map Ψ_B by

$$c = \Psi_B(a) .$$

To show that $\Psi_B \circ \Phi_B = id_{M_{1/2}}$ for all polynomials P_c with $c \in M_{1/2}$, apply first the construction from section 6 and 7 to obtain a polynomial Q_a with $a \in F_+$, then the construction from this section to obtain a polynomial $P_{c'}$. Following the construction we see that P_c and $P_{c'}$ are hybrid conjugated, therefore affine conjugated, hence $c = c'$. In other words

$$\Psi_B \circ \Phi_B = id_{M_{1/2}}$$

and Φ_B is injective.

Since Φ_B is continuous and injective and $M_{1/2}$ is compact, the mapping Φ_B defines a homeomorphism from $M_{1/2}$ onto a compact set F'_+ contained in F_+ . Since $M_{1/2}$ is full (i.e. $\mathbf{C} - M_{1/2}$ is connected) , then $M_{1/2}$ is simply connected in the sense of covering spaces (i.e. all the covering spaces of $M_{1/2}$ are trivial) . Therefore F'_+ is simply connected in the sense of covering spaces, hence full. It is easy to see that Φ_B induces a bijection between the Misiurewicz points of $M_{1/2}$ and those of F_+ . Since the Misiurewicz points of F_+ form a dense set in ∂F_+ , it follows that $F'_+ \supset \partial F_+$, and since F'_+ is full, we have $F'_+ \supset F_+$ and finally $F'_+ = F_+$. In other words Φ_B is surjective.

We could as well have shown, that $\Phi_B \circ \Psi_B = id_{F_+}$ by constructing a hybrid equivalence between Q_a and $Q_{a'}$ where $a' = \Phi_B(\Psi_B(a))$.

This ends the proof of theorem B .

III A VEIN IN M

10. The construction of a quasi-regular mapping of degree 2.

We start here the proof of theorem A.

As in part II we start from a quadratic polynomial P_c with $c \in M_{1/2}$. But this time our goal is to construct a quadratic polynomial $P_{c'}$ with $c' \in M_{1/3}$.

Therefore we first recall some features of the polynomials $P_{\tilde{c}}$ with $\tilde{c} \in M_{1/3}$. The fixed point $\alpha_{\tilde{c}}$ satisfies

$$\alpha_{\tilde{c}} = \gamma_{\tilde{c}}(\frac{1}{7}) = \gamma_{\tilde{c}}(\frac{2}{7}) = \gamma_{\tilde{c}}(\frac{4}{7})$$

and the other preimage $\alpha'_{\tilde{c}}$ of $\alpha_{\tilde{c}}$ satisfies

$$\alpha'_{\tilde{c}} = \gamma'_{\tilde{c}}(\frac{1}{14}) = \gamma_{\tilde{c}}(\frac{9}{14}) = \gamma_{\tilde{c}}(\frac{11}{14}) = -\alpha_{\tilde{c}} \, .$$

Fix a potential $\eta > 0$ and set

$$\widetilde{W}_i = \{ \, z \in \mathbf{C} \mid G_{\tilde{c}}(z) \leq \frac{\eta}{2^i} \, \}$$

and $\widetilde{W} = \widetilde{W}_0$. The 6 external rays

$$\mathcal{R}_{\tilde{c}}(\frac{1}{7}) \, , \mathcal{R}_{\tilde{c}}(\frac{2}{7}) \, , \mathcal{R}_{\tilde{c}}(\frac{4}{7}) \, , \mathcal{R}_{\tilde{c}}(\frac{1}{14}) \, , \mathcal{R}_{\tilde{c}}(\frac{9}{14}) \, , \mathcal{R}_{\tilde{c}}(\frac{11}{14})$$

together with $\alpha_{\tilde{c}}$ and $\alpha'_{\tilde{c}}$ decompose $\widetilde{W}$ into 5 compact subsets $\widetilde{V}$, $\widetilde{V}'^{(1)}$, $\widetilde{V}'^{(2)}$, $\widetilde{V}''^{(1)}$, $\widetilde{V}''^{(2)}$ as indicated on figure 10.1, where the parametervalue $\tilde{c}$ is chosen to be the center of the hyperbolic component $\Omega_{1/3}$.

Set $\widetilde{V}_i = \widetilde{V} \cap \widetilde{W}_i$ etc. The 4 sets

$$\widetilde{V}''^{(1)}_{i+1} \, , \, \widetilde{V}'^{(1)}_{i+1} \, , \, \widetilde{V}''^{(2)}_{i} \, , \, \widetilde{V}'^{(2)}_{i}$$

are homeomorphic with homeomorphisms which take points of $K_{\tilde{c}}$ into points of $K_{\tilde{c}}$.

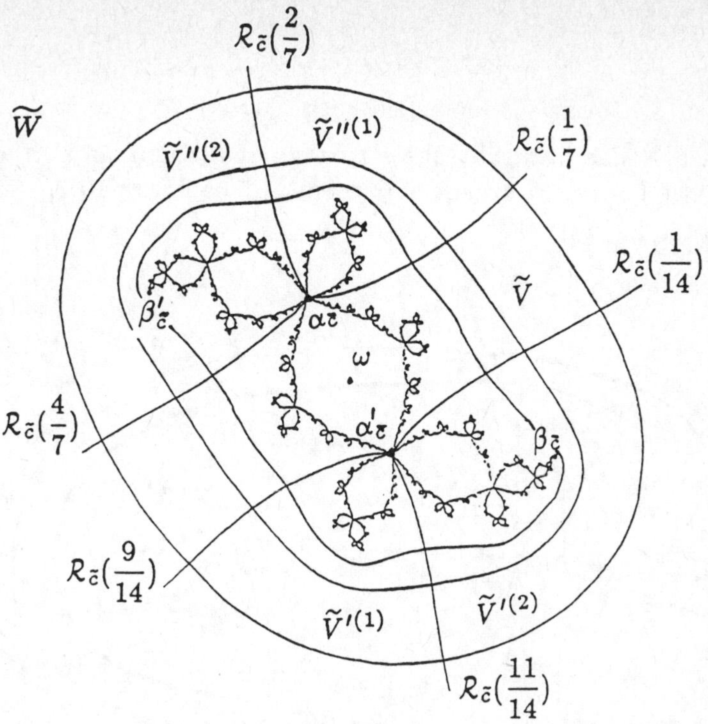

Figure 10.1.

Step (1) in the construction of $P_{c'}$ is to define a new space $\widehat{X}$ and a mapping f . Let P_c be given with $c \in M_{1/2}$ and let W , $S(\theta)$ etc. be as in section 5. Especially recall how W is decomposed into V , V' , V'' , see figure 5.1.

Motivated by the observation about the decomposition of $\widetilde{W}$ the space $\widehat{X}$ is obtained from W by cutting along $\mathcal{R}_c(\frac{2}{3})$ and open up such that we have 2 copies of $\mathcal{R}_c(\frac{2}{3}) \cap W$, then gluing in an extra copy of V'' in the opening. More formally, take 2 copies $V''^{(1)}$ and $V''^{(2)}$ of V'' , and for $x \in V''$ denote by $x^{(1)}$ and $x^{(2)}$ its images in $V''^{(1)}$ and $V''^{(2)}$ respectively. The space $\widehat{X}$ is obtained by taking the disjoint union

$$(V \cup V') \sqcup V''^{(1)} \sqcup V''^{(2)}$$

and identifying

$$\begin{array}{ll} z \text{ with } z^{(1)} & \text{for } z \in \mathcal{R}_c(\frac{1}{3}) \cap W \\ z \text{ with } z^{(2)} & \text{for } z \in \mathcal{R}_c(\frac{2}{3}) \cap W \\ z^{(2)} \text{ with } P(z)^{(1)} & \text{for } z \in \mathcal{R}_c(\frac{1}{3}) \cap W_1 \,. \end{array}$$

Let $\widehat{\mathcal{R}}$ denote the image of $\mathcal{R}_c(\frac{1}{3})^{(2)} \cap W$, which contains the image of $\mathcal{R}_c(\frac{2}{3})^{(1)} \cap W$. The identification along $\widehat{\mathcal{R}}$ is not equipotential, a point $z^{(2)}$ of potential s is identified with $P_c(z)^{(1)}$ of potential $2s$.

The space $\widehat{X}$ has naturally the structure of a Riemann surface with an **R**-analytic boundary except at 2 corners , see figure 10.2.

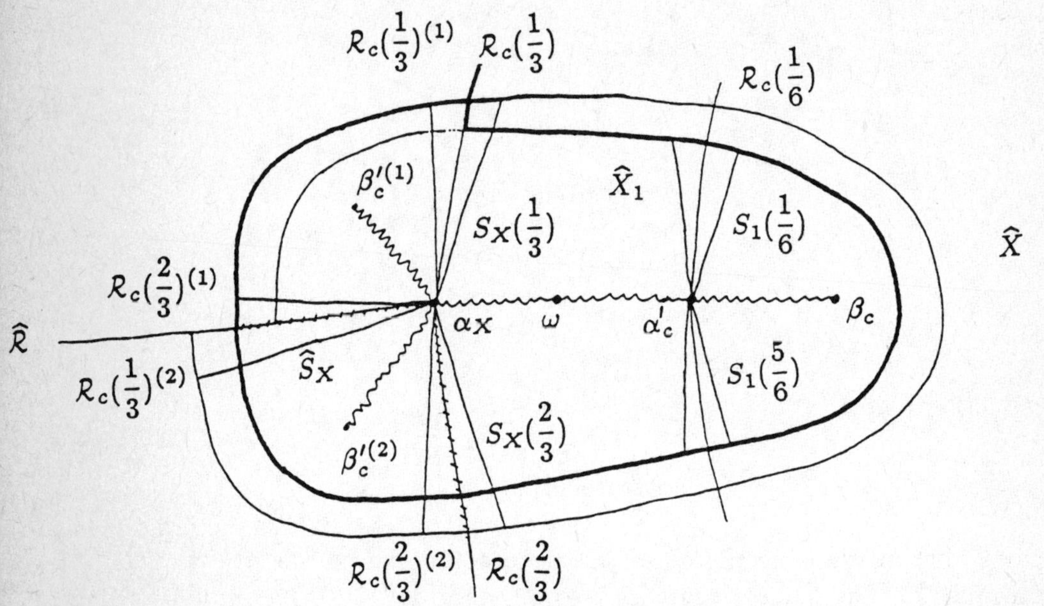

Figure 10.2.

Set

$$\widehat{X}_1 = V_1 \cup V_1' \cup V''^{(1)} \cup V_1''^{(2)}$$

and define

$$f : \widehat{X}_1 - \mathcal{R}_c(\frac{5}{6}) \;\to\; \widehat{X}$$

by

$$\begin{array}{lll} z & \mapsto P_c(z) & \text{for } z \in V_1' \\ z & \mapsto P_c(z)^{(1)} & \text{for } z \in V_1 \\ z^{(1)} & \mapsto z^{(2)} & \text{for } z \in V'' \\ z^{(2)} & \mapsto P_c(z) & \text{for } z \in V_1'' \,. \end{array}$$

This map is analytic in the interior of its domain of definition. But it can not be extended to $\mathcal{R}_c(\frac{5}{6})$, since when z tends towards this ray in V_1

and V_1' respectively , then its image tends towards $\widehat{\mathcal{R}}$ in $V''^{(1)}$ and $\mathcal{R}_c(\frac{2}{3})$ in V respectively.
This ends step (1) in the construction.

The second step , smoothing f to obtain a quasi-regular mapping g , is much harder.

We start by defining sectors in $\widehat{X}$, induced by the sectors in W . We shall use the following notation :

$$\alpha_X \qquad \text{for the image of } \alpha_c \ ,$$

$$S_X(\frac{1}{3}) \qquad \text{for the image of } (S(\frac{1}{3}) \cap V) \cup (S(\frac{1}{3}) \cap V'')^{(1)} \ ,$$

$$S_X(\frac{2}{3}) \qquad \text{for the image of } (S(\frac{2}{3}) \cap V) \cup (S(\frac{2}{3}) \cap V'')^{(2)} \ ,$$

$$\widehat{S}_X \qquad \text{for the image of } (S(\frac{2}{3}) \cap V'')^{(1)} \cup (S(\frac{1}{3}) \cap V'')^{(2)} \ ,$$

$$S_X = S_X(\frac{1}{3}) \cup \widehat{S}_X \cup S_X(\frac{2}{3}) \ .$$

Let $X \subset \widehat{X}$ be a subset with C^1-boundary, for which the boundary coincides with that of $\widehat{X}$ except in $V''^{(2)} \cup S_X(\frac{2}{3})$, where it is between the equipotentials at level η and $\frac{\eta}{2}$, and strictly between in $V''^{(2)} - \widehat{\mathcal{R}}$. Set

$$\Sigma = V'' - (S(\frac{1}{3}) \cup \{\alpha_c\} \cup S(\frac{2}{3})) \ ,$$

and

$$\widetilde{X}_1 = f^{-1}(X) \cup (\mathcal{R}_c(\frac{5}{6}) \cap V_1) \ ,$$

see figure 10.3 .

We shall change the mapping f to a quasi-regular mapping

$$g \colon X_1 \ \to \ X$$

such that g coincides with f outside the sector $S(\frac{5}{6}) \cap X_1$ and such that the restriction of g to $S(\frac{5}{6}) \cap X_1$ is a quasi- conformal homeomorphism onto $(\widehat{S}_X \cup \Sigma^{(2)} \cup S_X(\frac{2}{3})) \cap X$. As pointed out before we need to concentrate the non-holomorphy of g to regions where orbits pass at most once. Notice that $(S(\frac{5}{6}) - g^{-1}(\Sigma^{(2)})) \cap X_1$ is a region which orbits pass at most once, while the region $(S(\frac{5}{6}) \cap g^{-1}(\Sigma^{(2)})) \cap X_1$ is not.

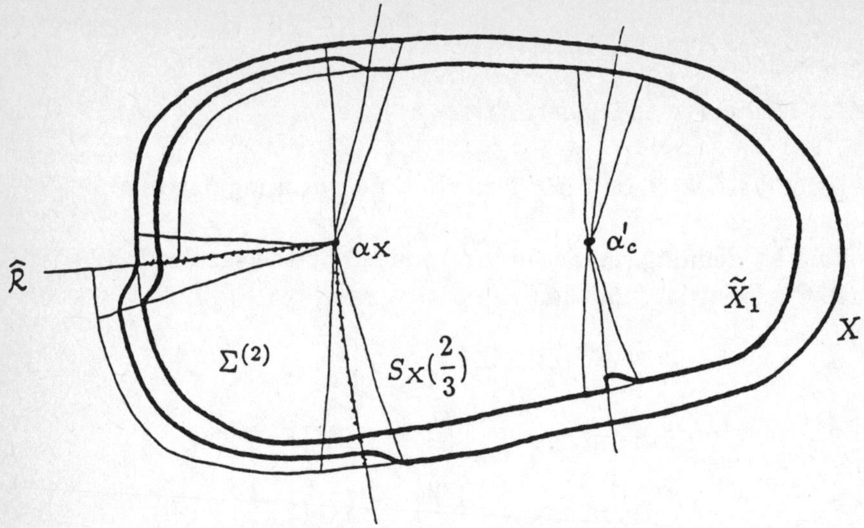

Figure 10.3.

The following proposition plays a role analogous to lemma 1 in section 6 . But the proof is more delicate , and it will occur in section 13 .

Recall that $\Omega_{1/2}$ denote the hyperbolic component of $\mathring{M}$, which is the satellite of the big cardioid of internal argument $\frac{1}{2}$.

PROPOSITION 1. *Suppose that* $c \in M_{1/2} - \overline{\Omega}_{1/2}$. *If we have chosen* η *sufficiently small and* q *sufficiently large (always with* $q\ \eta < \frac{1}{6}$ *), then there exists a subset* $X_1 \subset \mathring{X}$, *which coincides with* $\widetilde{X}_1$ *outside the sector* $S(\frac{5}{6})$ *and a quasi-regular mapping* $g\colon X_1 \to X$, *which coincides with* f *on* $\widetilde{X}_1 - S_1(\frac{5}{6})$, *and which is holomorphic on* $g^{-1}(\Sigma^{(2)})$.

In section 13 it will become clear why we can only complete step (2) in the surgery construction for polynomials P_c with c in $M_{1/2}$ but not in $\overline{\Omega}_{1/2}$.

REMARK.
In the two constuctions in part II the points $z \in X_1$ *for which* $g^n(z)$ *is defined for all* n *correspond directly to points in* K_c *or* K_a *respectively. This construction will be different. All points corresponding to points in* K_c *still satisfy that* $g^n(z)$ *is defined for all* n . *But also a lot of other points do: all the points which are eventually mapped onto* $(K_c \cap V'')^{(2)}$.

11. Opening modulus of a sector.

Again start from a quadratic polynomial P_c with $c \in M_{1/2}$. In a neighborhood of the fixed point α_c in W , the sectors $S(\frac{1}{3})$ and $S(\frac{2}{3})$ are both P_c^2-invariant sectors in a sense which we shall make precise below.

Let U be an open set in C and $F: U \to \mathbf{C}$ a holomorphic map which has a repulsive fixed point $\alpha \in U$; set $\rho = F'(\alpha)$. In a neighborhood of α we can find a *linearizing coordinate* for F , i.e. an isomorphism

$$\varsigma: U_1 \to \varsigma(U_1) \subset \mathbf{C}$$

where U_1 is a neighborhood of α in U such that

$$\varsigma(F(x)) = \rho\, \varsigma(x) \qquad \text{for} \qquad x \in U_1' = F^{-1}(U_1) \cap U_1 .$$

Let $\Delta \subset U_1$ be a neighborhood of α such that $\varsigma(\Delta)$ is a disc. Then

$$E = E_{F,\alpha} = \Delta - \{\alpha\}/(F)$$

is a Riemann surface of genus 1 , isomorphic to

$$\mathbf{C}/(\mathbf{Z}\ \log\rho \oplus \mathbf{Z}\ 2\pi i) .$$

Let ϖ denote the projection

$$\varpi: \Delta - \{\alpha\} \to E .$$

Let $A \subset E_{F,\alpha}$ be an annulus such that $S = \varpi^{-1}(A)$ is connected. Then we say that S is an *F-invariant sector* . And for any F-invariant sector S we shall denote the corresponding annulus by A_S . See figure 11.1. We call the modulus of the annulus $A = A_S$ the *opening modulus* of S *relative* to F . As a Riemann surface the annulus A (or $\overset{\circ}{A}$, if A is a closed annulus) is isomorphic to $B_h/\mathbf{Z}\,\ell$ for some h and ℓ where $B_h = \{z \mid 0 < \operatorname{Im} z < h\}$. The modulus of A is h/ℓ and two annuli are isomorphic if and only if they have the same moduli.

In W the sectors $S(\frac{1}{3})$, $S(\frac{2}{3})$ and Σ are P_c^2-invariant sectors in a neighborhood of α_c . The sectors $S(\frac{1}{3})$ and $S(\frac{2}{3})$ have the same opening modulus, μ , relative to P_c^2 . (The mapping P_c induces an isomorphism between the corresponding annuli $A_{S(1/3)}$ and $A_{S(2/3)}$.) The sectors $S_1(\frac{1}{6})$ and $S_1(\frac{5}{6})$ are $\widetilde{P}_c^2$-invariant sectors in a neighborhood

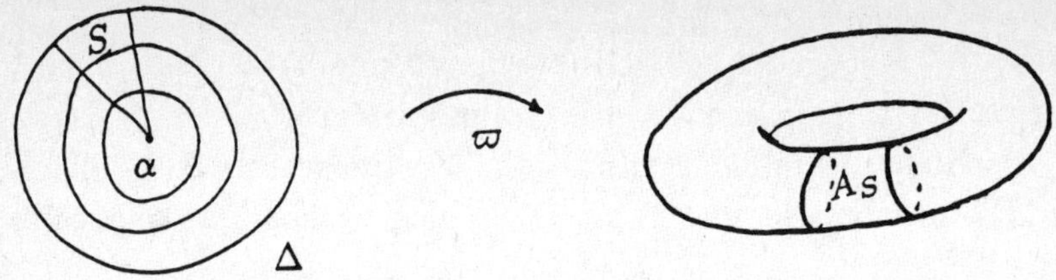

Figure 11.1.

of α'_c , where $\widetilde{P}_c = -P_c$. The sectors $S_1(\frac{1}{6})$ and $S_1(\frac{5}{6})$ have μ as opening modulus relative to $\widetilde{P}_c^2$. (The mapping $\widetilde{P}_c$ induces an isomorphism between the corresponding annuli $A_{S(1/6)}$ and $A_{S(5/6)}$. The mapping P_c induces an isomorphism from $A_{S(1/6)}$ onto $A_{S(1/3)}$.) In X the sectors $S_X(\frac{1}{3})$, $S_X(\frac{2}{3})$ and $\widehat{S}_X$ are f^3-invariant sectors in a neighborhood of α_X . The sectors $S_X(\frac{1}{3})$, $S_X(\frac{2}{3})$ and $\widehat{S}_X$ have μ as opening modulus relative to f^3 . (The mapping f induces isomorphisms between the corresponding annuli, and the mapping P_c induces an isomorphism from $A_{S(1/6)}$ onto $A_{S_X(1/3)}$.)

Let ν denote the opening modulus of Σ relative to P_c^2 . In X the sector $\Sigma^{(2)}$ is a f^3-invariant sector in a neighborhood of α_X . The opening modulus of $\Sigma^{(2)}$ relative to f^3 is ν .

We shall see in lemma 5 that if the opening modulus of $S_1(\frac{5}{6})$ relative to $\widetilde{P}_c^2$ is greater that the opening modulus of $\Sigma^{(2)}$ relative to f^3 then we can smoothen f to a quasi-regular mapping g .

LEMMA 5. *If $\mu > \nu$, then we can find X_1 and g satisfying the conditions in proposition 1.*

PROOF:
When $\mu > \nu$ we can find a closed annulus B with C^1-boundary contained in the interior of the annulus $A_{S_1(5/6)}$ and with modulus ν , the same as the modulus of the annulus A_Σ . Let

$$\psi : B \to A_\Sigma$$

be an isomorphism . The mapping ψ can be lifted to an isomorphism $\widetilde{\psi}$ from a neighborhood $S_B{}^\vee$ of α'_c in the sector S_B defined by B onto a neighborhood $\Sigma^\vee$ of α_c in Σ. Replacing $\widetilde{\psi}$ by $P_c^{2n}\circ\widetilde{\psi}$ for n sufficiently big, we may assume that

$$S_B{}^\vee \subset S_1(\frac{5}{6})$$

and

$$\Sigma^\vee = \{x \in \Sigma \mid x^{(2)} \in X\} .$$

In $S_1(\frac{5}{6})$ we join the boundary of $S_B{}^\vee$ in S_B with the boundary of W_1 outside $S_1(\frac{5}{6})$ by a C^1-curve disjoint with its images by $(\widetilde{P}_c^2)^{-n} : S_1(\frac{5}{6}) \to S_1(\frac{5}{6})$; in this way we bound a compact set $S^\vee(\frac{5}{6})$. Let $S^\vee(\frac{1}{3})$ be the set of $x \in S(\frac{1}{3})$ satisfying : $x^{(2)} \in X$ if $x \in V''$ and $P_c(x)^{(1)} \in X$ if $x \in V$. Define similarly $S^\vee(\frac{2}{3})$ as the set of $x \in S(\frac{2}{3})$ such that x or $x^{(2)}$ belongs to X, see figure 11.2.

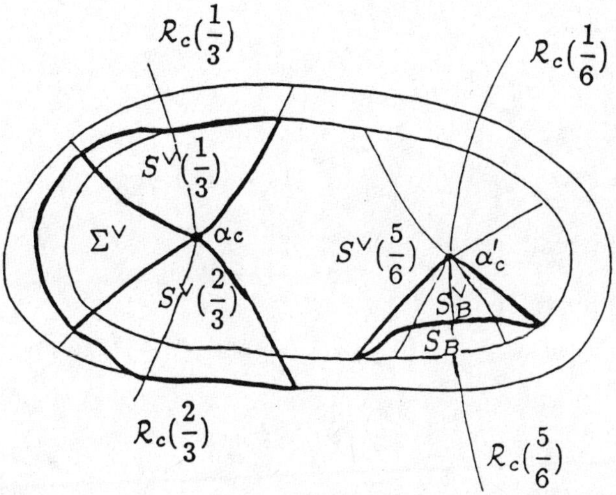

Figure 11.2.

Set

$$S^\vee = S^\vee(\frac{1}{3}) \cup \Sigma^\vee \cup S^\vee(\frac{2}{3}) ,$$

$$T_0^\vee = S^\vee - (P_c^2)^{-1}(S^\vee) ,$$

$$T_i^\vee = (P_c^2)^{-1}(T_{i-1}^\vee) \cap S^\vee ,$$

$$T_0^\vee(\frac{5}{6}) = S^\vee(\frac{5}{6}) - (\widetilde{P}_c^2)^{-1}(S^\vee(\frac{5}{6})) ,$$

$$T_i^\vee = (\widetilde{P}_c^2)^{-1}(T_{i-1}^\vee(\frac{5}{6})) \cap S^\vee(\frac{5}{6}) .$$

Choose a diffeomorphism

$$g_0^\vee : T_0^\vee(\frac{5}{6}) \rightarrow T_0^\vee$$

such that

$$g_0^\vee = \widetilde{\psi} \text{ on } T_0^\vee(\frac{5}{6}) \cap S_B{}^\vee ,$$

$$g_0^\vee \text{ is tangent to } P_c \text{ on } T_0^\vee(\frac{5}{6}) \cap \partial S_1(\frac{5}{6}) \cap V_1' ,$$

$$g_0^\vee \text{ is tangent to } \tau : x \mapsto -x \text{ on } T_0^\vee(\frac{5}{6}) \cap \partial S_1(\frac{5}{6}) \cap V_1 ,$$

$$P_c^2 \circ g_0^\vee = g_0^\vee \circ \widetilde{P}_c^2 \text{ on } I^\vee(\frac{5}{6}) = T_0^\vee(\frac{5}{6}) \cap T_1^\vee(\frac{5}{6}) .$$

Then define $g_i^\vee$ inductively by

$$P_c^2 \circ g_i^\vee = g_{i-1}^\vee \circ \widetilde{P}_c^2 .$$

The mapping

$$g^\vee : S^\vee(\frac{5}{6}) \rightarrow S^\vee$$

defined by

$$g^\vee(x) = g_i^\vee(x) \qquad \text{for} \qquad x \in T_i^\vee(\frac{5}{6})$$

is quasi-conformal.

Then we define

$$X_1 = (\widetilde{X}_1 - S_1(\frac{5}{6})) \cup S^\vee(\frac{5}{6}) ,$$

and

$$g : X_1 \rightarrow X$$

by

$$g(x) = \begin{cases} f(x) & \text{for } x \in \widetilde{X}_1 - S_1(\frac{5}{6}) \\ \iota \circ g^\vee(x) & \text{for } x \in S^\vee(\frac{5}{6}) , \end{cases}$$

where

$$\iota : S^\vee \rightarrow X$$

is defined by

$$\iota(x) = \begin{cases} P_c(x)^{(1)} & \text{for } x \in S^\vee(\frac{1}{3}) \cap V_1 \\ x^{(2)} & \text{for } x \in V_1'' \\ x & \text{for } x \in S^\vee(\frac{2}{3}) \cap V_1 . \end{cases}$$

The mapping g satisfies the conditions in proposition 1 . ∎

12. The size of the limbs of M.

Let Ω denote a hyperbolic component of $\overset{\circ}{M}$, i.e. a connected component corresponding to the existence of an attractive cycle. Denote by φ_Ω the conformal representation

$$\varphi_\Omega\colon \Omega \to D$$

which associates to each $c \in \Omega$ the multiplier of the attractive cycle. The inverse mapping φ_Ω^{-1} admits a continuous extention to the boundary, giving a parametrization

$$\gamma_\Omega\colon \mathbf{T} \to \partial\Omega .$$

For $t \in \mathbf{Q}/\mathbf{Z}$ the set $M - \{\gamma_\Omega(t)\}$ has 2 connected components (except if $\Omega = \Omega_0$ and if $t = 0$) . Let $M^\star_{\Omega,t}$ denote the component, which does not contain Ω , and set

$$M_{\Omega,t} = \overline{M^\star_{\Omega,t}} = M^\star_{\Omega,t} \cup \{\gamma_\Omega(t)\} .$$

We call $M_{\Omega,t}$ the *limb* of M *relative* to Ω of *internal argument* t .

YOCCOZ' THEOREM. *For each hyperbolic component, Ω , there exists a constant, C_Ω , such that for each $t = \frac{p}{q} \in \mathbf{Q}/\mathbf{Z}$, the (euclidean) diametre of $M_{\Omega,t}$ is $\leq \frac{C_\Omega}{q}$.*

COROLLARY.

$$M - \overline{\Omega} = \bigcup_{t\in \mathbf{Q}/\mathbf{Z}} M^\star_{\Omega,t} .$$

See [Y] and [L] .

13. Control of the opening modulus.

In this section we shall prove that if $c \in M_{1/2}$ is chosen <u>outside</u> the closure of the hyperbolic component $\Omega_{1/2}$, then it is always possible to find a slope $q(c)$ of the sectors $S(\theta)$ (and a potential $\eta(c)$ satisfying $q\,\eta < \frac{1}{6}$) such that the condition $\mu > \nu$ in lemma 5 is fulfilled. The existence of a quasi- regular mapping g as in proposition 1 follows immediately, and the topological surgery is completed but only for polynomials P_c with $c \in M_{1/2} - \overline{\Omega}_{1/2}$.

The opening moduli μ and ν are independent of the choice of η , they are entirely determined by the point $c \in M_{1/2}$ and by the slope q . The pair (q, η) is however subject to the inequality $q\,\eta < \frac{1}{6}$. The modulus μ is an increasing function of q , in fact

$$\mu = \frac{\arctan\,(2\pi q)}{\log 2} ,$$

while for c fixed ν is a decreasing function of q.

LEMMA 6. *For c fixed in $M_{1/2} - \overline{\Omega}_{1/2}$ the modulus ν tends to* 0 *when $q \to +\infty$.*

PROOF:
Let ς be a linearizing coordinate for P_c^2 in a neighborhood U_1 of the fixed point α_c

$$\varsigma : U_1 \;\to\; \varsigma(U_1)$$

such that

$$\varsigma(P_c^2(x)) = \rho\, \varsigma(x) \qquad \text{for} \qquad x \in U_1' = (P_c^2)^{-1}(U_1) \cap U_1 .$$

Let Δ_c be a neighborhood of α_c such that $\varsigma(\Delta_c)$ is a disc.

Under the hypothesis made on c , we have from Yoccoz' Corollary , that there exsists a $\tau \in \mathbf{Q}/\mathbf{Z} - \{0\}$ such that c is in $M_{1/2,\tau}$ the limb of M relative to $\Omega_{1/2}$ of internal argument τ .

Any polynomial P_c with $c \in M_{1/2,\tau}$ has the proporty that there exists a sequence of pinching points in the Julia set converging to α_c and such that each pinching point has external arguments in both $]\frac{1}{3}, \frac{1}{2}[$ and $]\frac{1}{2}, \frac{2}{3}[$.

For $c \in \Omega_{1/2}$ this is not the case. See figure 13.1 where c is the center of $\Omega_{1/2,1/2}$ and compare with figure 5.2 where c is the center of $\Omega_{1/2}$.

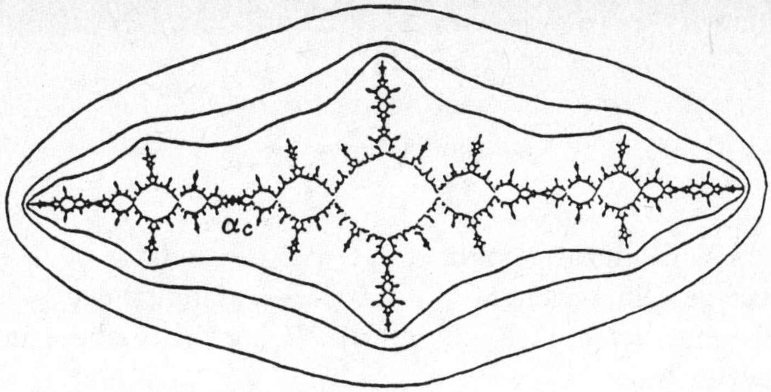

Figure 13.1.

Let x_0 be such a pinching point belonging to the same connected component of $K_c \cap \Delta_c$ as α_c and having at least 2 external arguments $\frac{1}{3}+u$ and $\frac{2}{3}-v$ with u and $v \in]0,\frac{1}{6}[$. Then the sequence of points $x_n \in \Delta_c$ satisfying

$$\varsigma(x_n) = \frac{1}{\rho}\,\varsigma(x_{n-1})$$

belong to the same component of $K_c \cap \Delta_c$ as α_c and

$$x_n = \gamma_c\Big(\frac{1}{3}+\frac{u}{4^n}\Big) = \gamma_c\Big(\frac{2}{3}-\frac{v}{4^n}\Big)\,.$$

For a given $s > 0$ let y'_n and y''_n respectively denote the points of potential $s/4^n$ on the rays $\mathcal{R}_c(\frac{1}{3}+\frac{u}{4^n})$ and $\mathcal{R}_c(\frac{2}{3}-\frac{v}{4^n})$ respectively. We can chose q such that the sectors $S_q(\frac{1}{3})$ and $S_q(\frac{2}{3})$ contain the points $y'_n(s)$ and $y''_n(s)$ for n sufficiently large. If necessary we reduce η such that $q\,\eta < \frac{1}{6}$.

Let E_c denote the complex torus $\Delta_c - \{\alpha_c\}/(P_c^2)$ (which does not depend on s) , and let $\varpi_c\colon \Delta_c - \{\alpha_c\} \;\to\; E_c$ denote the projection. Let A_Σ denote the annulus in E_c corresponding to the P_c^2-invariant sector Σ . When s tends to 0 , then the points $\varpi_c(y'_n(s))$ and $\varpi_c(y''_n(s))$ tend to $\varpi_c(x_n)$. The annulus A_Σ passes between these points and has its equator in a fixed homotopy class of loops in E_c . Provide E_c with a local euclidean metric, such that the geodesic in that class has lenght 1 , and denote by $\delta(s)$ the distance between $\varpi_c(y'_n(s))$ and $\varpi_c(y''_n(s))$ in this metric. Then

$$\text{mod }(A_\Sigma) = \frac{\pi}{2\ell}$$

where ℓ is the Poincaré lenght in $\overset{\circ}{A}_\Sigma$ of the equator of A_Σ . This equator has a euclidean lenght ≥ 1 and passes at a distance $\leq \delta/2$ from one of the points $\varpi_c(y'_n(s))$, $\varpi_c(y''_n(s))$. Using the Koebe-$\frac{1}{4}$-inequality, we see that

$$\ell \geq \frac{1}{2}\log(\frac{1}{\delta(s)}) ,$$

hence

$$\text{mod }(A_\Sigma) \leq \frac{\pi}{\log(1/\delta)} .$$

But $\delta(s) \to 0$ when $s \to 0$, therefore $\nu = \text{ mod }(A_\Sigma) \to 0$ when $s \to 0$. ∎

This ends step (2) in the construction.

Remark.
The annulus A_Σ and its modulus mod A_Σ depend continuously on $(c,q) \in M_{1/2} \times \mathbf{R}_+$. Therefore for each $c_0 \in M_{1/2} - \overline{\Omega}_{1/2}$ we can find a $q < +\infty$, such that the inequality $\nu < \mu$ is satisfied for all c in that neighborhood of c_0 . So we can also find a continuous function $c \mapsto q(c)$ of $M_{1/2} - \overline{\Omega}_{1/2}$ into $\mathbf{R}_+$ such that the inequality $\nu < \mu$ is satisfied for all $c \in M_{1/2} - \overline{\Omega}_{1/2}$ when we set $q = q(c)$.

14. Definition and continuity of the mapping Φ_A.

After having finished the topological surgery in section 13 for polynomials P_c with $c \in M_{1/2} - \overline{\Omega}_{1/2}$ we shall first complete it into holomorphic surgery and then define the mapping Φ_A .

The lemma 7 below is the analogue of lemma 2 in section 7 .

LEMMA 7. *Let $g\colon X_1 \to X$ be given as in proposition 1 .*
There exsits an almost complex structure σ on X , quasi-conformally equivalent to the standard structure σ_0 , such that $g^\star\sigma = \sigma$ and σ coincides with σ_0 on $X - (S^\vee(\frac{5}{6}) - S_B{}^\vee)$.

The complex structure is changed as in section 7 and 9 to an almost complex structure σ on X such that $g^\star\sigma = \sigma$. The essential point in the proof is that g is holomorphic except at $S^\vee(\frac{5}{6}) - S_B{}^\vee$ and that an orbit passes at most once the set $(S^\vee(\frac{5}{6}) - S_B{}^\vee)$: if x belongs to that set , then $g(x) \in S_X(\frac{2}{3}) \cup \widehat{S}_X$, hence $g^n(x) \in S_X$ for all n for which $g^n(x)$ is defined.

Let $\varphi\colon \overset{\circ}{X} \to U$ be a quasi-conformal homeomorphism which integrates σ and set

$$h = \varphi \circ g \circ \varphi^{-1}\colon U' \to U .$$

The mapping h is a polynomial-like mapping of degree 2 and it follows from the straightening theorem, that h is hybrid equivalent to a uniquely determined polynomial $P_{c'}$. Set

$$c' = \Phi_A(c) ;$$

it is clear that c' belongs to $M_{1/3}$.

This ends the construction in the dynamical plane. The uniquely determined polynomial $P_{c'}$ with $c' \in M_{1/3}$ is obtained from P_c with $c \in M_{1/2} - \overline{\Omega}_{1/2}$ through holomorphic surgery.

Thus we have defined a mapping

$$\Phi_A\colon M_{1/2} - \overline{\Omega}_{1/2} \to M_{1/3} .$$

By proceeding as in section 8 one can prove that Φ_A is continuous, except this time one can use proposition 7 , I.6 in [DH 2] directly.

We extend Φ_A to $M_{1/2}$ by defining

$$\Phi_A = \varphi_{\Omega_{1/3}}^{-1} \circ \varphi_{\Omega_{1/2}} \text{ on } \overline{\Omega}_{1/2} ,$$

where

$$\varphi_{\Omega_r} : \overline{\Omega}_r \to \overline{D}$$

is the homeomorphism, which associates to every $c \in \Omega_r$ the multiplier of the attractive cycle for P_c .

Thus the mapping Φ_A is continuous on $M_{1/2} - \overline{\Omega}_{1/2}$, and its restriction to $\overline{\Omega}_{1/2}$ is continuous . In order to prove that Φ_A is continuous on $M_{1/2}$ we still have to prove that

$$\Phi_A(c_n) \to \Phi_A(c) \qquad \text{when } c_n \to c$$

where

$$c_n \in M_{1/2} - \overline{\Omega}_{1/2} \qquad \text{and} \qquad c \in \partial\Omega_{1/2} .$$

The point c is then of the form $\gamma_{\Omega_{1/2}}(\theta)$. In the different cases we shall distinguish between θ rational or irrational.

Notice that it follows from the construction that for $\theta \in \mathbf{Q}/\mathbf{Z}$ the limb $M_{1/2,\theta}$ of M of internal argument θ relative to $\Omega_{1/2}$ is mapped by Φ_A into the limb $M_{1/3,\theta}$ of M of internal argument θ relative to $\Omega_{1/3}$.

a) Case where θ is irrational :

Each c_n is in a $M_{1/2,\theta_n}$ with $\theta_n \in \mathrm{Q}/\mathrm{Z}$, and the sequence θ_n tends to θ , in particular the denominator of θ_n tends to ∞ . Then $\Phi_A(c_n) \in M_{1/3,\theta_n}$, and we have $\gamma_{\Omega_{1/3}}(\theta_n) \to \Phi_A(c)$. From Yoccoz' theorem in section 12 we know that the diameter of $M_{1/3,\theta_n}$ tends to 0 , therefore $\Phi_A(c_n) - \gamma_{\Omega_{1/3}}(\theta_n) \to 0$. It follows that $\Phi_A(c_n) \to \Phi_A(c)$.

b) Case where θ is rational, but each c_n is in an $M_{1/2,\theta_n}$ with $\theta_n \neq \theta$:
That case is treated as the previous one.

c) Case where θ is rational and each $c_n \in M_{1/2,\theta}$:

Denote by $\Omega_{r,\theta}$ the component of $\overset{\circ}{M}$ attached to Ω_r at the point $\gamma_{\Omega_r}(\theta)$ and distinguish between the 2 cases :
c1) $c_n \in \overline{\Omega}_{1/2,\theta}$:
We notice, that $\Phi_A|_{\overline{\Omega}_{1/2,\theta}} = {\varphi_{\Omega_{1/3,\theta}}}^{-1} \circ \varphi_{\Omega_{1/2,\theta}}$.
c2) $c_n \in M_{1/2,\theta} - \overline{\Omega}_{1/2,\theta}$:
Each c_n is in $M_{1/2,\theta,\theta'_n}$ with $\theta'_n \to 0$ or 1 .
Then $\Phi_A(c_n) \in M_{1/3,\theta,\theta'_n}$. The conclusion is reached by applying Yoccoz' theorem from section 12 again.

Remark.
The mapping Φ_A is compatible with the tuning in the following sense : For any hyperbolic component Ω of $\overset{\circ}{M}$ let c_Ω denote the center of Ω . For $x \in M$ denote by $c_\Omega \perp x$ the *tuning* of c_Ω by x (cf. [D]). For every hyperbolic component Ω of $\overset{\circ}{M}_{1/2}$ we have

$$\Phi_A(c_\Omega \perp x) = (\Phi_A(c_\Omega)) \perp x \ .$$

In particular this can be applied to tuning of the center $c = -1$ of $\Omega_{1/2}$.

15. The injectivity of the mapping Φ_A.

To prove the injectivity of Φ_A and characterize its image we will construct a mapping Ψ_A from part of $M_{1/3}$ onto $M_{1/2}$, such that $\Psi_A \circ \Phi_A$ is the identity on $M_{1/2}$.

Starting from any quadratic polynomial $P_{\tilde{c}}$ with $\tilde{c} \in M_{1/3}$ we shall construct a quadratic polynomial P_c and characterize the polynomials $P_{\tilde{c}}$ for which we get $c \in M$. The construction develops in steps as before, parallel but different,

$$P_{\tilde{c}} \curvearrowright f \curvearrowright g \curvearrowright h \curvearrowright P_c .$$

In section 10 we defined the space X by adding a part of W. Since we want to make the opposite construction here we shall remove the part similar to what we added.

We shall use the notation from section 10 for $\widetilde{W}$, $\widetilde{V}$ etc. explained by figure 10.1.

We form the space $\widehat{X}$ from $\widetilde{V} \cup \widetilde{V}'^{(1)} \cup \widetilde{V}'^{(2)} \cup \widetilde{V}''^{(1)}$ by identifying $z \in \mathcal{R}_{\tilde{c}}(\frac{2}{7})$ with $P_{\tilde{c}}(z) \in \mathcal{R}_{\tilde{c}}(\frac{4}{7})$. The identification is not equipotential. We denote by $\widehat{R}$ the ray which is the image of $\mathcal{R}_{\tilde{c}}(\frac{2}{7})$ and $\mathcal{R}_{\tilde{c}}(\frac{4}{7})$. The space $\widehat{X}$ has naturally the structure of a Riemann surface with an R-analytic boundary except at 2 corners. Let $\widehat{X}_1$ denote the image of $\widetilde{V}_1 \cup \widetilde{V}_1'^{(1)} \cup \widetilde{V}_1'^{(2)} \cup \widetilde{V}_1''^{(1)}$ in $\widehat{X}$, and define

$$f : \widehat{X}_1 - \widetilde{V}_1'^{(1)} \to \widehat{X}$$

as the first return map of $P_{\tilde{c}}$, that is

$$f = \begin{cases} P_{\tilde{c}} & \text{on } \widetilde{V}_1 \cup \widetilde{V}_1'^{(2)} \\ P_{\tilde{c}}^2 & \text{on } \widetilde{V}_2''^{(1)} . \end{cases}$$

The map f is analytic in the interior of its domain of definition. But notice that f is not defined on $\widetilde{V}_1'^{(1)}$. This ends step (1) in the construction.

Fix a q such that $q\,\eta < \frac{1}{14}$ and define sectors with slope q. Denote by $\widehat{S}$ the image in $\widehat{X}$ of $(S(\frac{2}{7}) \cap \widetilde{V}''^{(1)}) \cup ((S(\frac{4}{7}) \cap \widetilde{V})$.

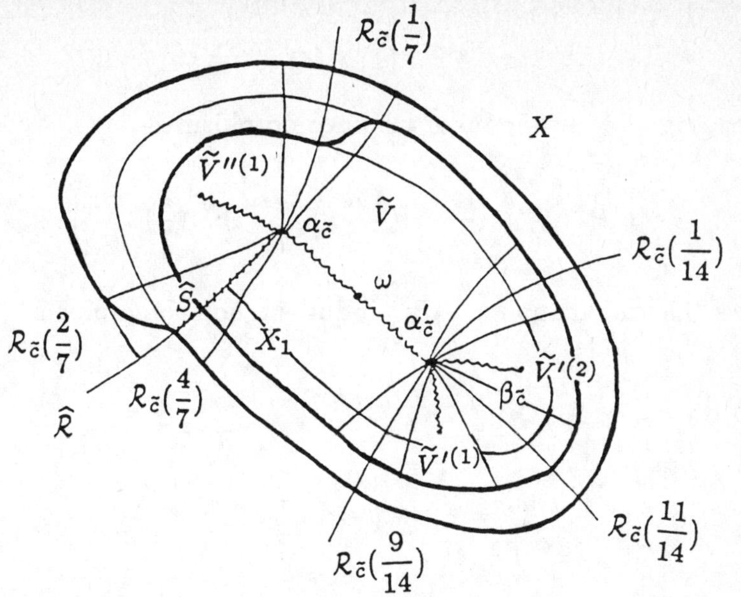

Figure 15.1.

We modify the boundary of $\widehat X$ in $\widehat S$ to obtain a subset $X \subset \widehat X$ with C^1-boundary, and we define X_1 by

$$X_1 \cap S(\frac{1}{7}) = f^{-1}(X) \cap S(\frac{1}{7})$$

$$X_1 - S(\frac{1}{7}) = \widehat X_1 - S(\frac{1}{7}) \,.$$

See figure 15.1.

Set

$$S^\vee = S_1(\frac{9}{14}) \cup \widetilde V_1'^{(1)} \cup S_1(\frac{11}{14}) - \{\alpha'_{\bar c}\} \,.$$

In the second step we shall obtain a quasi-regular mapping g which coincides with f on $X_1 - S^\vee$.

The polynomial $P_{\bar c}$ induces a homeomorphism from $S_{i+1}(\theta)$ onto $S_i(2\theta)$. In particular $P_{\bar c}^3$ induces a homeomorphism

$$P_{\bar c}^3 \colon S_{i+3}(\frac{k}{7}) \;\to\; S_i(\frac{k}{7}) \qquad \text{for } k = 1, 2, 4 \,.$$

Hence $\widehat S$ is an f^2-invariant sector. Set

$$\widehat T_0 = (\widehat S \cap X) - (f^2)^{-1}(\widehat S \cap X)$$

and

$$\widehat{T}_i = (f^2)^{-1}(\widehat{T}_{i-1}) \cap \widehat{S} \cap X .$$

The mapping $P_{\bar{c}}^3$ also induces a homeomorphism

$$P_{\bar{c}}^3 : S_{i+3}^{\vee} \to S_i(\frac{1}{7}) \cup V_i''^{(1)} \cup S_i(\frac{2}{7}) .$$

It follows that the mapping $-P_{\bar{c}}^3$ induces a homeomorphism

$$-P_{\bar{c}}^3 : S_{i+3}^{\vee} \to S_i^{\vee} .$$

Set

$$T_0^{\vee} = S^{\vee} - (-P_{\bar{c}}^3)^{-1}(S^{\vee})$$

and

$$T_i^{\vee} = (-P_{\bar{c}}^3)^{-1}(T_{i-1}^{\vee}) \cap S^{\vee} .$$

Choose a diffeomorphism

$$g_0 : T_0^{\vee} \to \widehat{T}_0$$

satisfying (1) and (2) as in section 6 with $\widetilde{P}_c^2$ replaced by $-P_{\bar{c}}^3$. Then define

$$g_i : T_i^{\vee} \to \widehat{T}_i$$

inductively by

$$f^2 \circ g_i = g_{i-1} \circ (-P_{\bar{c}}^3) .$$

Finally define $g : X_1 \to X$ similarly to g in section 6. This mapping g is a quasi-regular mapping. This ends step (2).

The complex structure is changed to an almost complex structure σ on X such that $g^*\sigma = \sigma$. Again the essential point in the proof is that g is holomorphic except at $S^{\vee}$ and that an orbit passes $S^{\vee}$ at most once : if $x \in S^{\vee}$, then $g^n(x) \in \widehat{S} \cup S(\frac{1}{7})$ for all n where $g^n(x)$ is defined. Therefore there exists an almost complex structure σ on X , quasi-conformally equivalent to the standard structure σ_0 , such that $g^*\sigma = \sigma$ and σ coincides with σ_0 on $X - \bigcup_n g^{-n}(S^{\vee})$.

Remark.

Not only is $S^\vee$ a region which orbits pass at most once, but if $x \in S^\vee$ then $g^n(x)$ is defined for only finitely many n. The set $K(g)$ of points $x \in X_1$ for which $g^n(x)$ is defined and belongs to X_1 for all n equals

$$K(g) = K_{\tilde{c}} - ((K_{\tilde{c}} \cap \overset{\circ}{\tilde{V}}{}''^{(2)}) \cup \bigcup_n g^{-n}(K_{\tilde{c}} \cap \overset{\circ}{\tilde{V}}{}'^{(1)}))\ .$$

Let $\varphi\colon \overset{\circ}{X} \to U$ be a quasi-conformal homeomorphism which integrates σ and set

$$h = \varphi \circ g \circ \varphi^{-1}\colon U' \to U\ .$$

The mapping h is polynomial-like of degree 2. The mapping h admits an hybrid equivalence to a quadratic polynomial P_c, hence a mapping

$$\widehat{\Psi}_A\colon \tilde{c} \mapsto c \qquad \text{from } M_{1/3} \text{ into } \mathbf{C}\ ,$$

there might be choices in the definition : c is uniquely determined only if $c \in M$. We have

$$\begin{aligned} c = \widehat{\Psi}_A(\tilde{c}) \in M &\iff 0 \in K_c \\ &\iff g^n(\omega) \in X \text{ for all } n \\ &\iff P_{\tilde{c}}^n(0) \notin K_{\tilde{c}} \cap \overset{\circ}{\tilde{V}}{}'^{(1)} \text{ for any n}\ . \end{aligned}$$

Let N denote the set

$$N = \{\, \tilde{c} \in M_{1/3} \mid (\forall n)\ P_{\tilde{c}}^n(0) \notin K_{\tilde{c}} \cap \overset{\circ}{\tilde{V}}{}'^{(1)} \,\}$$

and let

$$\Psi_A = \widehat{\Psi}_A|_N\colon N \to M\ .$$

One can verify that

$$\Psi_A(N) \subset M_{1/2}\ ,$$
$$\Phi_A(M_{1/2}) \subset N\ ,$$

and that if $c \in M_{1/2}$ and $c' = \Psi_A(\Phi_A(c))$, then there is a hybrid equivalence between P_c and $P_{c'}$, hence $c = c'$. In other words

$$\Psi_A \circ \Phi_A = id_{M_{1/2}}$$

and

$$\Phi_A \circ \Psi_A = id_N\ .$$

This ends the proof of theorem A.

It also implies the continuity of Ψ_A.

Remark.
It is only when we define the mapping Φ_A that we need to control the opening modulus. In the definition of Ψ_A , Φ_B and Ψ_B it is enough to choose diffeomorphisms. So maybe we could have avoided some difficulties by studying Ψ_A first and introducing Φ_A only to show the injectivity of Ψ_A . But since this study is intended to be generalized to all veins of M , there is no point in economizing in a way which might turn out to be specific for one particular vein.

We shall end this paper by a sequence of computer drawings : Figure 15.2 – 4 . In (a) is shown the Julia set of P_c , in (b) the Julia set of $P_{c'}$ with $c' = \Phi_A(c)$ and in (c) the Julia set of Q_a with $a = \Phi_B(c)$.

At the end of section 2 there is a description of how the dynamical behavior of P_c and Q_a on the filled in Julia sets are related.

From the construction in this section we can now give the similar description of how the dynamical behavior of P_c on the filled in Julia set K_c is related to the dynamical behavior of $P_{c'}$ with $c' = \Phi_A(c)$ on the filled in Julia set $K_{c'}$:
If we remove the part of $K_{c'}$ to 'the left ' of

$$\mathcal{R}_{c'}(\frac{2}{7}) \cup \{\alpha_{c'}\} \cup \mathcal{R}_{c'}(\frac{4}{7})$$

and everything which is eventually mapped onto $K_{c'} \cap \mathring{\tilde{V}}'^{(1)}$ and then change the map $P_{c'}$ to the first return map $f_{c'}$ on this amputated $K_{c'}$, then the dynamical behavior of $f_{c'}$ on that part of $K_{c'}$ is homeomorphically equivalent to the dynamical behavior of P_c on K_c .

In figures 15.2 – 4 the parameter value c is the center of different hyperbolic components : $\Omega_{1/2,1/2}$ and $\Omega_{1/2,1/3}$ respectively in figure 15.2 and 3. In figure 15.4 $c \in \mathbf{R}$ and the critical point has period 3 .

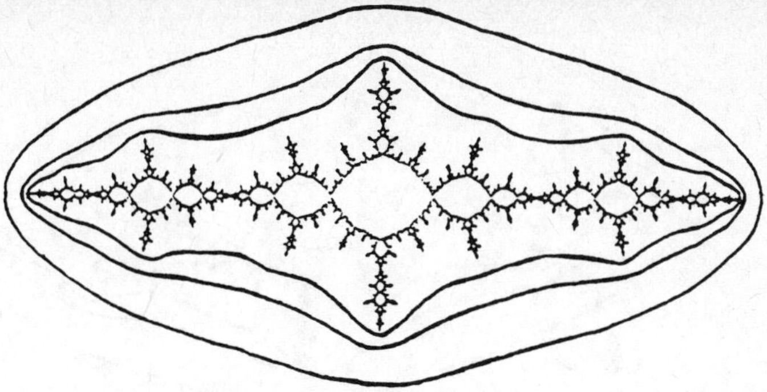

(a)

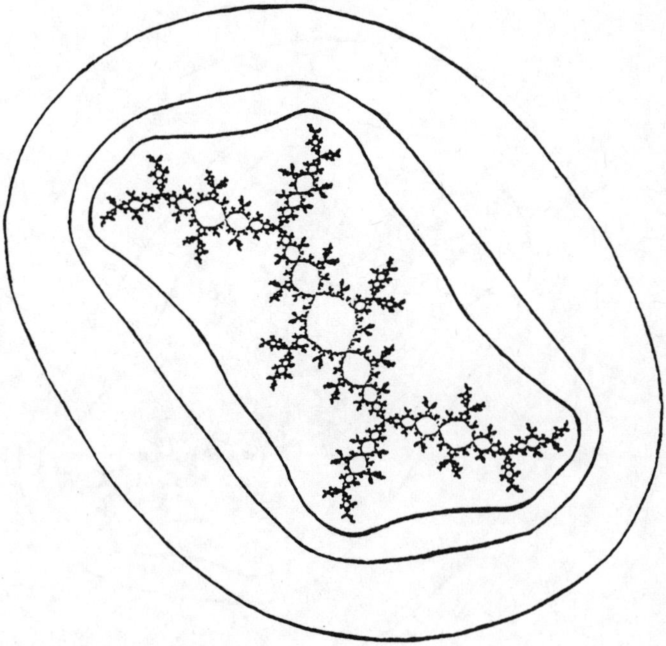

(b)

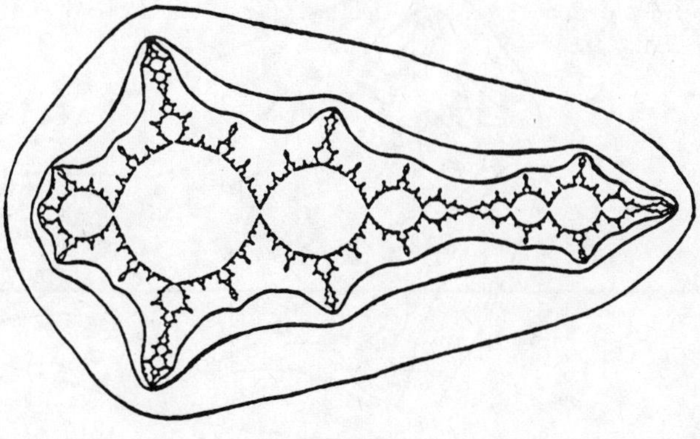

(c)

Figure 15.2.

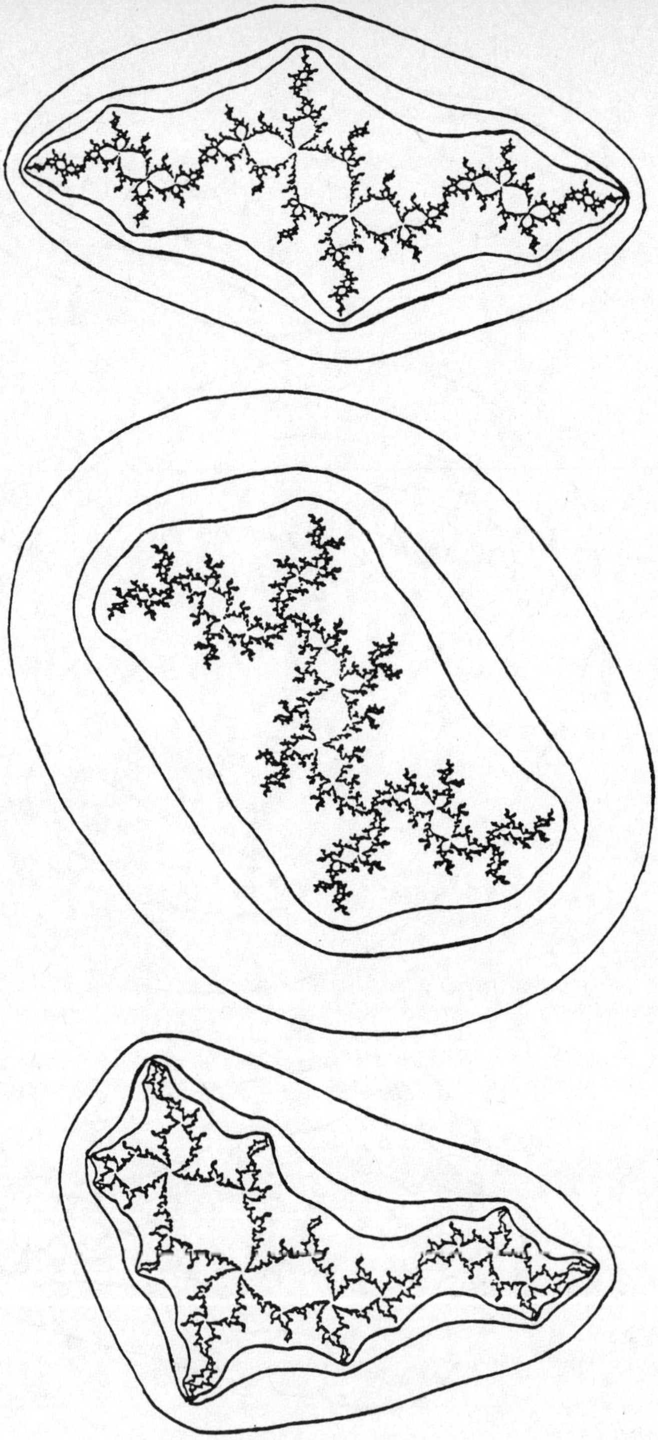

(a)

(b)

(c)

Figure 15.3.

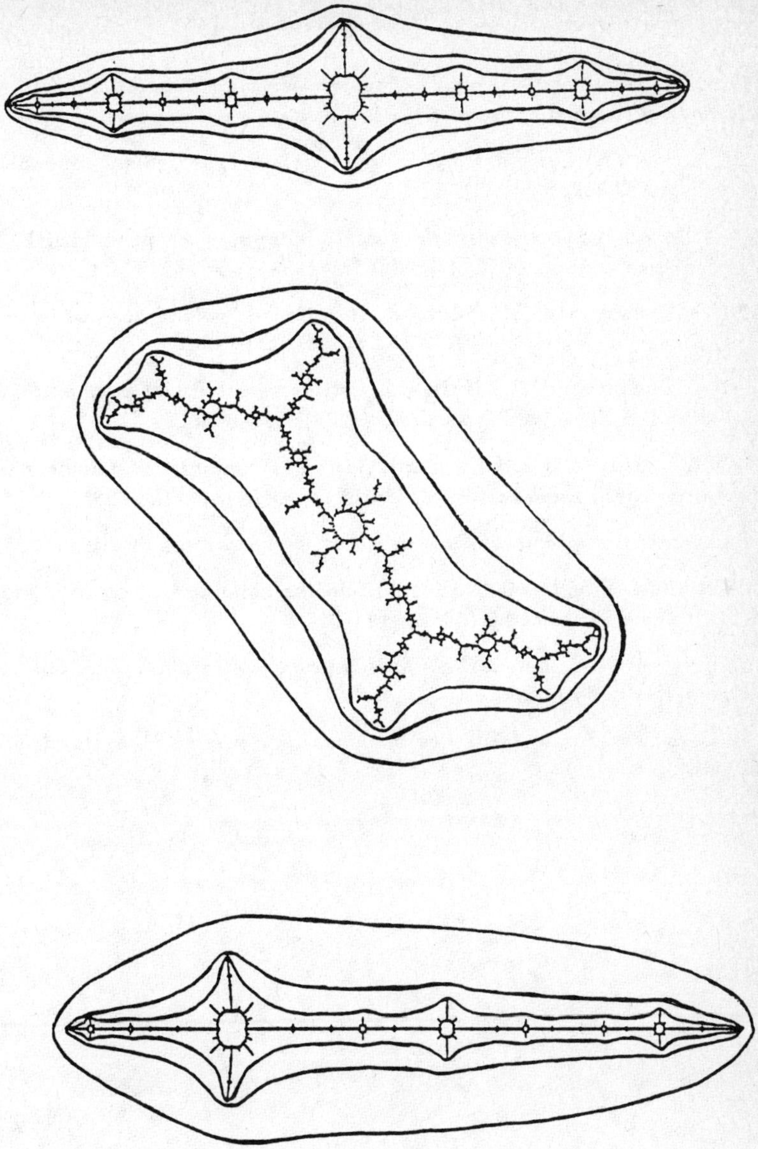

(a)

(b)

(c)

Figure 15.4.

References

[BH] B.Branner and J.H.Hubbard, *The iteration of cubic polynomials, Part I : The global topology of parameter space*, Acta Mathematica (to appear).

[D 1] A.Douady, *Systémes dynamiques holomorphes*, Sém. Bourbaki, exp. 599, Astérisque **105-106** (1983), 39–63.

[D 2] A.Douady, *Surgery on holomorphic mappings*, preprint MSRI (1986); french version: Proceedings of ICM (1986) (to appear).

[DH 1] A.Douady and J.H.Hubbard, *Itération de polynômes quadratiques complexes*, C.R. Acad. Sc. Paris, Série 1 **t.294** (1982), 123–126.

[DH 2] A.Douady and J.H.Hubbard, *On the dynamics of polynomial-like mappings*, Ann. Sc. E.S.N., 4éme Séries **t. 18** (1985), 287–343.

[DH 3] A.Douady and J.H.Hubbard, *Systémes dynamiques holomorphes I : Itération des polynômes complexes*, Publ. Math. Orsay (1984); *II*, (1985).

[L] P.Lavaurs, *Complément à une inégalité de Yoccoz*, manuscript (1986).

[MSS] R.Mañé, P.Sad and D.Sullivan, *On the dynamics of rational maps*, Ann. Sc. E.N.S., 4éme Séries **t. 16** (1982), 193–217.

[S] M.Shishikura, *On the quasi-conformal surgery of rational functions*, Ann. Sc. E.N.S., 4éme Séries **t. 20** (1987).

[Y] J.-C.Yoccoz, *Sur la taille des membres de l'ensemble de Mandelbrot*, manuscript (1986).

MAT-REPORT NO. 1987-05.

DICRITICALNESS OF A SINGULAR FOLIATION

F. Cano

The differential form $\omega = adx + bdy$ is classically said to be "dicritical" at the origin iff $x A + y B = 0$ (identically), where A, resp. B, is the homogeneous part of degree $r = \min(\text{order of } a, \text{order of } b)$, of a, resp. b. This is equivalent to say that after a blowing-up of the origin, then the exceptional divisor is not a leaf of the strict transform of the singular foliation given by ω. It is also equivalent to say that for each direction at the origin (unless a finite number of such directions) there is one integral curve of ω which is tangent to that direction.

The existence of a canonical desingularization for ω ([SE]) allows us to say that ω is "dicritical" at the origin if and only if the situation above occurs in one step of the desingularization process, and not only at the first step. With this new definition, dicriticalness is equivalent to the property of having infinitely many separatrices (which appear at the dicritical step of the desingularization).

This paper is mainly devoted to the study of the dicriticalness of a singular foliation in higher dimension.

Let us consider the integrable differential form given by Darboux:

$$\omega = (x^m z - y^{m+1})dx + (y^m x - z^{m+1})dy + (z^m y - x^{m+1})dz.$$

This form has no separatrices ([JOU]), but each non degenerate plane section is dicritical, hence it has infinitely many separatrices. Moreover, after making a blowing-up of the origin, then the exceptional divisor is not a leaf of the strict transform of the singular foliation given by ω.

Thus, we shall not consider the existence of infinitely many separatrices for a caracterization of the dicriticalness. We shall take two points of view: two-dimensional no degenerate sections and blowing-ups. Our main goal is to show that both approaches give rise to equivalent notions of dicriticalness.

From the viewpoint of two-dimensional sections, we shall need also singular ones. For these sections, dicriticalness consists, roughly speaking, in the existence of infinitely many separatrices.

Although there is not yet a desingularization in higher dimension for singular foliations (see [CA] for local results in the case of vector fields), we can give a reasonable concept of permissible center. Hence, decriticalness from this point of view will be to reach an exceptional divisor which is not a leaf of the strict transform of the singular foliation, after a finite number of permissible blowing-ups.

In order to compare both points of view, we shall need a result of "conditionated desingularization" for a singular surface with respect to a singular foliation of the ambient space (here we restrict ourselves to the three-dimensional case). This means that the surface can be desingularized by means of blowing-ups which are both permissible for the surface and for the singular foliation.

The origin of the word "dicritical" is in the works of Automme ([AU]) as J.F. Mattei has kindley communicated to me.

I am very gratefull to J.M. Aroca and D. Cervau for helpfull conversations during the preparation of this work.

§1. Adapted singular foliations

(1.1) Let X be a non singular complex analytic variety with dim $X = n \geq 2$ and let E be a normal crossings divisor of X. Here we introduce the singular foliations "adapted to E". If $E = \emptyset$, we obtain the classical definitions. We keep the divisor E in mind because we want to consider an intermediate step of a sequence of blowing-ups, E being the total exceptional divisor.

(1.2) Let $\Xi_X[E]$ be the sheaf of germs of vector fields which are tangent to E. Let us denote by $\Omega_X[-E]$ the sheaf of germs of meromorphic differential forms having logarithmic poles along E ([SA]). Then $\Xi_X[E]$ and $\Omega_X[-E]$ are mutually dual locally free $\mathcal{O}_X$-modules of rank n.

(1.3) Given a point $P \in X$, let $(x_1,\ldots,x_n)$ be a regular system of parameters of $\mathcal{O}_{X,P}$ such that E is locally given at P by $\Pi_{i \in A} x_i = 0$. Then $\Xi_{X,P}[E]$, resp. $\Omega_{X,P}[-E]$, has a basis given by

(1.3.1) $$\{x_i \partial/\partial x_i\}_{i \in A} \cup \{\partial/\partial x_i\}_{i \notin A} ,$$

resp.

(1.3.2) $$\{dx_i/x_i\}_{i \in A} \cup \{dx_i\}_{i \notin A}.$$

(1.4) Definition. A "singular foliation $\mathcal{F}$ over X adapted to E" is an invertible $\mathcal{O}_X$-submodule of $\Omega_X[-E]$ which satisfies the integrability condition

(1.4.1) $$\mathcal{F} \wedge d\mathcal{F} = 0$$

(d = exterior differential). We say that $\mathcal{F}$ is "saturated" if moreover

(1.4.2) $$\Omega_X[-E] / \mathcal{F} \text{ has no torsion}$$

(1.5) If

(1.5.1) $$\omega = \sum_{i\in A} a_i(dx_i/x_i) + \sum_{i\notin A} a_i dx_i$$

is a generator of $\mathcal{F}_P$, then (1.4.2) means that

(1.5.2) $$\text{g.c.d.}(a_i; i=1,..,n) = 1.$$

(1.6) Given a singular foliation $\mathcal{F}$ adapted to E, there is exactly one singular foliation Sat($\mathcal{F}$,E) adapted to E satisfying

(1.6.1) $$\mathcal{F} \subset \text{Sat}(\mathcal{F},E) \quad \text{and} \quad \text{Sat}(\mathcal{F},E) \text{ is saturated.}$$

We call Sat($\mathcal{F}$,E) the "saturation of $\mathcal{F}$ adapted to E". It is locally obtained by dividing (1.5.1) by (1.5.2).

(1.7) Let $E' \subset E$ be another normal crossings divisor of X. Then $\Omega_X[-E'] \subset \Omega_X[E]$. We define the "adaptation Adap($\mathcal{F}$,E;E') of $\mathcal{F}$ to E'" by

(1.7.1) $$\text{Adap}(\mathcal{F},E;E') = \mathcal{F} \cap \Omega_X[-E'].$$

Then Adap($\mathcal{F}$,E;E') is also a singular foliation over X adapted to E'. Moreover, if $\mathcal{F}$ is saturated, then Adap($\mathcal{F}$,E;E') is also saturated. In the case E' = $\emptyset$, we say that $\mathcal{T}$ = Adap($\mathcal{F}$,E;$\emptyset$) is the "singular foliation associated to $\mathcal{F}$". The leaves, separatrices, integrant factors, etc. of $\mathcal{F}$ are, by definition, the leaves, separatrices, integrant factors, etc. of $\mathcal{T}$.

(1.8) Definition. Let $\mathcal{F}$ be a singular foliation over X adapted to E. Let F be an irreducible component of E. We say that F is a "dicritical component of E for $\mathcal{F}$" iff F is not a leaf of Sat($\mathcal{F}$,E).

§2. Blowing-ups and numerical invariants

(2.1) Definition. Let $\mathcal{F}$ be a singular foliation over X adapted to E and let P be a point of X. The "adapted order $\nu(\mathcal{F},E;P)$ of $\mathcal{F}$ at P" is defined by

(2.1.1) $$\nu(\mathcal{F},E;P) = \max\{t;\ \mathcal{F}_P \subset m_P^t\, \Omega_{X,P}[-E]\},$$

where m_P = maximal ideal of $\mathcal{O}_{X,P}$. Equivalently , taking a generator ω as in (1.5.1), we have

(2.1.2) $$\nu(\mathcal{F},E;P) = \min(\nu_P(a_i)\ ; i=1,...,n)$$

where $\nu_P(a_i)$ is the m_P-adic order of $a_i \in \mathcal{O}_{X,P}$.

(2.2) The "adapted singular locus Sing($\mathcal{F}$,E)" is the set of the points P which $\nu(\mathcal{F};E;P) \geq 1$. It is an analytic set. If $\mathcal{T} = \mathrm{Adap}(\mathcal{F},E;\emptyset)$ we have

$$\mathrm{Sing}(\mathcal{T},\emptyset) \subset \mathrm{Sing}(\mathcal{F},E). \tag{2.2.1}$$

(2.3) Let $Y \subset X$ be a nonsingular analytic subvariety of X with $\mathrm{codim}_X Y \geq 2$ and having normal crossings with E. Let $\pi: X' \to X$ be the blowing-up centered at Y. Put $E' = \pi^{-1}(E \cup \{Y\})$. Let $\mathcal{F}$ be a saturated singular foliation over X adapted to E. Then there is a unique saturated singular foliation $\mathcal{F}'$ over X' adapted to E' such that

$$\mathcal{F}'|_{X'-\pi^{-1}(Y)} = \mathcal{F}|_{X-Y} \qquad (\text{vía } \pi). \tag{2.3.1}$$

(See [CA] for the case of vector fields)

(2.4) Definition. We say that $\mathcal{F}'$ is the "strict transform of $\mathcal{F}$ by π". For denoting the above situation we use the notation: $\pi: (X',E',\mathcal{F}') \to (X,E,\mathcal{F})$.

(2.5) Let $P \in Y$ be a point and let $p = (x_1,\ldots,x_n)$ be a regular system of parameters of $\mathcal{O}_{X,P}$ such that

$$E = (\prod_{i \in A} x_i = 0); \quad Y = (x_i = 0;\ i \in B) \tag{2.5.1}$$

locally at P. We shall say that p is "adapted to the pair (E,Y)". Let $P' \in \pi^{-1}(P) \subset \pi^{-1}(Y)$. Then there is an index $t \in B$ and scalars $\zeta_i \in \mathbb{C}$, $i \in B-\{t\}$, such that a certain regular system of parameters $p' = (x'_1,\ldots,x'_m)$ of $\mathcal{O}_{X',P'}$ is given by

$$\begin{aligned} x'_i &= x_i \qquad i \notin B \text{ or } i = t \\ x'_t(x'_i+\zeta_i) &= x_i \qquad i \in B-\{t\}. \end{aligned} \tag{2.5.2}$$

Assume that $\mathcal{F}_P$ is generated by ω as in (1.5.1). Let us define

$$\mu(\mathcal{F},E;Y,P) = \min(\{\nu_Y(a_i); i \notin B-A\} \cup \{\nu_Y(a_i)+1; i \in B-A\}) \tag{2.5.3}$$

Put $\mu = \mu(\mathcal{F},E;Y,P)$. Then a generator ω' of $\mathcal{F}'_{P'}$ is given by

$$\begin{aligned} \omega' = (1/x'_t)^{\mu}\Big[&\sum_{i \in A-B} a_i(dx'_i/x'_i) + \sum_{i \notin A \cup B} a_i dx'_i + \\ &+ \Big(\sum_{i \in B \cap A} a_i + \sum_{i \in B-A} x'_i(x'_i+\zeta_i)a_i\Big)(dx'_t/x'_t) + \\ &+ \sum_{\substack{i \in B \cap A-\{t\} \\ \zeta_i = 0}} a_i(dx'_i/x'_i) + \sum_{\substack{i \in B \cap A-\{t\} \\ \zeta_i \neq 0}} (a'_i/(x'_i+\zeta_i))dx'_i + \\ &+ \sum_{i \in B-A} x'_t a_i dx'_i\Big]. \end{aligned} \tag{2.5.4}$$

if $t \in A$. Analogous formula if $t \notin A$.

(2.6) Since $\mu(\mathcal{F},E;Y,P)$ does not depend on $P \in Y$, we always put $\mu(\mathcal{F},E;Y) = \mu(\mathcal{F},E;Y,P)$.

§3. Permissible centers

(3.1) Let $X,E,\mathcal{F},Y$ and $P \in Y$ be as above. Assume that $\mathcal{F}_P$ is generated by ω as in (1.5.1). Let us define the number $\rho(\mathcal{F},E;Y,P)$ by

(3.1.1) $$\rho(\mathcal{F},E;Y,P) = \min(\{\nu_P(a_i); i \notin B-A\} \cup \{\nu_P(a_i)+1; i \in B-A\}).$$

Compare with (2.5.3).

(3.2) Definition. In the above situation, we say that Y is "permissible for $\mathcal{F}$ at P" iff $\mathrm{Adap}(\mathcal{F},E;\emptyset)|_Y = 0$ (i.e., Y is an integral variety of $\mathcal{F}$) and

(3.2.1) $$\mu(\mathcal{F},E;Y) = \rho(\mathcal{F},E;Y,P).$$

We say that Y is "permissible for $\mathcal{F}$" iff it is permissible at each point $P \in Y$.

(3.3) Remarks. 1) If $Y = \{P\}$, a single point, then it is always permissible.

2) If Y is an integral variety of $\mathcal{F}$, then the set of points $P \in Y$ such that Y is not permissible at P is an analytic set $Z \subset Y$, $Z \neq Y$.

(3.4) Proposition. Let $\pi: (X',E',\mathcal{F}') \to (X,E,\mathcal{F})$ be the blowing-up with center Y and assume that Y is permissible at $P \in Y$. Then

a) $\pi^{-1}(P) \not\subset \mathrm{Sing}(\mathcal{F}',E')$.

b) For each point $P' \in \pi^{-1}(P)$ we have

(3.4.1) $$\nu(\mathcal{F}',E';P') \leq \nu(\mathcal{F},E;P).$$

Proof. It follows from (2.5.4) and the analogous formula for $t \notin A$.

(3.5) Remark. The property a) of (3.4) characterizes also the permissible centers.

(3.6) Proposition (stationary sequences). Let $X,E,\mathcal{F},Y$ and $P \in Y$ be as above. Assume that $\dim Y = 1$ and that Y is an integral variety of $\mathcal{F}$. Let us define an infinite sequence of blowing-ups

(3.6.1) $$\mathcal{S} = \{\pi(i):(X(i),E(i),\mathcal{F}(i)) \to (X(i-1),E(i-1),\mathcal{F}(i-1))\}_{i \geq 1}$$

inductively as follows: $(X(o),E(o),\mathcal{F}(o)) = (X,E,\mathcal{F})$, $P(o) = P$, $Y(o) = Y$; $\pi(i)$ is centered at the point $P(i-1)$, $i \geq 1$. $Y(i)$ is the strict transform of $Y(i-1)$ by $\pi(i)$, $i \geq 1$, $P(i)$ is the only point in $Y(i) \cap \pi(i)^{-1}(P(i-1))$. Then there is an index N such that for each $i \geq N$ then $Y(i)$ is permissible for $\mathcal{F}(i)$ at $P(i)$.

Proof. It can be done in a parallel way to the similar result for vector fields in [CA].

§4. Dicriticalness at the first infinitesimal neighborhood

(4.1) Definition. Let $\mathcal{H}$ be a singular foliation over X adapted to E. Let $\mathcal{F} = \mathrm{Sat}(\mathcal{H},E)$ and assume that E has no dicritical components for $\mathcal{F}$. Let us fix a point $P \in X$. We say that $\mathcal{H}$ is "strongly dicritical at the first infinitesimal neighborhood of P" iff the following property is satisfied:

(4.1.1) "Let $\pi: (X',E',\mathcal{F}') \to (X,E,\mathcal{F})$ be the blowing-up centered at P. Then $\pi^{-1}(P)$ is a dicritical component of E' for $\mathcal{F}'$."

We say that $\mathcal{H}$ is "dicritical at the first infinitesimal neighborhood of P " iff the following property is satisfied:

(4.1.2) "There is an open neighborhood $\mathcal{U}$ of P and a permissible center $Y \subset \mathcal{U}$ for $\mathcal{F}|_{\mathcal{U}}$, $P \in Y$, such that if $\pi: (X',E',\mathcal{F}') \to (\mathcal{U}, E \cap \mathcal{U}, \mathcal{F}|_{\mathcal{U}})$ is the blowing-up centered at Y, then $\pi^{-1}(Y)$ is a dicritical component of E' for $\mathcal{F}'$."

(4.2) Remark. $\mathcal{H}$ is strongly dicritical at the first infinitesimal neighborhood of P iff $\mathrm{Adap}(\mathcal{H},E;\emptyset)$ is.

(4.3) Proposition. Let $\mathcal{T}$ be a singular foliation over X (hence adapted to the empty divisor). Let $P \in X$ and let us fix an integer d, $2 \leq d \leq n$. Let $G(d;n)$ be the grassmannian of d-planes of T_PX. Then there is a nonempty Zariski open set $\mathcal{U}(\mathcal{T},d;P)$ of $G(d;n)$ such that $T \in \mathcal{U}(\mathcal{T},d;P)$ iff the following property is satisfied:

(4.3.1) "For each nonsingular analytic subvariety Z of X with $P \in Z$ and such that $T_PZ = T$, we have $\nu(\mathcal{T},\emptyset;P) = \nu(\mathcal{T}|_Z,\emptyset;P)$."

Proof. The condition $\nu(\mathcal{T},\emptyset;P) = \nu(\mathcal{T}|_Z,\emptyset;P)$ depends only on T and not on Z. Moreover, it is Zariski-open. Now, let us prove that it is a non empty condition. Taking a generator of $\mathcal{T}_P$ and its initial part, we have only to show that if

(4.3.2) $$W = \sum_{i=1}^{n} A_i(X_1,..,X_n)dX_i$$

is a homogeneous form of degree r, then there is a hyperplane L such that $W|_L \neq 0$. Let us reason by contradiction. Without loss of generality we can assume $A_n \neq 0$. Let us consider

(4.3.3) $$L = X_1 + \lambda X_2, \quad \lambda \in \mathbb{C}.$$

Then $W|_L = 0$ implies that

(4.3.4) $$A_n(-\lambda X_2, X_2, \ldots, X_n) = 0$$

for each $\lambda \in \mathbb{C}$. (Note $n \geq 3$). This is the desired contradiction.

(4.4) Remark. Assume that $\mathcal{F}$ is saturated. Then a similar result to (4.3) for $\mathcal{F}|_Z$ being saturated is not true. The obstruction to this is concentrated in the case $n = 3$ and $\mathcal{F}$ strongly dicritical in the first infinitesimal neighborhood. We can obtain $\mathcal{F}|_Z$ saturated if we allow no linear perturbations of T ("transversality theorem" in [MM]).

(4.5) Theorem. Let $\mathcal{F}$ be a singular foliation over X. Let P be a point of X and let us fix an integer d, $2 \leq d \leq n$. Fix an element $T \in \mathcal{U}(\mathcal{F}, d; P)$. Then the following statements are equivalent:

a) $\mathcal{F}$ is strongly dicritical at the first infinitesimal neighborhood of P.

b) For each nonsingular analytic subvariety Z of X with $P \in Z$, $T_P Z = T$, then $\mathcal{F}|_Z$ is strongly dicritical at the first infinitesimal neighborhood of P.

c) There is a nonsingular analytic subvariety Z of X with $P \in Z$, $T_P Z = T$, such that $\mathcal{F}|_Z$ is strongly dicritical at the first infinitesimal neighborhood of P.

Proof. The only non trivial fact is "c) implies a)". Assume that $\mathcal{F}_P$ is generated by $\omega = \sum a_i dx_i$. Let $r = \nu(\mathcal{F}, \emptyset; P)$ and denote A_i = homogeneous part of degree r of a_i, $i=1,\ldots,n$. Let us consider the integrable homogeneous form

(4.5.1) $$W = \sum_{i=1,\ldots,n} A_i dX_i .$$

Let $P(W) = \sum_{i=1,\ldots,n} A_i X_i$. Then $\mathcal{F}$ is strongly dicritical at the first infinitesimal neighborhood of P iff $P(W) = 0$. Without loss of generality we can assume

(4.5.2) $$T = (X_n = 0) \; ; \; A_1(X_1, \ldots, X_{n-1}, 0) \neq 0$$

(by induction we have only to consider the case $d = n-1$). Our assumption is

(4.5.3) $$P(W|_T) = \sum_{i=1}^{n-1} A_i(X_1, \ldots, X_{n-1}, 0) . X_i = 0.$$

Let us write

(4.5.4) $$A_i = \sum_{h=0}^{r} A_{i,h}.X_n^h \; ; \quad i = 1,\ldots,n.$$

(4.5.5) $$B_h = \sum_{i=1,\ldots,n-1} X_i A_{i,h} + A_{n,h-1}.$$

Where $A_{i,h}$ is homogeneous of degree r-h in the variables $X_1,\ldots,X_{n-1}$. Then P(W) = 0 is equivalent to

(4.5.6) $$B_s = 0 \quad \text{for each} \quad s = 0,1,\ldots,r+1$$

Let us prove (4.5.6) by induction on s. The case s=0 is (4.5.3). Assume $B_{s'}=0$ for $s' < s$ and let us prove $B_s = 0$. Let us write

(4.5.7) $$W = \sum_{h=0}^{r} X_n^h.W_h \; ; \; W_h = \sum_{i=1}^{n} A_{i,h} dX_i .$$

Then the integrability condition $W \wedge dW = 0$ implies

(4.5.8) $$\Big(\sum_{h=0}^{r} X_n^h W_h\Big) \wedge \Big(\sum_{h=0}^{r} hX_n^{h-1} dX_n \wedge W_h + X_n^h dW_h\Big) = 0$$

Looking at the coefficient of X_n^{s-1} in (4.5.8) we have

(4.5.9) $$0 = \sum_{m=0}^{s-1} H^{(m)}$$

(4.5.10) $$H^{(m)} = W_m \wedge dW_{s-m-1} + (s-m)W_m \wedge dX_n \wedge W_{s-m}.$$

Let $H^{(m,b)}$ be the coefficient of $dX_1 \wedge dX_b \wedge dX_n$ in $H^{(m)}$. Note that

(4.5.11) $$\sum_{m=0}^{s-1} H^{(m,b)} = 0.$$

We have that

(4.5.12) $$H^{(m,b)} = A_{1,m}(\partial A_{n,s-m-1}/\partial X_b) - A_{b,m}(\partial A_{n,s-m-1}/\partial X_1) +$$
$$+ A_{n,m}((\partial A_{b,s-m-1}/\partial X_1) - (A_{1,s-m-1}/\partial X_b)) +$$
$$+ (s-m)(A_{b,m}\, A_{1,s-m} - A_{1,m}\, A_{b,s-m}).$$

Let us denote

$$G^{(m)} = \sum_{b=1}^{n-1} X_b H^{(m,b)} \tag{4.5.13}$$

Note that

$$0 = \sum_{m=0}^{s-1} G^{(m)}. \tag{4.5.14}$$

We have that

$$\begin{aligned} G^{(m)} = {} & A_{1,m}\Big(\sum_{b=1}^{n-1} X_b(\partial A_{n,s-m-1}/\partial X_b)\Big) - \\ & - \Big(\sum_{b=1}^{n-1} X_b A_{b,m}\Big)(\partial A_{n,s-m-1}/\partial X_1) + \\ & + A_{n,m}\Big(\sum_{b=1}^{n-1} X_b(\partial A_{b,s-m-1}/\partial X_1) - \sum_{b=1}^{n-1} X_b(\partial A_{1,s-m-1}/\partial X_b)\Big) + \\ & + (s-m)\Big(A_{1,s-m}\sum_{b=1}^{n-1} X_b A_{b,m} - A_{1,m}\sum_{b=1}^{n-1} X_b A_{b,s-m}\Big). \end{aligned} \tag{4.5.15}$$

Note that by the Euler's Identity and the induction assumption we have that

$$\sum_{b=1}^{n-1} X_b(\partial A_{n,s-m-1}/\partial X_b) = (r-s+m+1)A_{n,s-m-1}. \tag{4.5.16}$$

$$\sum_{b=1}^{n-1} X_b(\partial A_{1,s-m-1}/\partial X_b) = (r-s+m+1)A_{1,s-m-1} \tag{4.5.17}$$

$$\sum_{b=1}^{n-1} X_b A_{b,m} = -A_{n,m-1}. \tag{4.5.18}$$

$$\sum_{b=1}^{n-1} X_b(\partial A_{b,s-m-1}/\partial X_1) = -A_{1,s-m-1} - \partial A_{n,s-m-2}/\partial X_1. \tag{4.5.19}$$

Hence, substituing in (4.5.15), we have

$$\begin{aligned} G^{(m)} = {} & (r-s+m+1)A_{1,m}A_{n,s-m-1} - (r-s+m+2)A_{1,s-m-1}A_{n,m} + \\ & + A_{n,m-1}(\partial A_{n,s-m-1}/\partial X_1) - A_{n,m}(\partial A_{n,s-m-2}/\partial X_1) - \\ & - (s-m)A_{1,s-m}A_{n,m-1} - (s-m)A_{1,m}\sum_{b=1}^{n-1} X_b A_{b,s-m}. \end{aligned} \tag{4.5.20}$$

Hence

$$0 = \sum_{m=0}^{s-1} G^{(m)} = \sum_{t=0}^{s-1} (2t-s)A_{1,t}A_{n,s-1-t} - \tag{4.5.21}$$

$$- \sum_{t=1}^{s} t A_{1,t} A_{n,s-1-t} - \sum_{t=1}^{s-1} (s-t) A_{1,t} \sum_{b=1}^{n-1} X_b A_{b,s-t} -$$

$$- s A_{1,0} \sum_{b=1}^{n-1} X_b A_{b,s} .$$

By induction hypothesis we have

(4.5.21) $$\sum_{b=1}^{n-1} X_b A_{b,s-t} = -A_{n,s-t-1} , \quad t=1,\ldots,s-1.$$

Hence

(4.5.22) $$0 = \sum_{m=0}^{s-1} G^{(m)} = -s A_{1,0}(A_{n,s-1} + \sum_{b=1}^{n-1} X_b A_{b,s}) = -s A_{1,0} B_s .$$

Since $s A_{1,0} \neq 0$ by (4.5.2), we have $B_s = 0$.

(4.6) Proposition. Let $\mathcal{F}$ be a saturated singular foliation over X adapted to E. Let $Y \subset X$ be a nonsingular analytic subvariety of X having normal crossings with E. Let $F \subset E$ be the normal crossings divisor consisting on the components of E which don not contain Y. Let $\mathcal{T} = \mathrm{Adap}(\mathcal{F},E;F)$. Then

a) Y is permissible for $\mathcal{F}$ adapted to E iff Y is permissible for $\mathcal{T}$ adapted to F.

b) Assume that Y is permissible for $\mathcal{F}$ adapted to E and let $\pi: X' \to X$ be the blowing-up of X centered at Y. Denote by $\mathcal{F}'$, resp. $\mathcal{T}'$, the strict transforms of $\mathcal{F}$, resp. $\mathcal{T}$. Then $\pi^{-1}(Y)$ is a dicritical component for $\mathcal{F}'$ iff it is a dicritical component for $\mathcal{T}'$.

Proof. b) is immediate, since $\mathcal{F}'$ and $\mathcal{T}'$ have the same associated singular foliation. a) is straigh-forward from the definitions.

(4.7) Remark. Generically, a point of Y is not in F. Hence the above Proposition justifies the "non adapted" statement of the following theorem.

(4.8) Theorem. Let $\mathcal{T}$ be a saturated singular foliation over X (hence adapted to the empty divisor). Let $Y \subset X$ be a permissible center for $\mathcal{T}$ and let $\pi: (X',E',\mathcal{T}') \to (X,\emptyset,\mathcal{T})$ be the blowing-up with center Y (hence $\pi^{-1}(Y) = E'$). Put $r = \nu(\mathcal{T},\emptyset;Y) = \min\{\nu(\mathcal{T},\emptyset;P); P \in Y\}$ and $\mu = \mu(\mathcal{T},\emptyset;Y)$. Then:

a) If $\mu = r+1$, then there is an analytic subspace $Z \subset Y$, $Z \neq Y$ satisfying the following property: if $P \in Y-Z$, then $\mathcal{T}$ is strongly dicritical in the first infinitesimal neighborhood of P iff E' is a dicritical component of $\mathcal{T}'$.

b) If $\mu = r+1$, $P \in Y$, $\nu(\mathcal{T},\emptyset;P) = r$) and E' is a dicritical component of $\mathcal{T}'$, then $\mathcal{T}$ is strongly dicritical at the first infinitesimal neighborhood of P.

c) If $\mu = r$, then E' is a dicritical component for $\mathcal{T}'$. Moreover, we have the following property:

(4.8.1) "For each generic germ $Z \supset Y \ni P$ of nonsingular analytic subvariety Z of X such that $\dim Z = \dim Y+1$, then $\mathcal{H} = \mathrm{Sat}(\mathcal{T}|_Z, \emptyset)$ is not singular and Y Y is not a leaf of $\mathcal{H}$".

Proof. a) We may assume that $\nu(\mathcal{T},\emptyset;P) = r$ for each $P \in Y$. Moreover, we can take Z such that if $P \in Y-Z$ then there is a regular system of parameters $(x_1,\ldots,x_n)$ of $\mathcal{O}_{X,P}$ and a generator $\omega = \sum a_i dx_i$ of $\mathcal{T}_P$ satisfying the following properties

(4.8.2) $$Y = (x_1 = \ldots = x_m = 0)$$

(4.8.3) "Let $b = \sum_{i=1,\ldots,m} x_i a_i$. Then $\mathrm{In}^{r+1}(b) = 0$ iff $\mathrm{In}_Y^{r+1}(b) = 0$". Where $\mathrm{In}^{r+1}(b)$, resp. $\mathrm{In}_Y^{r+1}(b)$, denotes the homogeneous part of degree r+1 of b, resp. the homogeneouos part of degree r+1 of b in the indeterminates $x_1,\ldots,x_m$. Now, by (2.5.4), E' is a dicritical component for $\mathcal{T}'$ iff $\mathrm{In}_Y^{r+1}(b) = 0$, hence iff $\mathrm{In}^{r+1}(b)=0$. But since $\mu = r+1$ and $\nu(\mathcal{T},\emptyset;P) = r$, we have

(4.8.4) $$\nu(\mathcal{T}|_{x_{m+1}=\ldots=x_n=0}, \emptyset;P) = r.$$

Then, by Theorem (4.5), $\mathrm{In}^{r+1}(b) = 0$ iff $\mathcal{T}$ is strongly dicritical at the first infinitesimal neighborhood of P.

b) It is enough to remark that $\mathrm{In}_Y^{r+1}(b) = 0$ implies that $\mathrm{In}^{r+1}(b) = 0$.

c) From (2.5.4) we know that E' is a dicritical component for $\mathcal{T}'$. Now, take a regular system of parameters $(x_1,\ldots,x_n)$ of $\mathcal{O}_{X,P}$ such that $Y = (x_1 = \ldots = x_m = 0)$ and $Z = (x_2 = \ldots = x_m = 0)$. Let $\omega = \sum a_i dx_i$ be a generator of $\mathcal{T}_P$. Since Z is generic and there is an index $j > m$ with $\nu_Y(a_j) = r$, we can assume that

(4.8.5) $$a_j(x_1,0,\ldots,0) \neq 0.$$

Since $\nu_Y(a_i) \geq r$ for each $i=1,\ldots,n$, then $\mathcal{T}|_Z$ is generated at P by

(4.8.9) $$\omega|_Z = a_1|_Z dx_1 + \sum_{i=m+1,\ldots,n} a_i|_Z dx_i = x_1^r \cdot \omega'$$

and ω' is not singular by (4.8.5) and by $\nu_Y(a_j) = r$. Moreover, $x_1 = 0$ is not a leaf of ω'.

§ 5. Foliation induced in a singular surface

(5.1) In the sequel we shall assume that $n = \dim X = 3$.

(5.2) Definition. Let $\mathcal{F}$ be a saturated singular foliation over X adapted to E. Assume that E does not have dicritical components for $\mathcal{F}$. Let $S \subset X$ be a surface. We say that S is "admissible" for $\mathcal{F}$ if it is not a leaf of $\mathcal{F}$ and each component of Sing S is an integral variety of $\mathcal{F}$. If S is admissible for $\mathcal{F}$, we say that the "adapted restriction $(\mathcal{F},E)|_S$ "is a pair $(\mathcal{T}|_{S^o},T)$ where $\mathcal{T}$ is the singular foliation associated to $\mathcal{F}$, S^o = S-Sing S, and

(5.2.1) $$T = (S \cap E) \cup \text{Sing } S \cup \text{Sing}(\mathcal{T}|_{S^o},T).$$

We say that T is the "adaptation locus of $\mathcal{F}$ to S ".

(5.3) Remarks. 1) T is an analytic subspace $T \subset S$. $T \neq S$.

2) $\mathcal{T}|_{S^o}$ is not necessarily saturated.

(5.4) Definition. We say that $(\mathcal{F},E)|_S = (\mathcal{T}|_{S^o},T)$ is "dicritical" iff there is an irreducible component T_1 of T satisfying one of the following conditions

a) There is a point $P \in T_1$ and infinitely many germs of curve $\Gamma \subset S$ at P such that Γ-T is a leaf of $\mathcal{T}|_{S^o}$.

b) $\dim T_1 = 1$ and except for a set of isolated points, for each point $P \in T_1$ there is a germ of curve $\Gamma \subset S$ at P, $\Gamma \subset T$, such that Γ-T is a leaf of $\mathcal{T}|_{S^o}$.

(5.5) Proposition. Let $\mathcal{F}$,E,S be as above. Assume that Y is a permissible center for $\mathcal{F}$ and let $\pi: (X',E',\mathcal{F}') \to (X,E,\mathcal{F})$ be the blowing-up center at Y. Assume that $\pi^{-1}(Y)$ is not a dicritical component for $\mathcal{F}'$. Then

a) The strict transform S' of S by π is admissible for $\mathcal{F}'$.

b) $(\mathcal{F},E)|_S$ is dicritical iff $(\mathcal{F}',E')|_{S'}$ is dicritical.

Proof. a) is obvious. For b), assume that $(\mathcal{F},E)|_S$ is dicritical. Let T_1 be as in (5.4), since leaves have good behaviour under blowing-up, the infinitely many germs of curve of a) or b) of (5.4) produce infinitely many germs of curve in X' which are adherent to a certain irreducible component of $\pi^{-1}(T_1) \cap S'$, which is necessarily in the adaptation locus of $(\mathcal{F}',E')|_{S'}$. Conversely, assume that $(\mathcal{F}',E')|_{S'}$ is dicritical and let T_1' be as in (5.4). It is enough to show that $\pi(T_1') \subset T$ = = adaptation locus of $(\mathcal{F},E)|_S$. The only obstruction to this is that $\pi(T_1')$ = = a point of Y-T. But this is not possible since if $P \in$ Y-T, then $\mathcal{T}|_{S^o}$ is not singular and hence it has only one integral branch at P.

(5.6) Remark. The theorem of desingularization of singular foliations in a two-dimensional non singular space Z asserts that after finitely many blowing-ups we have a singular foliation $\mathcal{F}^{\sim}$ over $Z^{\sim}$ and an exceptional divisor $E^{\sim}$ of $Z^{\sim}$ such that all the singularities of $\mathcal{F}^{\sim}$ are "simple" (eigenvalues without resonnances, see [MM] or [CM]). Moreover, the original foliation $\mathcal{F}$ has infinitely many separatrices at the considered singular point iff $E^{\sim}$ has at least one dicritical component for $\mathcal{F}^{\sim}$. (See [SE], or [CA2] for an adapted proof of the theorem of desingularization). Now, assume that S has normal crossings with E and that the adaptation locus T of $\mathcal{F}$ to S is exactly $T = E \cap S$ (hence a normalcrossings divisor on S). Then it is easy to verify that $\mathrm{Sat}(\mathcal{F}|_S, T)$ is well defined and T has a dicritical component for $\mathrm{Sat}(\mathcal{F}|_S, T)$ iff $(\mathcal{F}, E)|_S$ is dicritical. Hence, in this case, Definition (5.4) corresponds to the classical situation of dicriticalness . The following result shows that we can always reach $T = E \cap S$.

(5.7) Proposition. Let $\mathcal{F}$ be a singular foliation over X adapted to E. Let S be an admissible surface for $\mathcal{F}$. Then there is a finite sequence of permissible blowing-ups

(5.7.1) $\mathcal{S} = \{\pi(i): (X(i), E(i), \mathcal{F}(i)) \rightarrow (X(i-1), E(i-1), \mathcal{F}(i-1))\}_{i=1,..,N}$

which satisfies the following properties:

a) $(X(o), E(o), \mathcal{F}(o), S(o)) = (X, E, \mathcal{F}, S)$.

b) $\pi(i)$ is centered at $Y(i-1)$, which is also a permissible center for $S(i-1)$, $i=1,...,N$.

c) $S(i)$ = strict transform of $S(i-1)$ by $\pi(i)$, $i=1,...,N$.

d) The adaptation locus $T(N)$ of $\mathcal{F}(N)$ to $S(N)$ satisfies that $T(N) = S(N) \cap E(N)$.

Proof. It is enough to show that we can make permissible all the one-dimensional components of T(= adaptation locus of $\mathcal{F}$ to S) both for $\mathcal{F}$ and S, after finitely many blowing-ups of points. But this follows from Proposition (3.6) and the analogous results for surfaces (see e.g. [SG]).

§6. Conditionated desingularization of a surface

(6.1) This paragraph is devoted to the proof of the following:

Theorem. (Conditionated desingularization). Assume that dim X = 3. Let $\mathcal{F}$ be a saturated singular foliation over X adapted to E. Let $S \subset X$ be an admissible surface for $\mathcal{F}$. Then we have two possibilities:

1st Possibility. There is a finite sequence of permissible blowing-ups

(6.1.1) $= \{\pi(i): (X(i), E(i), \mathcal{F}(i)) \rightarrow (X(i-1), E(i-1), \mathcal{F}(i-1))\}_{i=1,..,N}$

satisfying the following properties:

a) $(X(o),E(o),\mathcal{F}(o),S(o)) = (X,E,\mathcal{F},S)$.

b) $\pi(i)$ is centered at $Y(i-1)$, which is both permissible for $\mathcal{F}(i-1)$ and for $S(i-1)$. $i=1,\dots,N$.

c) $S(i)$ = strict transform of $S(i-1)$ by $\pi(i)$. $i=1,\dots,N$.

d) The adaptation locus $T(N)$ of $\mathcal{F}(N)$ to $S(N)$ is $T(N) = E(N) \cap S(N)$.

e) $S(N)$ is non singular and has normal crossings with $E(N)$.

2nd Possibility. There is a sequence of permissible blowing-ups like (6.1.1) such that $E(N)$ has a dicritical component for $\mathcal{F}(N)$.

(6.2) Remarks. 1) In order to simplify we shall consider only the case "X is a germ of analytic variety", i.e. we desingularize over a small enough neighborhood of a point $P \in X$. This assumption has been tacitily made in other statements, namely Proposition (5.7). Anyway, the situation of Theorem (6.1) is not local at each step. By example, if we blow-up P, then the exceptional divisor is a projective plane.

2) We shall detail here only the difficult part of the local control of the singularity. The global strategy which allows to prove the Theorem may be taken as in [CA3] and we do not include it. (See [OGC], Appendix for more details in the surface case).

3) The 2nd Possibility corresponds to the case that $\mathcal{F}$ is "dicritical" (to be defined in §7). Hence we can always desingularize S in the non dicritical case.

(6.3) By Proposition (5.7) we can assume that the adaptation locus $T(i)$ of $\mathcal{F}(i)$ to $S(i)$ satisfy $T(i) = E(i) \cap S(i)$ for each $i=0,1,\dots,N$.

(6.4) Proposition. Let $X,E,\mathcal{F}$ and S be as in (6.1). Let $P \in S$. Let us denote

(6.4.1) $$\mathrm{Inv}(\mathcal{F},E,S;P) = (\nu,d,r,\mu)$$

where $\nu = \nu_P(S)$, d = dimension of the strict tangent space of Hironaka of S at P (see [HI]), $r = \nu(\mathcal{F},E;P)$, $\mu = \mu(\mathcal{F},E;\{P\})$. Let $Y \subset X$ be a permissible center both for $\mathcal{F}$ and S and let $\pi\colon (X',E',\mathcal{F}',S') \to (X,E,\mathcal{F},S)$ be the blowing-up centered at Y. Let $P' \in \pi^{-1}(P)$. Assume that $\pi^{-1}(Y)$ is not a dicritical component of E' for $\mathcal{F}'$. Then

(6.4.2) $$\mathrm{Inv}(\mathcal{F},E,S;P) \geq \mathrm{Inv}(\mathcal{F}',E',S';P') = (\nu',d',r',\mu')$$

for the lexicographical order.

Proof. The general results for surfaces (e.g. [OGC]) and the Proposition (3.4) allows us to show only that if $r' = r$ then $\mu' \leq \mu$. But this follows from a computation over the equations (2.5.4), the fact that $\pi^{-1}(Y)$ is not a dicritical component of E' for $\mathcal{F}'$ and an application of Theorem (4.5).

(6.5) Remark. If $d = 0$, then $\nu' < \nu$ and the invariant $Inv(\mathcal{F},E,S;P)$ decreases strictly with one quadratic blowing-up If $d = 1$, we generate a stationary sequence by quadratic blowing-ups (if the invariant does not decrease) and it produces a curve permissible both for $\mathcal{F}$ and S. Blowing-up this curve the invariant decreases strictly. Hence, let us make the "local control" only in the "difficult case" $d = 2$.

(6.6) Let $Z \subset X$ be a subvariety which has maximal contact with S at P (see e.g. [AHV]). Let us consider the condition

(6.6.1) "Z is transversal to each component of E at P and it has normal crossings with E at P".

Let us say, without details, that (6.6.1) may be reached by desingularization procedures which are similar to the ones we are going to present here. Once (6.6.1) is reached, it is persistent if $Inv(\mathcal{F},E,S;P)$ does not decrease strictly. Hence, let us assume (6.6.1) for the sequel.

(6.7) In the above situation, we can choose a regular system of parameters $p = (x,y,z)$ of $\mathcal{O}_{X,P}$ satisfying the following conditions:

(6.7.1)
a) $(x=0) \subset E$.
b) $E = (x=0)$ or $E = (xy=0)$.
c) $Z = (z=0)$ has maximal contact with S.

Let us call "adecuate" such a regular system of parameters p. Let us now fix an element $g \in \mathcal{O}_{X,P}$, we can write

(6.7.2) $$g = \sum g_{hij} x^h y^i z^j .$$

For a fixed integer $t > 0$, let us define the "characteristic polygon $\Delta^t(g;p)$ of g with respect to p" by

(6.7.3) $$\Delta^t(g;p) = [[\{(h,i)/(t-j); g_{hij} \neq 0,\ t > j\}]] \subset \mathbb{R}^2_+$$

where [[...]] means "positively convex hull" (see [OGC], Appendix).

(6.8) Definition. Let f be a local equation of S at P, let ω be a generator of $\mathcal{F}_P$ and let $p = (x,y,z)$ be an adecuate regular system of parameters. We define the "characteristic polygon $\Delta(\mathcal{F},E,S;P;p) = \Delta$" by

(6.8.1) $\Delta = [[\Delta^{\nu}(f;p) \cup \Delta^{\mu}(\omega(x\partial/\partial x);p) \cup \Delta^{\mu}(\omega(y\partial/\partial y);p) \cup \Delta^{\mu}(\omega(z\partial/\partial z);p)]]$,

where $\mathrm{Inv}(\mathcal{F},E,S;P) = (\nu,d,r,\mu)$. The "main vertex" of Δ will be the vertex of lowest abscissa. We shall denote its coordinate by $(\alpha,\beta) = (\alpha(\mathcal{F},E,S;P;p),\beta(\mathcal{F},E,S;P,p))$.

(6.9) Remark. Like in the case of surfaces, $(x=z=0)$, resp. $(y=z=0)$, is a permissible center both for $\mathcal{F}$ and S iff all the points of Δ have abscissa, resp. ordinate, bigger or equal than one. (See e.g. [OGC], Appendix).

(6.10) Proposition. In the above situation, let $\pi: (X',E',\mathcal{F}',S') \to (X,E,\mathcal{F},S)$ be a blowing-up centere at Y, where Y satisfies the following properties:

a) If neither $(x=z=0)$ nor $(y=z=0)$ are both permissible for $\mathcal{F}$ and S, then $Y = \{P\}$.

b) $Y = (x=z=0)$ or $Y = (y=z=0)$ otherwise, Y being both permissible for $\mathcal{F}$ and S.

Assume that $P' \in \pi^{-1}(P)$ is a point such that

(6.10.1) $$\mathrm{Inv}(\mathcal{F}',E',S';P') = \mathrm{Inv}(\mathcal{F},E,S;P).$$

Then there is an adecuate regular system of parameters $p' = (x',y',z')$ of $\mathcal{O}_{X',P'}$ such that

(6.10.2) $$\beta' = \beta(\mathcal{F}',E',S';P';p') \leq \beta = \beta(\mathcal{F},E,S;P;p).$$

Moreover, the inequality in (6.10.2) is strict except may be for the cases $Y=(x=z=0)$ or $Y = \{P\}$ and P' corresponds to a direction which is transversal to $x=0$.

Proof. Since we have (6.10.1), we can find $p' = (x',y',z')$ given by one of the following transformations

(6.10.3)
$$\begin{array}{llll} (T\text{-}1,\zeta) : & x=x'; & y = (y'+\zeta)x'; & z = z'x'. \\ T\text{-}2 : & x=x'y'; & y = y'; & z = z'y'. \\ T\text{-}3 : & x=x'; & y = y'; & z = z'x'. \\ T\text{-}4 : & x=x'; & y = y'; & z = z'y'. \end{array}$$

where $(T\text{-}1,\zeta)$ and T-2 are for the case $Y = \{P\}$, T-3 is for the case $Y = (x=z=0)$ and T-4 is for the case $Y = (y=z=0)$. The "vertical stability" of the maximal contact [AHV] assures that p' is adecuate. On the other hand, the characteristic polygon of (6.8) has a similar behaviour to the one of Hironaka in ([OGC],Appendix) and hence we can apply the same kind of control. (See [CA] and [CA3] for related situations in the case of vector fields).

(6.11) The Proposition (6.10) provides the local control. In fact, the only conflictive case is the appearence of a stationary sequence for the invariant β. In this case, only a finite number of $Y = (x=0)$ succesive is possible (e.g. by looking at the movement of Δ). Hence the existence of infinitely many situations $Y = \{P\}$ and P' corresponding to a transversal direction to x=0, provides the existence of a pemissible curve Γ having the points P' as infinitely near points. By a change of parameters which does not modify β, we can assume that $\Gamma = (y=z=0)$ and then blowing-up Γ the invariant β decreases strictly. The globalization of this strategy may be made as in [CA3]. This ends the proof of Theorem (6.1).

§7. Dicritical singular foliations

(7.1) Definition. Let $\mathcal{F}$ be a saturated singular foliation over X adapted to E. We say that $\mathcal{F}$ is "dicritical" iff there is a dicritical component of E for $\mathcal{F}$ or there is a finite sequence of permissible blowing-ups

(7.1.1) $\mathcal{S} = \{\pi(i): (X(i),E(i),\mathcal{F}(i)) \rightarrow (X(i-1),E(i-1),\mathcal{F}(i-1))\}_{i=1,..,N}$

with $(X(o),E(o),\mathcal{F}(o)) = (X,E,\mathcal{F})$, such that E(N) has a dicritical component for $\mathcal{F}(N)$.

(7.2) Remark. At least for the case dim X = 3, the above Definition is equivalent to the fact of having a finite sequence of "local permissible blowing-ups" such that the last exceptional divisor has a dicritical component. The reason is that the "local blowing-ups" may be made to be "global" by means of Proposition (3.6).

(7.3) The rest of this paragraph is devoted to the proof of the following:

Theorem. Let X be a germ of nonsingular analytic variety of dim X = 3. Let $\mathcal{F}$ be a saturated singular foliation over X adapted to a normal crossings divisor E of X. Then the following two statements are equivalent

a) $\mathcal{F}$ is dicritical.

b) There is an admissible surface $S \subset X$ such that the restriction $(\mathcal{F},E)|_S$ is dicritical.

(7.4) Let us prove first "a) implies b) ".

(7.5) Assume first that there is a dicritical component F of E for $\mathcal{F}$. Then it is enough to take $S \subset X$ such that $S \cap F$ is not a separatrix of $\mathcal{F}|_S$ (S nonsingular), since $S \cap F$ gives the needed component of the adaptation locus.

(7.6) Assume now that there is no dicritical components of E for $\mathcal{F}$. Let us take a sequence $\mathcal{S}$ like in (7.1.1) of minimal length = N. Let $E_N(N)$ be the exceptional divisor of $\pi(N)$. Let $Z(N) = (\pi(1)\circ\ldots\circ\pi(N))^{-1}(P(0))$, where $P(0)$ = origin of $X(0) = (\mathbb{C}^3, 0)$. Then (3.4) a) allows us to assure that there is a point $Q \in Z(N) \cap E_N(N)$ such that

(7.6.1) $\qquad \nu(\mathcal{F}(N), E(N); Q) = 0 \quad$ and $\quad e(E(N), Q) \leq 2.$

Let us make a local study at Q. Assume first that $e(E(N), Q) = 1$ and let (x, y, z) be a regular system of parameters of $\mathcal{O}_{X(N),Q}$ such that $E(N)=(x=0)$ locally at Q. Since $E_N(N)$ is a dicritical component, we can write a generator of $\mathcal{F}(N)_Q$ as follows

(7.6.2) $$w = a'x(dx/x) + dy + cdz.$$

Hence w is also a generator of $\mathcal{T}(N)_Q$, where $\mathcal{T}(N)$ = singular foliation associated to $\mathcal{F}(N)$. We see easily that if $S(N)$ is an admissible surface for $\mathcal{F}(N)$ which is not singular at Q and $T_Q S(N)=(z=0)$, then $(\mathcal{F}(N), E(N))|_{S(N)}$ is dicritical. Let S be the image of $S(N)$ by $\pi(1)\circ\ldots\circ\pi(N)$, then $(\mathcal{F},E)|_S$ is dicritical. Assume now that $e(E(N), Q)=2$ and let (x, y, z) be a regular system of parameters of $\mathcal{O}_{X(N),Q}$ such that $E(N)=(xy=0)$ and $E_N(N)=(x=0)$ locally at Q. We can write a generator of $\mathcal{F}(N)_Q$ in one of the two following ways

(7.6.3) $$w = a'x(dx/x) + dy/y + cdz$$

(7.6.4) $$w = a'x(dx/x) + b(dy/y) + dz.$$

In the case (7.6.3), a generator of $\mathcal{T}(N)_Q$ is given by yw and we can reason as in the case (7.6.2). Assume we have the case (7.6.4). We have two possibilities

(7.6.5) $$(x = y = 0) \not\subset (b = 0).$$

(7.6.6) $$(x = y = 0) \subset (b = 0).$$

In the case (7.6.5), by choosing if necessary another Q, we can assume that b is a unit and we can reason as in cases (7.6.2) and (7.6.3). Assume we have (7.6.6). Then $(x=y=0)$ is a permissible curve at Q. Without loss of generality we can assume that it is a globally permissible curve. Let $\pi(N+1)\colon X(N+1) \longrightarrow X(N)$ be the blowing-up centered at $(x=y=0)$. Then the exceptional divisor is also a dicritical component for $\mathcal{F}(N+1)$. Since $\pi(N+1)^{-1}(Q)$ is generically contained in only one component of $E(N+1)$, we reduce ourselves to the case $e(E(N), Q)=1$.

(7.7) By (7.6) above, in order to prove "a) implies b)" of (7.3), it is enough to prove the following

<u>Proposition.</u> Let $X(0) = (\mathbb{C}^3, 0)$ and let

(7.7.1) $$\mathcal{S} = \{\pi(i)\colon X(i) \longrightarrow X(i-1)\}_{i=1,\ldots,N}$$

be a sequence of blowing-ups such that if $E(0)=\emptyset$, then $E(i)=\pi(i)^{-1}(E(i-1)\ Y(i-1))$, where $Y(i-1)$ is the center of $\pi(i)$, $i=1,\ldots,N$, and $Y(i)$ has normal crossings with $E(i)$ (Hence each $E(i)$ is a normal crossings divisor of $X(i)$). Let us fix a point Q of $E(N)$ and a plane $T \subset T_Q X(N)$ which is not tangent to the components of $E(N)$. Then there is a closed surface $S(N) \subset X(N)$ satisfying the following properties

a) $Q \in S(N)$, $S(N)$ is not singular at Q and $T_Q S(N)=T$.

b) $\mathrm{Sing}S(N) \subset E(N)$.

(7.8) Lemma. Let $\Gamma \subset (\mathbb{C}^3, 0)$ be a germ of analytic irreducible curve. Then there is a surface S with isolated singularity such that $\Gamma \subset S \subset (\mathbb{C}^3, 0)$.

Proof. Let $x=t^n$, $y=t^m\phi(t)$, $z=t^p\Psi(t)$, with $\phi(0)\neq 0\neq\Psi(0)$, be an irreducible Puiseux's parametrization of Γ with $n<m<p$. Then two plane projections of Γ are given by

(7.8.1) $$f(x,y) = y^n + a_1(x)y^{n-1} + \dots + a_n(x) = 0$$
$$g(x,y) = z^n + b_1(x)z^{n-1} + \dots + b_n(x) = 0$$

with $\nu(a_i(x), b_i(x)) \geq i+1$. We shall show that $\mu f + \lambda g = 0$ has isolated singularity for generic (μ, λ). We have

(7.8.2) $$\partial(\mu f + \lambda g)/\partial y = \mu\partial f/\partial y$$
$$\partial(\mu f + \lambda g)/\partial z = \lambda\partial g/\partial z.$$

Now, for generic (μ, λ), the tangent cones of $\mu f + \lambda g = 0$, $\partial(\mu f + \lambda g)/\partial y = 0$ and $\partial(\mu f + \lambda g)/\partial z = 0$ do not contain a common line and hence $\mu f + \lambda g = 0$ has isolated singularity.

(7.9) (Proof of (7.7)). We know that $X(N)$ is a closed analytic subspace of $(\mathbb{C}^3,0)\times \mathbb{P}^M$ for a certain M. Thus, we can take a hypersurface H of $(\mathbb{C}^3,0)\times \mathbb{P}^M$ such that $Q \in H$ and $T_Q H \cap T_Q X(N) = T$. Let $S_1(N) = H \cap X(N)$. Then $S_1(N)$ satisfies (7.7) a). Let S_1 be the image of $S_1(N)$ by $\pi(1)\circ\dots\circ\pi(N)$ and let $f=0$ be a local equation of S_1. Then $S_1(N)$ is the strict transform of S_1 (provided that H does not contain any component of $E(N)$).

Before continuing with the proof of (7.7), let us remark the following property

(7.9.1) "We can find a sequence of infinitely near points P_0=origin, P_1, P_2, P_3, P_4 of the origin of $(\mathbb{C}^3,0)$ such that if $S \subset (\mathbb{C}^3,0)$ is a surface passing through these points, then multiplicity of S at P_0 is strictly bigger than multiplicity of S at P_4."

(For proving (7.9.1), take P_3 in the corner of the exceptional divisor of the transformations and take P_4 corresponding to a direction tangent to the third exceptional divisor). Now, by (7.9.1), we can take a germ of analytic curve $\Gamma(N)$ at Q such that for each surface $S_2(N) \supset \Gamma(N)$, then

(7.9.2) $$\text{multiplicity of } S_2(N) \text{ at } Q \gg \text{multiplicity of } S_1 \text{ at } Q = 1.$$

By Lemma (7.8) we have a surface with isolated singularity $S_2 \supset \Gamma = (\pi(1) \dots \pi(N))(\Gamma(N))$. Let $g=0$ be a local equation of S_2. By the Theorem of Bertini (see e.g. [ZA]), the generic surface S of the linear system $\mu f + \lambda g = 0$ has isolated singularity. Let $S(N)$ be the strict transform of S by $\pi(1)\circ\dots\circ\pi(N)$. By (7.9.2) we have that $T_Q S(N) = T_Q S_1(N)$. This ends the proof of (7.7).

(7.9.10) Let us prove now "b) implies a)" of the theorem (7.3).

(7.11) Let us reason by contradiction, assuming that $\mathcal{F}$ is not dicritical. Once applied Theorem (6.1), we can assume that $X,\mathcal{F},E$ and S satisfy the following conditions

a) $X\subset(\mathbb{C}^3,0)\times P^M$ is a nonsingular analytic subvariety of dimension three.

b) $S\subset X$ is a nonsingular surface which has normal crossings with E.

c) The adaptation locus T of $\mathcal{F}$ to S is $T=S\cap E$.

d) $(\mathcal{F}, E)|_S$ is dicritical.

Let $\mathcal{H}=\mathrm{Sat}(\mathcal{F}|_S;T)$. Applying the theorem of desingularization in dimension two (see e.g. [SE] or [CA2]), we can reach after finitely many blowing-ups of points the following additional conditions

e) There is a dicritical component D of T for $\mathcal{H}$.

f) There is a point $P\in D$ which is a simple corner for $\mathcal{H}$. (See [CLNS]; for this blow-up enough number of times following the direction given by D).

(7.12) Proposition. In the above situation, after finitely many permissible blowing-ups we may also assume that

g) $\mu(\mathcal{F},E;\{P\}) = 0$.

Proof. Take a regular system of parameters $p=(x,y,z)$ of $\mathcal{O}_{X,P}$ satisfying

$$E = (xy=0);\ S = (z=0);\ D = (x=z=0) \tag{7.12.1}$$

Let $\Delta = \Delta(\mathcal{F},E;P,p)$ be defined by

$$\Delta = [[\Delta^{\mu}(\omega(x\partial/\partial x);p) \cup \Delta^{\mu}(\omega(y\partial/\partial y)) \cup \Delta^{\mu}(\omega(z\partial/\partial z))]] \tag{7.12.2}$$

where $\mu = \mu(\mathcal{F},E,\{P\})$ and ω is a generator of $\mathcal{F}_P$. (Compare with (6.8.1)). Now, let us take priorities as in Proposition (6.10) for choosing the centers. Let us consider over P the point given by D and S. Locally, the situation repeats, and a control of Δ as in (6.10) allows us to have $\mu' < \mu$ after finitely many steps. Hence we reach $\mu= 0$.

(7.13) Assume we have a)-g) above and $p = (x,y,z)$ satisfying the conditions (7.12.1). Since $\mu(\mathcal{F},E;\{P\}) = 0$, we can choose a generator ω of $\mathcal{F}_P$ satisfying one of the following equalities:

$$\omega = (dx/x) + b(dy/y) + c\,dz \tag{7.13.1}$$

$$\omega = a(dx/x) + (dy/y) + c\,dz \tag{7.13.2}$$

where x, resp. y, does not divide a, resp. b.

(7.14) Assume first we have (7.13.1). Then $\mathcal{H}_P$ is generated by

$$\omega|_S = (dx/x) + b(x,y,0)(dy/y) \tag{7.14.1}$$

but this contradicts the fact that (x=z=0) = D is a dicritical component of T for $\mathcal{H}$. Hence assume that we have (7.13.2). Then $\mathcal{H}_P$ is generated by

(7.14.2) $$\omega|_S = a(x,y,0)(dx/x) + (dy/y).$$

Since D is a dicritical component for $\mathcal{H}$, we have

(7.14.3) $$a(x,y,0) = xa'(x,y).$$

On the other hand, a coordinate change $y \to u.y$ where $u(0,0,0) = 1$ allows us to assume c=0 an hence

(7.14.4) $$\omega = adx/x + dy/y$$

(this is a formal computation). Then the integrability condition $\omega \wedge d\omega=0$ implies

(7.14.5) $$\partial a/\partial z = 0.$$

Hence $a(x,y,z) = a(x,y,0) = xa'(x,y)$ and then x=0 is a dicritical component of E for $\mathcal{F}$. This is the desired contradiction. This ends the proof of Theorem (7.3).

REFERENCES

|AHV| AROCA-HIRONAKA-VICENTE. "Infinitely near singular points. The theory of maximal contact. Desingularization Theorems". Mem. Mat. Inst. Jorge Juan nº 28,29,30. (1974).

|AU| AUTONNE. "Sur d'equation différentielle du premier ordre". J. Ecole Polytechnique c. 63,64, 1983.

|CA| CANO. "Desingularization stratefies for a three-dimensional vector field". Lecture Notes in Math. nº 1259. Springer-Verlag. (1978).

|CA2| ——. "Desingularization of plane vector fields". Transact. of the A.M.S. 296 1. (1986). 83/93.

|CA3| ——. "Local and global results on the desingularization of three-dimensional vector fields". Colloque sur les singularités des équations différentielles. Dijon 1985. To appear in Asterisque.

|CLNS| CAMACHO-LINS NETO-SAD. "Topological invariants and equidensingularization for holomorphic vector fields". J. of Diff. Geom. 20. (1984) 143/174

|CM| CERVEAU-MATTEI. "Formes holomorphes intégrables singuliéres". Asterisque 97. (1982).

|HI| HIRONAKA. "Resolution of singularities of an algebraic variety over a field of characteristic zero". Ann. of Math.79. (1964).

|JOU| JOUANOLOU. "Equations de Pfaff algébriques". Lecture Notes in Math. nº 708 Springer-Verlag (1979).

|MM| MATTEI-MOUSSU. "Holononie et intégrales premières". Ann. Sci. Ec. Norm. Sup. 4º serie t.13. (1980). 469/523.

|OGC| ORBANZ-GIRAUD-COSSART. "Desingularization of surfaces (Appendix: Bowdoin College Seminar of Hironaka)". Lecture Notes in Math nº 1101. (1985).

|SA| SAITO. "Theory of logarithmic differential forms and logarithmic vector fields". J. of Fac. Sci. Tokyo. Vo. 27. nº 2. (1980).

|SE| SEIDENBERG. "Reduction of singularities of the differential equation Ady = Bdx". Ann. J. of Math. (1968). 248/269.

|SG| SANCHEZ-GIRALDA. "Caracterisations des variétés permises d'une hypersurface algébrique". C.R. Ac. Sci. París. L. 285 (1977).

|ZA| ZARISKI. "The theorem of Bertini on the variable singular points of a linear system of varieties". Trans. A.M.S.. 45. (1944). pp 472/542.

F. Cano.
Dpto. Algebra y Geometría.
Facultad de Ciencias.
47005 Valladolid.
SPAIN.

INVARIANT MANIFOLDS AND A PREPARATION LEMMA FOR LOCAL HOLOMORPHIC FLOWS AND ACTIONS

Marc Chaperon

Centre de mathématiques,
U.A. n° 169 du C.N.R.S.,
Ecole Polytechnique
91128 PALAISEAU Cedex.

1. INTRODUCTION

(1.1) Notation and definitions.

Let n denote a fixed positive integer, and let $\underline{\underline{d}}$ be the Lie algebra of those germs at $0 \in \mathbb{C}^n$ of holomorphic vector fields which vanish at the origin. Call two elements X and Y of $\underline{\underline{d}}$ conjugate if there exists a germ $h : (\mathbb{C}^n,0) \circlearrowleft$ of a holomorphic diffeomorphism such that $h^*Y = X$. For each $X \in \underline{\underline{d}}$, let X^1 stand for the linear part $dX(0) \in gl(n,\mathbb{C})$ of X.

THROUGHOUT THE SEQUEL, WE DENOTE BY S A DIAGONALISABLE ELEMENT OF $gl(n,\mathbb{C})$. An S-vector field is an element X of $\underline{\underline{d}}$ such that S is the semi-simple part of X^1 (thus, every $X \in \underline{\underline{d}}$ is an S-vector field for a unique S). An S-normal form is an S-vector field of the form(*) $S + N$, where N is polynomial and $[S, N] = 0$ (since $S + N$ is an S-vector field, the linear part N^1 is nilpotent, and $[S, N^1] = 0$).

(1.2) Strongly invariant manifolds.

Define a strongly invariant manifold (s.i.m.) of S to be a subspace of $\mathbb{C}^n$ which is the unstable subspace E_a^- of the vector field aS, viewed as an element

(*) We shall not distinguish between polynomial vector fields and their germs at 0.

of gl(2n, $\mathbb{R}$) , for some $a \in \mathbb{C}$; in other words, E_a^- is the direct sum of those eigenspaces of S associated to eigenvalues c with $\mathcal{R}e(ac) > 0$ - in particular, the s.i.m.'s of S are complex vector subspaces of $\mathbb{C}^n$, and there is but a finite number of them. Moreover, every S-normal form S+N is tangent to every s.i.m. W of S at each of its points (proof : if $W = E_a^-$, then it is the set of those $v \in \mathbb{C}^n$ such that $e^{saS} v$ tends to 0 when $s \in \mathbb{R}$ tends to $-\infty$; now, for each $v \in E_a^-$, the commutation relation $[S, N] = 0$ yields $e^{saS} N(v) = N(e^{saS} v)$ for every $s \in \mathbb{R}$, hence $\lim_{s \to -\infty} e^{saS} N(v) = N(0) = 0$, hence $N(v) \in E_a^-$).

The reader is referred to Appendix 1 for further information on s.i.m.'s.

THROUGHOUT THE SEQUEL, $\mathcal{V}$ DENOTES THE UNION OF THE S.I.M.'S OF S .

(1.3) Statement of the Preparation Lemma.

The following result plays a crucial role in the classification of elements of $\underline{\underline{d}}$ ([Ch 86a]) :

PREPARATION LEMMA. - For every $m \in \mathbb{N}$, each S-vector field is conjugate to an S-vector field Z having m^{th} order contact with an S-normal form along $\mathcal{V}$, and the conjugacy can be chosen tangent to the identity at 0 .(*)

The reader is referred to Appendix 2 for a more precise statement and a generalisation to holomorphic action-germs.

Here is an important particular case of the Preparation Lemma : define S to be in the Poincaré domain when the convex hull of its spectrum in $\mathbb{C}$ does not contain 0 . Clearly, S is in the Poincaré domain if and only if $\mathbb{C}^n$ itself is one of its s.i.m.'s ; therefore, our Preparation Lemma includes the Poincaré-Dulac Theorem : if S is in the Poincaré domain, every S-vector field is conjugate to an S-normal form.

Aknowledgements. I have to thank Spyros Pnevmatikos for his kind invitation to participate in the Heraklion Symposium in 1983, for the Preparation Lemma was obtained

---

(*) In particular, Z is tangent to every s.i.m. of S "at each of its points".

during this meeting - and, therefore, announced in [Ch 85].

I also wish to thank the organisers of the present Symposium for their kind invitation, which enabled me to enjoy Mexico and its mathematicians.

2. PROOF OF THE PREPARATION LEMMA.

(2.1) Algebraic background : formal normal forms.

We let $\underline{\underline{D}}$ denote the group of all germs $(\mathbb{C}^n,0)\circlearrowleft$ of holomorphic diffeomorphisms. We denote by $\mathcal{E}$ the complex algebra of all germs at $0\in\mathbb{C}^n$ of holomorphic complex functions, by $\mathfrak{m}=\{f\in\mathcal{E} : f(0)=0\}$ its maximal ideal and by $\mathcal{E}_k$ the algebra $\mathcal{E}/\mathfrak{m}^{k+1}$ for each non-negative integer k .

Each $g\in\underline{\underline{D}}$ (resp. $X\in\underline{\underline{d}}$) induces an automorphism g^* (resp. a derivation X^*) of algebra $\mathcal{E}$, given by $g^*f=f\circ g$ (resp. $X^*f=L_X f$), hence, for each k , an automorphism g_k^* (resp. a derivation X_k^*) of $\mathcal{E}_k$. Clearly, g_k^* (resp. X_k^*) is determined by and determines the k-jet g_k (resp. X_k) of g (resp. X) at 0 : the space $\underline{\underline{D}}_k$ (resp. $\underline{\underline{d}}_k$) of k-jets at 0 of elements of $\underline{\underline{D}}$ (resp. $\underline{\underline{d}}$) is the quotient of $\underline{\underline{D}}$ (resp. $\underline{\underline{d}}$) by the kernel of the group (resp. Lie algebra) (anti-) morphism $g\mapsto g_k^*$ (resp. $X\mapsto X_k^*$). We obviously have

PROPOSITION 1. - The mapping $g\mapsto g^*$ (resp. $X\mapsto X^*$) is a bijection of $\underline{\underline{D}}$ (resp. $\underline{\underline{d}}$) onto the space of automorphisms (resp. derivations) of algebra $\mathcal{E}$. Therefore, the same statement holds true when $\underline{\underline{D}}$, $\underline{\underline{d}}$, $\mathcal{E}$ are replaced by $\underline{\underline{D}}_k$, $\underline{\underline{d}}_k$, $\mathcal{E}_k$ respectively. ■

Here comes the key result of this paragraph :

PROPOSITION 2. - For each k and each $g_k\in\underline{\underline{D}}_k$ (resp. $X_k\in\underline{\underline{d}}_k$), the semi-simple part of g_k^* (resp. X_k^*) as a linear endomorphism is an algebra automorphism (resp. a derivation) of $\mathcal{E}_k$.

Proof. Let $\mathfrak{m}_k=\mathfrak{m}/\mathfrak{m}^{k+1}$ denote the maximal ideal of $\mathcal{E}_k$, and let $F_c=\bigcup_{m\in\mathbb{N}}\operatorname{Ker}(g_k^*-c\,\mathrm{Id})^m$, $c\in\mathbb{C}$. Since $g_k^*\mathfrak{m}_k^j=\mathfrak{m}_k^j$, there exist complex numbers

c_j and vectors $x_j \in F_{c_j}$, $1 \leqslant j \leqslant n$, such that the projections $x_1^1, \dots, x_n^1$ of $x_1, \dots, x_n$ form a basis of the complex vector space $\mathfrak{m}_k/\mathfrak{m}_k^2$, alias $L(\mathbb{C}^n,\mathbb{C})$. As $(x^p : p \in \mathbb{N}^n , 0 \leqslant |p| \leqslant k)$ clearly is a basis of the complex vector space $\mathcal{E}_k$, our assertion about g_k in Proposition 2 stems from the following

LEMMA. - Given complex numbers c , c' , the two relations $z \in F_c$ and $z' \in F_{c'}$ imply $zz' \in F_{cc'}$.

The proof is by induction on $(d_c(z), d_{c'}(z'))$, where $d_c(z) = \min\{m \in \mathbb{N} : (g_k^* - c\,\mathrm{Id})^m z = 0\}$. If $d_c(z)$ or $d_{c'}(z')$ equals 0 , then $zz' = 0$. In the remaining case, we have $g_k^* z = cz + z_1$ and $g_k^* z' = cz' + z_1'$, with $z_1 \in F_c$, $z_1' \in F_{c'}$, $d_c(z_1) < d_c(z)$ and $d_{c'}(z_1') < d_{c'}(z')$; since $g_k^*(zz') = g_k^*(z)\, g_k^*(z') = cc'\, zz' + c\, zz_1' + c'\, z' z_1 + z_1 z_1'$, using induction, we obtain $(g_k^* - cc'\,\mathrm{Id})\, zz' \in F_{cc'}$, hence our Lemma. □

As the case of X_k is similar, this proves Proposition 2. ■

Call $g_k \in \underline{\underline{D}}_k$ (resp. $X_k \in \underline{\underline{d}}_k$) <u>semi-simple</u> when g_k^* (resp. X_k^*) is a diagonalisable endomorphism of $\mathcal{E}_k$. For each $g \in \underline{\underline{D}}$ (resp. $X \in \underline{\underline{d}}$) and each k , if $g^1 = dg(0)$ (resp. $X^1 = dX(0)$) denotes the linear part of g (resp. X), the <u>linear part</u> of g_k (resp. X_k) is $g_k^1 := (g^1)_k$ (resp. $X_k^1 := (X^1)_k$). Clearly, if g_k (resp. X_k) is semi-simple, so is its linear part.

For each $h \in \underline{\underline{D}}$ and each k , the inner automorphism $g \mapsto h^* g := h^{-1} \circ g \circ h$ (resp. $X \mapsto h^* X$) of $\underline{\underline{D}}$ (resp. $\underline{\underline{d}}$) induces an automorphism $g_k \mapsto h_k^*\, g_k = h_k^{-1} \circ g_k \circ h_k$ (resp. $X_k \mapsto h_k^* X_k$) of $\underline{\underline{D}}_k$ (resp. $\underline{\underline{d}}_k$), determined by h_k .

PROPOSITION 3. - For each non-negative integer k , every semi-simple $s_k \in \underline{\underline{D}}_k$ (resp. $\sigma_k \in \underline{\underline{d}}_k$) is linearisable : there exists $h_k \in \underline{\underline{D}}_k$, with $h_k^1 = \mathrm{Id}_k^1$, such that $h_k^* s_k = s_k^1$ (resp. $h_k^* \sigma_k = \sigma_k^1$).

<u>Proof</u>. As s_k^* is diagonalisable and $s_k^* \mathfrak{m}_k = \mathfrak{m}_k$, there exist eigenvectors $y_1, \dots, y_n$ of s_k^* whose canonical projections $y_1^1, \dots, y_n^1$ form a basis of $\mathfrak{m}_k/\mathfrak{m}_k^2$, alias $L(\mathbb{C}^n,\mathbb{C})$; in other words, $(y_1^1, \dots, y_n^1)$ is a system of complex linear coordinates on $\mathbb{C}^n$. Let $x_j = (y_j^1)_k$ and let c_j be the eigenvalue of s_k^*

associated to y_j for $1 \leq j \leq n$. Clearly, we have $s_k^{1*} x_j = c_j x_j$, $1 \leq j \leq n$. Therefore, the element h_k of $\underline{\underline{D}}_k$ defined by $h_k^* y_j = x_j$, $1 \leq j \leq n$, which satisfies $h_k^1 = Id_k^1$, is such that $h_k^*(s_k^* y_j) = h_k^*(c_j y_j) = c_j x_j = s_k^{1*} x_j = s_k^{1*}(h_k^* y_j)$ for every j, hence $s_k \circ h_k = h_k \circ s_k^1$, as required. The case of $\underline{\underline{d}}_k$ is similar. ∎

FORMAL PREPARATION LEMMA. – For each positive integer k, every S-vector field X is conjugate to an S-vector field which has k^{th} order contact at 0 with an S-normal form of degree k. Moreover, the conjugacy can be chosen polynomial of degree k and tangent to the identity at 0.

Proof. Let σ_k denote the semi-simple part of X_k, and let ν_k be its nilpotent part. Proposition 3 provides an $h_k \in \underline{\underline{D}}_k$ with $h_k^1 = Id_k^1$ and $h_k^* \sigma_k = \sigma_k^1$, hence (as "the linear part of the semi-simple part equals the semi-simple part of the linear part") $h_k^* \sigma_k = S_k$. Let h (resp. N) be that polynomial element of $\underline{\underline{D}}$ (resp. $\underline{\underline{d}}$) of degree k which admits h_k (resp. $h_k^* \nu_k$) as its k-jet at 0. Clearly, $S+N$ is an S-normal form, and $h^* X$ has k^{th} order contact with it at 0. ∎

A more general version of this classical lemma is proved in Appendix 2. For the convenience of the reader, we shall now explain "what S-normal forms look like" :

NOTATION. – Let $(x_1, \ldots, x_n)$ be a system of complex linear coordinates on $\mathbb{C}^n$ in which S is diagonal, i.e. $S^* x_j = c_j x_j$, $c_j \in \mathbb{C}$, $1 \leq j \leq n$. For each $p \in \mathbb{N}^n$, we denote by $p_1, \ldots, p_n$ its coordinates, and let $|p| = \sum p_j$ and $x^p = x_1^{p_1} \ldots x_n^{p_n}$. Given $X \in \underline{\underline{d}}$, the notation $X = \sum a_j \partial_j$ means that $X^* x_j = a_j$ for $1 \leq j \leq n$ (for example, $S = \sum c_j x_j \partial_j$). The following results are obvious :

PROPOSITION 4. – For every S-normal form $S+N$, the polynomial vector field $N - N^1$ is a $\mathbb{C}$-linear combination of the monomials $x^p \partial_j$ with $p \in P_j := \{p \in \mathbb{N}^n : |p| > 1 \text{ and } \sum p_k c_k = c_j\}$ and $1 \leq j \leq n$. ∎

PROPOSITION 5. – Let $\mathbb{N}^n$ be equipped with the ordering $\leq$ defined by " $p \leq q$ if and only if $p_j \leq q_j$ for every j " ; then, for each $A \subset \mathbb{N}^n$, the minimal set min A of A is finite, and every $p \in A$ satisfies $p \geqslant q$ for some $q \in \min A$. Therefore, if $P_o = \{p \in \mathbb{N}^n \setminus \{0\} : \sum p_k c_k = 0\}$ and if $P_1, \ldots, P_n$ are as in Proposition 4, then, for $0 \leqslant j \leqslant n$, each element of P_j can be written - in a non-unique fashion in general - as the sum of one element of $\min P_j$ and finitely many elements of $\min P_o$.

In particular, if P_o is empty - which of course is the case when S is in the Poincaré domain -, then each P_j, being equal to $\min P_j$, is finite - and, therefore, the centraliser of S in $\underline{\underline{d}}$ is a finite dimensional complex vector subspace, generated by the centraliser of S in $gl(n,\mathbb{C})$ and the monomials $x^p \partial_j$, $p \in P_j$, $1 \leqslant j \leqslant n$. Thus, in this case, the space of all S-normal forms is finite dimensional. ■

COROLLARY. - For every positive integer k, the set of those $S \in gl(n,\mathbb{C})$ such that the only S-normal form of degree k is S itself is an open and dense subset of $gl(n,\mathbb{C})$, the complementary subset of which has codimension one.

This comes at once from Proposition 4. ■

(2.2) Key lemmas.

HYPOTHESES AND NOTATION. - Let $T \in GL(n,\mathbb{C})$ be semi-simple, with eigenvalues $a_1, \ldots, a_n$ (repeated according to their multiplicities), let W denote a T-invariant complex vector subspace of $\mathbb{C}^n$, and let

$$K = \max |a_j| \quad , \quad L = \max |a_j|^{-1} .$$

We assume

$$c := \max \left\{ |a_j|^{-1} : a_j \in \operatorname{Spec}(T|_W) \right\} < 1 \text{ , hence } K > 1 .$$

Define a function s of $\mathbb{N}$ into itself by

$$s(k) = \begin{cases} k + 1 + \left[\dfrac{\operatorname{Log} K + k \operatorname{Log} L}{-\operatorname{Log} c} \right] & \text{for } K L^k \geqslant 1 \\ \left[(\operatorname{Log} K)/(-\operatorname{Log} L) \right] + 1 & \text{for } K L^k < 1 , \end{cases} \tag{1}$$

where $[x]$ is the greatest integer $\leqslant x$. An element f of $\underline{\underline{D}}$ is called a

<u>T -diffeomorphism</u> if T is the semi-simple part of its linear part f^1.

LEMMA A. - Let f be a T-diffeomorphism, preserving the germ of W at 0. For every $k \in \mathbb{N}$, every $q \geq s(k)$ and every $g \in \underline{\underline{D}}$, if f and g have q^{th} order contact at 0, then there exists $h \in \underline{\underline{D}}$ with the following two properties :

(i) $h^* g$ and f have k^{th} order contact along W (which is therefore invariant by $h^* g$).

(ii) h has $(q-1)^{th}$ order contact with the identity at 0.

Moreover, the restrictions $(D^j h)|_W$, $0 \leq j \leq k$, are uniquely determined by conditions (i)-(ii).

REMARK. - Since f preserves W, condition (i) involves only the restrictions $(D^j h)|_W$, $0 \leq j \leq k$. Moreover, one can replace (ii) by

(ii)' For $0 \leq j \leq k$, $(D^j h)|_W$ has $(q-j-1)^{th}$ order contact with $(D^j \mathrm{Id})|_W$ at 0 (negative contact means no contact is required) :

indeed, if (ii)' is satisfied, one can easily modify h so that (ii) is satisfied and $(D^j h)|_W$ remains the same for $0 \leq j \leq k$ (<u>see</u> the proof of Proposition (2.4) below). In other words, Lemma A is not a statement about h itself -introduced for notational convenience-, but a statement about its kth order jet along W.

LEMMA B. - Given f as in Lemma A and $k \in \mathbb{N}$, let $Z \in \underline{\underline{d}}$ satisfy the following two conditions :

(i) Z has $(s(k)-1)^{th}$ order contact with 0 at 0.

(ii) Z and $f_* Z$ have k^{th} order contact along W.

Then, for $0 \leq j \leq k$, $(D^j Z)|_W = 0$.

REMARK. - As in Lemma A and for the same reason, one can weaken (i) into

(i)' For $0 \leq j \leq k$, $(D^j Z)|_W$ has $(s(k)-j-1)^{th}$ order contact with 0 at 0.

Proof. We shall use the following

FACT. - There exists a hermitian norm on $\mathbb{C}^n$ such that, denoting by $A \mapsto |A|$ the induced norm on $gl(n,\mathbb{C})$, if we replace K, L and c by $K_o := |f^1|$, $L_o := |(f^1)^{-1}|$ and $c_o := |(f^1|_W)^{-1}|$ respectively in (1), we obtain the same $s(j)$ as before for $0 \leqslant j \leqslant k$, and c_o is less than 1.

Indeed, let $(x_1, \dots, x_n)$ be a system of $\mathbb{C}$-linear coordinates on $\mathbb{C}^n$ in which f^1 is in Jordan normal form. Clearly, for every positive ε, there exist positive constants $k_1, \dots, k_n$ with the following property : if we endow $\mathbb{C}^n$ with the hermitian norm $v \mapsto (\sum k_j |x_j(v)|^2)^{1/2}$, then the norms of $f^1 - T$ and $(f^1)^{-1} - T^{-1}$ are less than ε. □

Thus, denoting again by f a representative of f, the hypotheses of [Ch 86, section (4.2.2)] are satisfied (with $\Sigma = \{0\}$ and $Q = \mathbb{C}^n$, equipped with the above norm). In the particular case we are considering, we can use the following two facts:

(a) Given a closed ball B centered at $0 \in W$, the space of those continuous maps $B \to \mathbb{C}^j$ which are holomorphic in $\mathring{B}$ is closed for (any norm finer than) the norm of uniform convergence.

(b) Given such a map F and a non-negative integer m, the m-jet of F at 0 vanishes if and only if $|v|^{-m-1} F(v)$ is uniformly bounded for $v \in B \setminus \{0\}$.

Because of (a)-(b), the proofs of Lemma A and Lemma B are contained in those of [Ch 86, (4.2.2)], Théorème 1 and Théorème 2 respectively(*). The uniqueness statement at the end of Lemma A stems from that of the fixed point in [Ch 86, (4.2.1)] and (a)-(b). ∎

(2.3) First consequence.

NOTATION. - Let $W_1, \dots, W_w$ denote those s.i.m.'s of S which are maximal for

---

(*) For instance, for each small enough B as in (a)-(b), the restriction $j_B^k h$ to B of the kth order jet of h is obtained as the limit of $j_B^k(g^m \circ f^{-m})$ when m tends to $+\infty$, for a finer norm than that of uniform convergence.

inclusion, and let $M = \{W_j \cap W_m : 1 \leqslant j \leqslant m \leqslant w\} \setminus \{0\}$. For each $W \in M$, there exists $a \in \mathbb{C} \setminus \{0\}$ such that the hypotheses of (2.2) are satisfied by W and $T = e^{aS}$. We let s_W denote the minimum of all functions s associated to such T 's by (1).

COROLLARY. - Given $W \in M$, let X be an S -vector field, tangent to W (*). For every positive integer k and every integer $q \geqslant s_W(k)$, if $Y \in \underline{\underline{d}}$ has q^{th} order contact with X at 0 , then there exists $h \in \underline{\underline{D}}$ with the following two properties :

(i) $h^* Y$ and X have k^{th} order contact along W .

(ii) The identity has $(q-1)^{\text{th}}$ order contact with h at 0 .

Moreover, the restrictions $(D^j h)|_W$, $0 \leqslant j \leqslant k$, are uniquely determined by (i)-(ii).

<u>Proof</u>. Let $\underline{\underline{a}} \in \mathbb{C} \setminus \{0\}$ be such that the function s associated to $T = e^{\underline{\underline{a}}S}$ by (1) satisfies $s(k) = s_W(k)$. Clearly, the time $\underline{\underline{a}}$ values f and g of the flows of X and Y respectively, fulfil the hypotheses of Lemma A, and the h we are looking for must satisfy conditions (i)-(ii) of Lemma A, hence uniqueness and part (ii) of our Corollary.

Conversely, given h as in Lemma A, we shall see that $Z = X - h^* Y$ fulfils the hypotheses of Lemma B, hence our result. As h has $(q-1)^{\text{th}}$ order contact with Id at 0 , our hypothesis on Y implies that Z satisfies Lemma B (i). Moreover,

$$(2) \qquad f_* X = X \quad \text{and} \quad g_* Y = Y \text{ , hence } (h^* g)_* (h^* Y) = h^* Y \text{ .}$$

Now, since f and $h^* g$ have k^{th} order contact along W , it follows that $h^* Y$ and $f_* h^* Y$ have $(k-1)^{\text{th}}$ order contact along W . Therefore, as s is non-decreasing and k is positive, Lemma B implies that Z has $(k-1)^{\text{th}}$ order contact with 0 along W . In particular, $h^* Y$ is tangent to W ; thus, (2) and Lemma A (i) show that Z and $f_* Z$ have k^{th} order contact along W , hence our result by Lemma B. ■

NOTE. - The above Corollary holds true if $k=0$, and can be proven without Lemma B.

(*) meaning that so is some representative of X at each point of W in its domain.

However, the latter is an interesting rigidity result, which can be used in various situations -for another example, see Appendix 2 below.

The idea of both Lemma A and Lemma B goes back to S. Sternberg - see [N].

(2.4) Proof of the Preparation Lemma.

Let $W_1, \ldots, W_w$, M and s_W, $W \in M$, be as in (2.3).

- If $M = \emptyset$, the only s.i.m. of S is $\{0\}$, and the Preparation Lemma is just the Formal Preparation Lemma (2.1).

- If M consists of one element W, then $\mathcal{V} = W$; by the Formal Preparation Lemma, for each positive integer m, every S-vector field X_o is conjugate to an S-vector field Y having $s_W(m)^{th}$ order contact with an S-normal form X at 0. Since X is tangent to W, Corollary (2.3) shows that Y, hence X_o, is conjugate to an S-vector field having m^{th} order contact with X along $\mathcal{V}$. Moreover, the conjugacy is tangent to the identity, hence the Preparation Lemma in this case.

- If M has several elements, let $r : \mathbb{N}^* \circlearrowleft$ be defined by

$$r(m) = \max \{s_W(2m) : W \in M\} .$$

As in the case when M has one element, the Preparation Lemma stems from the Formal Preparation Lemma and the following

PROPOSITION. - Under the above hypotheses, let X be an S-vector field, tangent to every s.i.m. of S (for example an S-normal form). For every positive integer m, if $Y \in \underline{\underline{d}}$ has $r(m)^{th}$ order contact with X at 0, then it is conjugate to an S-vector field having m^{th} order contact with X along $\mathcal{V}$, and the conjugacy can be chosen tangent to the identity at 0.

Proof. Let $(x_1, \ldots, x_n)$ be a system of $\mathbb{C}$-linear coordinates in which S is diagonal. For $1 \leq j \leq w$, let I_j be the subset of $\{1, \ldots n\}$ such that $W_j = \bigcap_{k \in I_j} x_k^{-1}(0)$. Set $p_I = \sum_{k \in I} p_k$, $p \in \mathbb{N}^n$, $I \subset \{1, \ldots, n\}$. Every $g \in \underline{\underline{D}}$ can be identified with a convergent power series $\sum_{p \in \mathbb{N}^n} a_p x^p$ with coefficients $a_p \in \mathbb{C}^n$; then, for $1 \leq j \leq w$

and $k\in\mathbb{N}$, $\sum_{p_{I_j}\leq k} a_p x^p$ is the only element of $\underline{\underline{D}}$ which has k^{th} order contact with g along W_j and -as a power series- does not contain any x^p with $p_{I_j}>k$.

Therefore, under the hypotheses of our Proposition, Corollary (2.3) contains the following assertion : for $1\leq j\leq w$, there exists exactly one $h_j\in\underline{\underline{D}}$ of the form $h_j=\sum_{p_{I_j}\leq 2m} a_{j,p} x^p$, such that conditions (i)-(ii) of Corollary (2.3) are satisfied with $(W,k,h,q)=(W_j,2m,h_j,r(m))$. Thus, for $1\leq j<j'\leq w$, conditions (i)-(ii) of Corollary (2.3) are fulfilled with $W=W_j\cap W_{j'}$, $k=2m$, $q=r(m)$, $h=h_j$ AND $h=h_{j'}$. Therefore, by the uniqueness part of Corollary (2.3), h_j and $h_{j'}$ have $2m^{th}$ order contact along $W_j\cap W_{j'}$, i.e. $a_{j,p}=a_{j',p}$ for $p_{I_j\cup I_{j'}}\leq 2m$, hence

$$a_{j,p}=a_{j',p} \quad \text{if } p\in\mathbb{N}^n \text{ satisfies } p_{I_j}\leq m \text{ AND } p_{I_{j'}}\leq m .$$

If $h\in\underline{\underline{D}}$ is defined by

$$h=\sum_{j=1}^{w}\ \sum_{\begin{cases} p_{I_t}>m \text{ for } 1\leq t<j \\ p_{I_j}\leq m\end{cases}} a_{j,p} x^p ,$$

it follows that, for $1\leq j\leq w$, h and h_j have m^{th} order contact along W_j ; therefore, since X is tangent to W_j and has $2m^{th}$ order contact with h_j^*Y along W_j , it has m^{th} order contact with h^*Y along W_j , hence our result (it is easily checked that h is tangent to the identity at 0). ■

REFERENCES.

[Ch 85] M. CHAPERON, Differential geometry and dynamics : two examples, in Singularities and Dynamical Systems, S.N. Pnevmatikos ed., Elsevier B.V. (North Holland), 1985, 187-207.

[Ch 86] M. CHAPERON, Géométrie différentielle et singularités de systèmes dynamiques, Astérisque 138-139 (1986).

[Ch 86a] M. CHAPERON, C^k-conjugacy of holomorphic flows near a singularity, Pub. Math. I.H.E.S. 64 (1987), 143-183.

[N] E. NELSON, Topics in dynamics, Part I, Flows, Princeton (1970).

APPENDIX 1. MORE ON STRONGLY INVARIANT MANIFOLDS.

(A1.1) Strongly invariant manifolds of $X \in \underline{\underline{d}}$.

If (f^t) denotes the complex flow of X (in particular, $t \mapsto f^t$ is a group morphism $\mathbb{C} \to \underline{\underline{D}}$), a s.i.m. of X is a germ W at 0 of a holomorphic submanifold of $\mathbb{C}^n$ which is the (strict) unstable manifold of f^a for some $a \in \mathbb{C}$. Clearly, the tangent space $W^1 = T_0W$ is the unstable subspace of e^{aX^1} , hence a s.i.m. of X^1 - and of its semi-simple part S . The following consequence of the unstable manifold theorem is a particular case of [Ch 86, (4.4.1), Proposition 3] :

PROPOSITION. - X is tangent to each of its s.i.m.'s (in other words, strongly invariant manifolds are invariant manifolds). Conversely, for each s.i.m. W^1 of X^1 , there exists exactly one germ W of submanifold of $\mathbb{C}^n$ with $T_0W = W^1$ to which X is tangent. Moreover, W is a s.i.m. of X . ∎

COROLLARY. - There exists an element h of $\underline{\underline{D}}$ which sends every s.i.m. of X onto the germ of its tangent space at 0 .

Proof. By the Preparation Lemma, there exists $h \in \underline{\underline{D}}$, tangent to the identity at 0 , such that $h_* X$ is tangent to every s.i.m. of X^1 (indeed, the s.i.m.'s of X^1 are those of its semi-simple part S , to which every S -normal form is tangent). Thus, for each s.i.m. W^1 of X^1 , X is tangent to $W = h^{-1}(W^1)$ and $T_0W = W^1$, hence our result, by the above Proposition. ∎

(A1.2) Topological characterisation of strongly invariant manifolds.

Two elements X and Y of $\underline{\underline{d}}$ are C°-conjugate if there exists a germ at 0 of a local homeomorphism of $\mathbb{C}^n$, having 0 as a fixed point, sending the complex flow of X onto that of Y (this makes sense if we take representatives ; see [Ch 86a] for a formal definition).

PROPOSITION. - Let X and Y be two elements of $\underline{\underline{d}}$ such that 0 lies neither in the spectrum of X^1 , nor in that of Y^1 , and let (f^t) and (g^t) denote their respective complex flows. If X and Y are C°-conjugate, there exists $h \in \underline{\underline{D}}$ such that, for every $a \in \mathbb{C}$, the image by h of the unstable manifold of f^a is that of g^a. In particular, h (resp. h^{-1}) sends every s.i.m. of X (resp. Y) onto a s.i.m. of Y (resp. X).

Proof. Let S and T denote the semi-simple parts of X^1 and Y^1 respectively. By Corollary (A1.1) (or the Preparation Lemma), we may assume that, for each a $\mathbb{C}$, the (strict) unstable manifold of f^a (resp. g^a) is the germ at 0 of the unstable subspace V_a (resp. W_a) of e^{aS} (resp. e^{aT}). Moreover (taking coordinates in which S and T respectively are diagonal), we may assume that each V_a (resp. W_a) is a coordinate subspace of $\mathbb{C}^n$ for the standard coordinates. As S and T do not have 0 in their spectrum, the set H of those $a \in \mathbb{C}$ for which both e^{aS} and e^{aT} are hyperbolic automorphisms of $\mathbb{C}^n$ is the complementary subset of finitely many lines, and there is a finite subset A of H such that every V_b (resp. W_b) with $b \in H$ is of the form V_a (rep. W_a) for some $a \in A$.

As a C°-conjugacy between two hyperbolic germs of diffeomorphisms exchanges their unstable manifolds, we have the following : if k denotes a C°-conjugacy between X and Y, it sends the germ at 0 of V_a onto that of W_a for every $a \in A$. Therefore, since the V_a's and the W_a's are coordinate subspaces, the topological invariance of dimension implies the following : there exists an $L \in GL(n,\mathbb{C})$ (obtained by a permutation of the standard coordinates of $\mathbb{C}^n$) such that $L(V_a) = W_a$ for every $a \in A$. Now, for every $b \in \mathbb{C}^n$, there exist $a,a' \in A$ such that $V_b = V_a \cap V_{a'}$ and $W_b = W_a \cap W_{a'}$, hence $L(V_b) = W_b$. ■

Call S weakly hyperbolic if the closed line segment between two of its eigenvalues never contains $0 \in \mathbb{C}$. Let us mention the following straight-forward consequence of [Ch 86a, Complex Isolating Block Lemma, (vi)] :

THEOREM. - If S is weakly hyperbolic, then, for every S-vector field X, the union $\mathcal{W}$ of the s.i.m.'s of X can be characterised as "the union of those leaves of the foliation-germ defined by X which are adherent to 0". More precisely, denoting again by X and $\mathcal{W}$ representatives of X and $\mathcal{W}$ respectively, every neighbourhood of 0 in $\mathbb{C}^n$ contains an open neighbourhood U of 0 in the domain of X such that, if $\mathcal{F}_U$ is the foliation of $U \setminus \{0\}$ by complex curves everywhere tangent to X, then $\mathcal{W} \cap U$ is the union of those leaves of $\mathcal{F}_U$ which are adherent to 0. ■

In particular, the topological type of $\mathcal{W}$ is not only a C°-conjugacy invariant (see the above Proposition), but a C°-equivalence invariant -for a definition, see for example [Ch 86a].

APPENDIX 2. FURTHER RESULTS.

(A2.1) A precised version of the Preparation Lemma.

The following result is needed in [Ch 86a, Theorem 4] :

THEOREM. - If $S_o \in gl(n,\mathbb{C})$ is weakly hyperbolic and semi-simple, then, for each positive integer m, there exist an integer q and an open neighbourhood U of S_o in $gl(n,\mathbb{C})$ such that, if S lies in U, the S-normal forms in the Preparation Lemma can be chosen of degree q.

Proof. Since S_o is weakly hyperbolic, either it is in the Poincaré domain, in which case our theorem is true with $q = s_{\mathbb{C}^n}(m)$ (notation of (2.3), with $S:=S_o$), or it has several maximal s.i.m.'s, in which case we can take $q = r(m)$ (notation of Proposition (2.4), with $S:=S_o$). Indeed, our weak hyperbolicity asumption implies that those s.i.m.'s of S which are maximal for inclusion depend analytically on S in a neighbourhood of S_o. Therefore, it is easily checked that, if S is close enough to S_o, each $s_W(m)$ and $s_W(2m)$ in (2.3) is less than or equal to the corresponding number with $S := S_o$; thus, the argument in (2.4) yields our theorem. ■

(A2.2) Actions.

Let S be a $\mathbb{C}$-linear representation of $\mathbb{C}^k$ in $\mathbb{C}^n$, i.e. a morphism of the abelian Lie algebra $\mathbb{C}^k$ into $gl(n,\mathbb{C})$; assume that each $S(t)$, $t \in \mathbb{C}^k$, is semi-simple. An S _-action_ is a complex Lie Algebra morphism $Y:\mathbb{C}^k \to \underline{\underline{d}}'$ such that, for each $t \in \mathbb{C}^k$, $Y(t)$ is an $S(t)$ -vector field. A _s.i.m. of_ S is a complex vector subspace of $\mathbb{C}^n$ which is the unstable subspace of $S(t)$ for some t. Clearly, _each_ $S(t)$ _is tangent to every s.i.m. of_ S. IN THE SEQUEL, WE DENOTE BY $\mathcal{V}$ THE UNION OF THE S.I.M. 'S OF S.

Two S -actions Y and Z are _conjugate_ if there exists $h \in \underline{\underline{D}}$ such that $h^* Z(t) = Y(t)$ for every t. An S -action Y _has_ m^{th} _order contact with an_ S _-normal form_ X _of degree_ $q>0$ _along some closed subset_ $V \ni 0$ _of_ $\mathbb{C}^n$ if there exists a mapping $X : \mathbb{C}^k \to \underline{\underline{d}}$ such that the following three properties hold for every $t \in \mathbb{C}^k$:

(i) $Y(t)$ has m^{th} order contact with $X(t)$ along V.

(ii) $X(t)$ is an $S(t)$ -normal form of degree q, commuting with every $S(s)$.

(iii) The Lie bracket $[X(t),X(s)]$ has q^{th} order contact with 0 at 0 for every s (in fact, this implies (b). Note that we do _not_ require X to be an S -action).

By (ii), _each_ $X(t)$ _is tangent to every s.i.m. of_ S. Therefore, the following extension of the Preparation Lemma does provide information :

THEOREM. - For each positive integer m, there exists a positive integer q such that every S -action Z is conjugate to an S -action having m^{th} order contact with an S -normal form of degree q along $\mathcal{V}$, and the conjugacy can be chosen tangent to the identity at 0.

The proof is the same as before : first, a Formal Preparation Lemma tells us that, _for each integer_ q, Z _is conjugate to an_ S _-action_ Y _having_ q^{th} _order contact with an_ S _-action of degree_ q _at_ 0 (just use the approach of (2.1) and the following fact : given two endomorphisms A and A' of a finite dimensional complex vector space, with semi-simple parts s and s' respectively, if A commutes with A', then A, A', s, s' commute). There exists a <u>finite</u> subset T of $\mathbb{C}^k$ such that every s.i.m. of S is contained in the unstable manifold W_t of $S(t)$ for some $t \in T$. By Lemma A, and the argument of (2.4), if q is large enough, then there exists $h \in \underline{\underline{D}}$, having $(q-1)^{th}$ order contact with the identity at 0, such that, for every $t \in T$, $h^*Y(t)$ has m^{th} order contact with $X(t)$ along W_t. Therefore, Lemma B and the remark following it imply that, for each s, $h^*Y(s)$ and $X(s)$ have m^{th} order contact along every W_t. As our notion of a large enough q depends only on S, this proves our theorem. ∎

STABILITE DES V-VARIETES KAHLERIENNES

par

Aziz EL KACIMI ALAOUI

Résumé : On montre que toute petite déformation d'une V-variété kählérienne compacte est encore kählérienne.

MOTS CLES : V-variété, feuilletage, opérateur différentiel, déformation.

Dans [7] K. Kodaira et D.C. Spencer ont démontré un résultat de stabilité pour les variétés kahlériennes compactes par des méthodes de déformation du spectre d'un opérateur fortement elliptique auto-adjoint.

Le but de cette note est de montrer que ce théorème reste vrai pour les V-variétés [11] kählériennes.

Nous aurons besoin de plusieurs résultats de [7] dans le cadre équivariant par l'action d'un groupe de Lie compact. Nous reprendrons d'autre part certains théorèmes de décomposition (notamment pour les complexes de de Rham et de Dolbeault) sur les V-variétés qu'on trouvera déjà dans [1][9] et [2].

Sauf mention expresse du contraire ou précision toutes les structures considérées seront supposées de classe C^{∞}.

Dans toute la suite M sera une variété compacte, G un groupe de Lie compact et E un fibré vectoriel complexe de rang N' muni d'une métrique hermitienne h.

Soit $\phi : G\times M \to M$ une action de G sur M. On dira que ϕ est <u>localement libre</u> si tout point $y \in M$ a un groupe d'isotropie discret ; donc fini. Une telle action définit un feuilletage compact $\mathcal{F}$ (i.e. toutes les feuilles sont compactes). Plus précisément toute feuille L_y est difféomorphe au quotient de G par G_y. Notons

n la codimension de F et soient Ω une boule ouverte de R^n et G_o un sous-groupe fini de $O(n)$ agissant sur une variété compacte L_o (difféomorphe à G). On définit une action de G_o sur $L_o \times \Omega$ par

$$g_o'(g_o, z) = (g_o g_o'^{-1}, g_o' z)$$

Le feuilletage défini par les fibres de la projection $L_o \times \Omega \to \Omega$ est invariant par cette action et définit donc un feuilletage F_o sur le quotient $L_o \times_{G_o} \Omega$ qui constitue le modèle local de F i.e. si L est une feuille, il existe un voisinage V de cette feuille dans M tel que la restriction de F à V est équivalent à F_o sur $L_o \times_{G_o} \Omega$.

Dans toute la suite M sera munie d'une action localement libre de G. On confondra cette action avec le feuilletage compact F qu'elle définit.

1. *Décomposition spectrale équivariante.*

Considérons la variété $V = L_o \times_{G_o} \Omega$ et soit $\eta : G_o \to C^\infty(\Omega, SU(N'))$ une application vérifiant $\eta(g_o g_o')(z) = \eta(g_o)(g_o' z) \circ \eta(g_o')(z)$ pour tous $g_o, g_o' \in G_o$ et $z \in \Omega$. En posant $g_o'(g_o, z, v) = (g_o g_o'^{-1}, g_o' z, \eta(g_o')(z)v)$ on définit une action libre de G_o sur $L_o \times \Omega \times \mathbb{C}^{N'}$. Le quotient est un fibré de base $V = L_o \times_{G_o} \Omega$ et de fibre $\mathbb{C}^{N'}$ appelé <u>fibré admissible</u> (cf [4]).

1.1. <u>Définition</u>. On dira que E est <u>admissible</u> si sa restriction à tout ouvert $V \simeq L_o \times_{G_o} \Omega$ est un fibré admissible.

Un fibré admissible est feuilleté [4]. L'action de G sur M se relève alors en une action sur E pour laquelle on peut supposer que la métrique hermitienne h est invariante. On dira alors que E est un fibré <u>hermitien admissible</u>.

Soit E un tel fibré. L'action de G induit une action sur l'espace $C^\infty(E)$ des sections de E donnée par

$$(g\alpha)(y) = g\cdot\alpha(y\, g^{-1})$$

pour tout point $y \in M$, toute section $\alpha \in C^\infty(E)$ et tout $g \in G$.

1.2. <u>Définition</u>. Une section α de E est dite <u>invariante</u> si elle vérifie $g\cdot\alpha = \alpha$ pour tout $g \in G$.

L'ensemble $C^{\infty}_G(E)$ des sections invariantes de E est un module sur l'anneau $A_G(M)$ des fonctions G-invariantes (ou basiques) i.e. constantes sur les feuilles de F.

Considérons maintenant deux sections α et β de E. On pose :

$$\langle\alpha,\beta\rangle = \int_M h(\alpha(y),\beta(y))dy$$

où dy est la mesure canonique associée à une métrique riemannienne G-invariante sur M. On définit ainsi sur $C^{\infty}(E)$ un produit scalaire $\langle\,,\,\rangle$ G-invariant pour lequel $C^{\infty}_G(E)$ est un sous-espace fermé.

Soit D un opérateur différentiel agissant sur $C^{\infty}(E)$. On dira que D est <u>invariant</u> s'il commute à l'action de G sur $C^{\infty}(E)$. Dans ce cas il préserve $C^{\infty}_G(E)$. Dans un système de coordonnées locales $y = (x,z)$ adaptés à F, D s'écrit :

$$(D\alpha)^i(y) = \sum_{s=0}^{\ell} \sum_{j=1}^{N'} P^{ij}_s(z\ ;\frac{\partial}{\partial y_1},\ldots,\frac{\partial}{\partial y_m})\ \alpha^j(y) \qquad (*)$$

où $\alpha(y) = (\alpha^1(y),\ldots,\alpha^{N'}(y))$ est l'écriture locale d'une section α de E et P^{ij}_s un polynôme homogène de degré s en $\frac{\partial}{\partial y_1},\ldots,\frac{\partial}{\partial y_m}$ à coefficients des fonctions qui ne dépendent que de la coordonnée transverse z. Supposons que D est fortement elliptique et auto-adjoint pour $\langle\,,\,\rangle$. On a alors le

1.3. <u>Théorème</u>. L'espace vectoriel $H_G(E) = \mathrm{Ker}(C^{\infty}_G(E)\xrightarrow{D} C^{\infty}_G(E))$ est de dimension finie et on a une décomposition orthogonale $C^{\infty}_G(E) = H_G(E) \oplus \mathrm{Im}\, D$.

D'où l'on déduit le

1.4. <u>Théorème de décomposition spectrale</u>. L'opérateur D a un système de sections propres invariantes (e_h) formant une base hilbertienne du complété $L^2_G(E)$ de $C^{\infty}_G(E)$ dans $L^2(E)$. Pour tout $\alpha \in L^2_G(E)$, on a :

$$\alpha = \sum_{h=1}^{\infty} \langle\alpha,e_h\rangle\ e_h$$

et la série converge en norme L^2. D'autre part, $\alpha \in C^{\infty}_G(E)$ si et seulement si pour tout entier r on a :

$$\sum_{h=1}^{\infty} \lambda_h^{2r}|\langle\alpha,e_h\rangle|^2 < +\infty$$

où λ_h est la valeur propre réelle associée à e_h. En plus on a $\lambda_1 \leq \lambda_2 \leq \ldots \leq \lambda_h \leq \ldots$ et

$$\lim_{h \to +\infty} \lambda_h = +\infty$$

Ces deux théorèmes découlent immédiatement de leurs analogues classiques (cf [6] ou [12] par exemple) en utilisant le fait que D commute à l'action de G sur E.

2. *Application aux formes basiques.*

Notons $T\mathcal{F}$ le fibré tangent à $\mathcal{F}$, $\nu\mathcal{F} = TM/T\mathcal{F}$ et $\iota : TM \to \nu\mathcal{F}$ la projection canonique. On a une suite exacte

$$0 \to T\mathcal{F} \longrightarrow TM \xrightarrow{\iota} \nu\mathcal{F} \longrightarrow 0$$

On peut réaliser $\nu\mathcal{F}$ comme un sous-fibré admissible ν de TM à l'aide d'une section de ι. On a alors une décomposition en somme directe

$$TM = T\mathcal{F} \oplus \nu$$

Une section du fibré $E^r = \Lambda^r \nu_{\mathbb{C}}^*$, où $\nu_{\mathbb{C}} = \nu \otimes_R \mathbb{C}$, est appelée r-forme différentielle (à valeurs dans $\mathbb{C}$) semi-basique. Si en plus, elle est G-invariante alors elle est dite basique. On notera $(\Omega^*(M/\mathcal{F}),d)$ le complexe des formes basiques de $\mathcal{F}$.

Sur $C^\infty(E^*)$ la différentielle d se décompose en une somme

$$d = d_{\mathcal{F}} + \bar{d}$$

où $d_{\mathcal{F}}$ est la différentielle extérieure le long des feuilles et $\bar{d}$ un opérateur invariant de différentiation dans la direction de ν et dont la restriction à $\Omega^*(M/\mathcal{F})$ est égale à d.

La métrique hermitienne G-invariante sur $\nu_{\mathbb{C}}$ permet de définir un opérateur

$$\bar{*} : C^\infty(E^r) \to C^\infty(E^{n-r})$$

de manière analogue au cas classique.

Soient $Q_1,\ldots,Q_N$ les champs fondamentaux de l'action de G sur M. Notons $\theta^1,\ldots,\theta^N$ les 1-formes duales et $\chi = \theta^1 \wedge \ldots \wedge \theta^N$ la forme caractéristique de $\mathcal{F}$.

La forme différentielle $\alpha \wedge \bar{*} \beta \wedge \chi$ est de degré $m = \dim M$. On pose :

$$<\alpha,\beta> = \int_M \alpha \wedge \bar{*} \beta \wedge \chi$$

On définit ainsi un produit scalaire $< , >$ sur $C^\infty(E^r)$. Soit d'autre part $\bar{\delta} : C^\infty(E^r) \to C^\infty(E^{r-1})$ défini par $\bar{\delta} = (-1)^r \bar{*}^{-1} \bar{d}\bar{*}$.

2.1. *Proposition*. Soient $\alpha \in C^\infty(E^{r-1})$ et $\beta \in C^\infty(E^r)$. On a :

$$<\bar{d}\alpha,\beta> = <\alpha,\bar{\delta}\beta>$$

Démonstration : On a :

$$\bar{d}(\alpha \wedge \bar{*} \beta \wedge \chi) = d(\alpha \wedge \bar{*} \beta \wedge \chi)$$

pour des raisons de degré évidentes. D'autre part

$$\bar{d}(\alpha \wedge \bar{*} \beta \wedge \chi) = \bar{d}\, \alpha \wedge \bar{*} \beta \wedge \chi + (-1)^{r-1}\, \alpha \wedge \bar{d}\, \bar{*}\, \beta \wedge \chi + (-1)^{n-1}\, \alpha \wedge * \beta \wedge \bar{d}\, \chi.$$

Comme F est défini par une action localement libre de groupe compact on peut choisir sur TF une métrique riemannienne de telle sorte que les feuilles soient des sous-variétés minimales de M. Quitte à renormer les champs $Q_1,\ldots,Q_N$ on peut alors supposer que χ est relativement fermée (cf [10]). Ce qui implique $\bar{d}\chi = 0$. Par définition de $\bar{\delta}$ on a $\bar{d}\bar{*} = (-1)^r\, \bar{*}\, \bar{\delta}$. D'où finalement

$$d(\alpha \wedge \bar{*} \beta \wedge \chi) = \bar{d}\alpha \wedge \bar{*} \beta \wedge \chi - \alpha \wedge \bar{*}\, \bar{\delta}\, \beta \wedge \chi$$

En intégrant les deux membres sur M on obtient :

$$0 = <\bar{d}\alpha,\beta> - <\alpha,\bar{\delta}\beta>$$

C.Q.F.D.

2.2. Corollaire. L'opérateur $\bar{\Delta} = \bar{d}\bar{\delta} + \bar{\delta}\bar{d}$ est auto-adjoint.

Soient maintenant $Q_1,\ldots,Q_N$ les opérateurs différentiels d'ordre 1 définis par les champs fondamentaux. On pose

$$Q = \sum_{i=1}^{N} Q_i \cdot \bar{Q}_i$$

où $\bar{Q}_i$ est le conjugué complexe de Q_i. On obtient alors un opérateur invariant sur $C^\infty(E^r)$ dont il est facile de voir qu'il est auto-adjoint. En posant

$$L = \bar{\Delta} + Q$$

on obtient un opérateur fortement elliptique (cf [2]) et auto-adjoint. Sa restriction à $\Omega^r(M/F)$ n'est rien d'autre que le laplacien basique tel qu'il est décrit par exemple dans [3] ou [9]. D'après le théorème 1.3, on a

2.3. <u>Théorème</u>. L'espace vectoriel $\mathcal{H}^r(M/F) = \mathrm{Ker}(\Omega^r(M/F) \to \Omega^r(M/F))$ est de dimension finie et on a une décomposition orthogonale

$$\Omega^r(M/F) = \mathcal{H}^r(M/F) \oplus \mathrm{Im}\,\bar{\Delta} = \mathcal{H}^r(M/F) \oplus \mathrm{Im}\,\bar{d} \oplus \mathrm{Im}\,\delta.$$

On en déduit que la cohomologie basique $H^r(M/F)$ de F s'identifie à $\mathcal{H}^r(M/F)$ qui est l'espace des r-formes harmoniques basiques. Ce théorème a été établi par plusieurs auteurs [1] [9] et [3].

On peut remarquer que $H^*(M/F)$ est canoniquement isomorphe à la cohomologie de de Rham de la V-variété $B = M/F$ (cf [9]). Donc si F est transversalement orientable, $H^n(M/F) \neq 0$ et dans ce cas $H^*(M/F)$ vérifie la dualité de Poincaré.

Supposons maintenant que F est transversalement holomorphe. Le fibré $\nu_{\mathbb{C}}$ se décompose alors sous la forme

$$\nu_{\mathbb{C}} = \nu^{(1,0)} \oplus \nu^{(0,1)}$$

où $\nu^{(1,0)}$ et $\nu^{(0,1)}$ sont les sous-fibrés propres de l'automorphisme feuilleté J de la structure complexe sur $\nu_{\mathbb{C}}$. On en déduit une décomposition

$$\Lambda^r \nu_{\mathbb{C}}^* = \bigoplus_{p+q=r} \Lambda^p\, \nu^{(1,0)*} \otimes \Lambda^q\, \nu^{(0,1)*}$$

Une section du fibré $E^{pq} = \Lambda^p\, {}^{(1,0)*} \otimes \Lambda^q\, \nu^{(0,1)*}$ est appelée forme différentielle <u>semi-basique de type</u> (p,q). Si en plus elle est G-invariante, on dira qu'elle est <u>basique de type</u> (p,q). On notera $\Omega^{pq}(M/F)$ l'espace des formes différentielles basiques de type (p,q).

L'opérateur $\bar{d}$ se décompose en une somme

$$\bar{d} = \partial + \bar{\partial}$$

respectivement de type (1,0) et (0,1). La restriction de $\bar{\partial}$ à $\Omega^{pq}(M/F)$ permet de définir un complexe différentiel

$$0 \to \Omega^{p0}(M/F) \xrightarrow{\bar{\partial}} \Omega^{p1}(M/F) \xrightarrow{\bar{\partial}} \ldots \xrightarrow{\bar{\partial}} \Omega^{pn}(M/F) \to 0$$

dont l'homologie notée $H^{p*}(M/F)$ sera appelée la <u>cohomologie de Dolbeault basique</u> de F.

On note $\bar{\mathfrak{F}}$ l'adjoint de $\bar{\partial}$ pour < , > et on pose $\bar{\Delta}' = \bar{\partial}\bar{\mathfrak{F}} + \bar{\mathfrak{F}}\bar{\partial}$. En procédant alors comme pour le complexe de de Rham basique on obtient le

2.4. Théorème. L'espace vectoriel $\mathcal{H}^{pq}(M/F) = \mathrm{Ker}(\Omega^{pq}(M/F) \xrightarrow{\bar\Delta'} \Omega^{p,q}(M/F))$ est de dimension finie et on a une décomposition orthogonale

$$\Omega^{pq}(M/F) = \mathcal{H}^{pq}(M/F) \oplus \mathrm{Im}\,\bar\partial \oplus \mathrm{Im}\,\bar\delta$$

D'où l'on déduit $H^{pq}(M/F) \simeq \mathcal{H}^{pq}(M/F)$. Comme d'autre part, F est transversalement holomorphe il est transversalement orientable : donc $H^{2n}(M/F) \neq 0$. D'après [2] $H^{**}(M/F)$ vérifie la dualité de Serre i.e.

$$H^{pq}(M/F) \simeq H^{n-p,n-q}(M/F)$$

pour tout $p,q = 0,\dots,n$.

2.5. Cas où F est transversalement kählerien.

Notons toujours h la métrique hermitienne G-invariante sur $\nu_{\mathbb{C}}$ et posons $\omega(\cdot,\cdot) = h(J\cdot,\cdot)$. On obtient ainsi une forme basique de type $(1,1)$. On dira que F est transversalement kählerien si ω est fermée. Pour un tel feuilletage on a les propriétés suivantes (cf. [2]).

i) $\bar\Delta = 2\bar\Delta'$

ii) $H^{pq}(M/F) = H^{qp}(M/F)$

iii) $H^{r}(M/F) = \bigoplus_{p+q=r} H^{pq}(M/F)$

iv) $\bar\Delta\omega^p = 0$ pour tout $p = 0,1,\dots,n$ où $\omega^p = \underbrace{\omega \wedge \dots \wedge \omega}_{p \text{ fois}}$;

d'où $H^{p,p}(M/F) \neq 0$ pour tout $p = 0,1,\dots,n$.

3. *Déformation équivariante des opérateurs fortement elliptiques.*

Soient T une boule de R^d et E un fibré hermitien admissible de rang N' sur $M\times T$ qui est en fait une famille $(E_t)_{t\in T}$ de fibrés admissibles au-dessus de M.

On considère une famille $(D_t)_{t\in T}$ d'opérateurs différentiels invariants d'ordre $\ell = 2\ell'$ agissant respectivement sur $C^\infty(E_t)$. Dans un système de coordonnées locales $y = (x,z)$ adaptées à F, D_t a la forme (*) où P_s^{ij} dépend de z et de t.

3.1. Définition. On dira que D_t dépend différentiablement de t si les coefficients du polynôme P_s^{ij} dépendent différentiablement de z et de t.

On supposera comme dans [7] que $\mathcal{E}$ est de la forme $\mathcal{E} = E\times T$ où E est un fibré hermitien admissible sur M i.e les fonctions de transitions de $\mathcal{E}$ sont indépendantes de t.

Ceci étant on considère une famille $(D_t)_{t\in T}$ d'oépratеurs différentiels d'ordre $\ell = 2\ell'$ dépendant différentiablement de t, invariants, fortement elliptiques et auto-adjoints agissant sur $C^\infty(E)$.

Soient U un ouvert de M trivialisant E et $(y_1,\dots,y_m)$ un système de coordonnées sur U. Pour tout multiindice $s = (s_1,\dots,s_m)$ on notera $|s| = s_1+\dots+s_m$ sa longueur et D^s l'opérateur $\dfrac{\partial^{|s|}}{\partial y_1^{s_1}\dots\partial y_m^{s_m}}$. On posera alors

$$||\alpha||_r^2 = \sum_U \sum_{i=1}^{N'} \sum_{|s|=0}^{r} \int_U |D^s\,\alpha^i(y)|^2\, dy_1\dots dy_m$$

pour tout $\alpha\in C^\infty(E)$ et tout entier $r = 0,1,\dots$ où U parcourt un recouvrement de M. Le complété $H_r(E)$ de $C^\infty(E)$ pour cette norme $||\ ||_r$ est le $r^{\text{ème}}$ espace de Sobolev des sections de E. On a $C^\infty(E) = \bigcap_{r\geq 0} H_r(E)$.

3.2. <u>Proposition</u>. Si $D_t : C_G^\infty(E)\to C_G^\infty(E)$ est surjectif et vérifie l'inégalité

$$||D_t\alpha||_o \geq C||\alpha||_o$$

pour tout $\alpha\in C_G^\infty(E)$ où C est une constante positive indépendante de t, alors D_t^{-1} dépend différentiablement de t.

La démonstration est la même que celle de la proposition 1 de [6] p.51.

Soit maintenant W un domaine borné de $\mathbb{C}$ contenant l'origine et dont le bord Γ est une courbe C^∞. Notons $\mathcal{H}_G(\Gamma)_t$ le sous-espace de $C_G^\infty(E)$ engendré par les sections propres $e_h(t)$ de D_t dont les valeurs propres correspondantes sont contenues dans W et $P_t(\Gamma)$ la projection orthogonale

$$P_t(\Gamma) : C_G^\infty(E)\longrightarrow \mathcal{H}_G(\Gamma)_t$$

Si $\alpha = \sum_{h=1}^{\infty} \langle\alpha,e_h(t)\rangle\, e_h(t)\in C_G^\infty(E)$, alors

$$P_t(\Gamma)\alpha = \sum_{\lambda_h(t)\in W} \langle\alpha,e_h(t)\rangle\, e_h(t).$$

On pose d'autre part

$$K_t(\Gamma)\alpha = \sum_{\lambda_h(t)\notin W} \frac{<\alpha, e_h(t)>}{\lambda_h(t)} e_h(t)$$

On a clairement

$$\alpha = D_t K_t(\Gamma)\alpha + P_t(\Gamma)\alpha$$

On notera $H_{G,t} = \{\alpha \in C_G^\infty(E) \;/\; D_t\alpha = 0\}$, P_t la projection orthogonale $C_G^\infty(E) \to H_{G,t}$ et K_t l'opérateur de Green associé à D_t défini par :

$$K_t\alpha = \sum_{\lambda_h(t)\neq 0} \frac{<\alpha, k(t)>}{\lambda_h(t)} e_h(t)$$

pour $\alpha \in C_G^\infty(E)$.

On a le

3.3. <u>Théorème</u>. Pour h fixé, $\lambda_h(t)$ est une fonction continue de t.

<u>Démonstration</u> : On considère $\lambda_h(t)$ comme valeur porpre de D_t opérant sur $C^\infty(E)$. C'est une fonction continue de D_t d'après [6] p.342. Soient $t_o \in T$ et $e_h(t_o)$ une section propre invariante associée à $\lambda_h(t_o)$. Pour tout $t \in T$ soit $e'_h(t)$ une section propre (non nécessairement invariante) associée à $\lambda_h(t)$. On pose

$$e_h(t) = \int_G g\cdot e'_h(t)\, d\mu(g)$$

où μ est une mesure de Haar normalisée sur G. On obtient ainsi une section invariante.

Si t est suffisamment voisin de t_o, alors $e_h(t)$ est voisine de $e_h(t_o)$, donc non nulle. D'autre part, on a

$$D_t\, e_h(t) = D_t \int_G g\cdot e'_h(t)\, d\mu(g)$$

$$D_t\, e_h(t) = \int_G g\cdot D_t\, e'_h(t)\, d\mu(g)$$

$$D_t\, e_h(t) = \int_G g\cdot \lambda_u(t)\, e'_h(t) d\mu(g)$$

$$D_t\, e_h(t) = \lambda_h(t)\, e_h(t)$$

Ainsi quand t est voisin de t_o, $\lambda_h(t)$ est voisine de $\lambda_h(t_o)$ en tant que valeur propre associée à une section propre invariante.

De cette proposition et en suivant la démarche de [6] p.343 on déduit le

3.4. Théorème. La dimension de $H_{G,t}$ est une fonction semi-continue supérieurement en t.

De même, on a :

3.5. Théorème. [6] . Si $\dim H_{G,t}$ est indépendante de t, les opérateurs K_t et P_t dépendent différentiablement de t.

La démonstration utilise la proposition 3.2 et la continuité de $\lambda_h(t)$. Elle est la même que celle du théorème 7.4 de [6] p.343.

4. *Déformation des V-variétés kählériennes.*

Soit B un espace topologique séparé. Une uniformisation locale de B est la donnée d'un ouvert U de B et d'un triplet $(\tilde{U},\Gamma,\varphi)$ où

i) $\tilde{U}$ est un ouvert de $\mathbb{R}^n$;

ii) Γ un groupe fini de difféomorphismes de $\tilde{U}$;

iii) φ est une application de $\tilde{U}$ dans U qui vérifie $\varphi \circ \gamma = \varphi$ pour tout $\gamma \in \Gamma$ et induit un homéomorphisme de $\tilde{U}/\Gamma$ sur U.

4.1. Définition. [11]. On dira que B est une V-variété de dimension n si B admet un recouvrement ouvert $\{U\}$ auquel est associée une famille $\{(\tilde{U},\Gamma,\varphi)\}$ d'uniformisations locales telles que si $(\tilde{U},\Gamma,\varphi)$ et $(\tilde{U}',\Gamma',\varphi')$ sont deux éléments de cette famille, alors il existe une injection de $(\tilde{U},\Gamma,\varphi)$ dans $(\tilde{U}',\Gamma',\varphi')$ i.e. un difféomorphisme $f : \tilde{U} \to \tilde{U}'$ tel que $f \circ \gamma = \gamma' \circ f$ et $\varphi = \varphi' \circ \gamma$ pour tout $\gamma \in \Gamma$ et tout $\gamma' \in \Gamma$.

On définit de manière analogue une V-variété complexe de dimension n en remplaçant ouvert de $\mathbb{R}^n$ par ouvert de $\mathbb{C}^n$ et difféomorphisme par application biholomorphe.

Les objets géométriques sur B tels que fibré vectoriel, opérateur différentiel, métrique riemannienne etc... correspondent à ceux définis localement sur U invariants par Γ.

L'espace des feuilles d'un feuilletage riemannien compact est une V-variété [8]. En fait toute V - variété s'obtient de cette manière. En particulier, si B est une V-variété complexe compacte de dimension n on a la

4.2. <u>Proposition</u> [5]. B est l'espace des feuilles d'une action localement libre du groupe $G = SO(2n)$ sur une variété compacte M.

On dira que B est <u>kählerienne</u> si le feuilletage F défini par cette action est transversalement kählérien.

On désigne toujours par T une boule de $\mathbb{R}^d$.

4.3. <u>Définition</u>. Une déformation de B est la donnée d'une submersion $\mathcal{B} \xrightarrow{\pi} T$, où $\mathcal{B}$ est une V-variété, telle que :

i) $B_t = \pi^{-1}(t)$ est une V-variété complexe compacte de dimension n ;

ii) $B_o = B$.

Il est démontré dans [5] que B ne se déforme pas différentiablement i.e. pour tout t, il existe un difféomorphisme de B_t sur B. Par contre la structure complexe sur B_t dépend de t en général. D'autre part par la proposition 4.2.3 de [5] toute déformation de B définit une déformation (dans l'espace des feuilletages compacts hermitiens i.e. riemanniens et transversalement holomorphes) du feuilletage F donné par 4.2.

4.4. <u>Définition</u>. Une déformation de F est une submersion

$$\mathcal{M} \xrightarrow{q} T$$

où $\mathcal{M}$ est une variété munie d'un feuilletage de même dimension que F induisant sur chaque fibre $M_t = q^{-1}(t)$ un feuilletage F_t compact hermitien de codimension n tel que $F_o = F$ sur $M_o = M$.

Dans toute la suite la déformation de F que l'on considérera sera celle qui provient de la déformation de B.

4.5. <u>Théorème</u>. Si $B = B_o$ est kählerienne alors il existe $\varepsilon > 0$ tel que pour tout $t \in T$ vérifiant $|t| < \varepsilon$, la V-variété B_t est kählerienne.

De manière équivalente si le feuilletage $F = F_o$ est transversalement kählerien, alors pour t suffisamment petit F_t admet une structure transverse kählerienne. C'est cette assertion que nous allons prouver en appliquant les résultats des paragraphes 1,2,3 et en adaptant la démarche de K. Kodaria et D.C. Spencer [7] au cas basique.

Démonstration du théorème 4.5. Considérons l'opérateur d'ordre 4 :

$D_t = \partial_t \bar{\partial}_t \bar{\delta}_t \delta_t + \bar{\delta}_t \delta_t \partial_t \bar{\partial}_t + \bar{\delta}_t \partial_t \delta \bar{\partial}_t + \delta_t \bar{\partial}_t \bar{\delta}_t \partial_t + \bar{\delta}_t \bar{\partial}_t + \delta_t \partial_t$ dont il est facile de vérifier qu'il est auto-adjoint et tel que

$$D_o = \Delta'_o \Delta'_o + \bar{\delta}_o \bar{\partial}_o + \delta_o \partial_o$$

et qu'on a

$$H_t^{pq} = \mathrm{Ker}(D_t : \Omega^{pq}(M_t/F_t) \to \Omega^{pq}(M_t/F_t)) = \mathrm{Ker}\ \partial_t \cap \mathrm{Ker}\ \bar{\partial}_t \cap \mathrm{Ker}\ \bar{\delta}_t \delta_t$$

Cet opérateur peut aussi être vu comme la restriction à $\Omega^{pq}(M_t/F_t) = C_G^\infty(E_t^{p,q})$ de l'opérateur

$$L_t = D_t + Q_t^2$$

qui est fortement elliptique auto-adjoint pour le produit scalaire $< >$ sur $C^\infty(E_t^{pq})$. Ici $Q_t = \Sigma\, Q_{ti} \bar{Q}_{ti}$ où $(Q_{ti})_{i=1 \ldots \mathrm{Dim}\, G}$ sont les champs fondamentaux de l'action de $G = SO(2n)$ qui définit F_t.

Soit $Z_t^{pq} = \{\alpha \in \Omega^{pq}(M_t/F_t)/d\alpha = 0\}$. La démonstration de la proposition 7 de [7] reste encore valable dans le cas basique qui nous intéresse ici. Elle donne

$$Z_t^{pq} = \partial_t \bar{\partial}_t \Omega^{p-1,q-1}(M_t/F_t) \oplus H_t^{pq} .$$

La variété B_t ne se déforme pas différentiablement [5]. Donc $b_o^2 = b_t^2 = \dim H^2(M_t/F_t) = \dim H^2(B_t,\mathbb{C})$ est indépendant de t. Un raisonnement analogue à celui de [7] p.73-74 montre qu'il existe $\varepsilon > 0$ tel que $\dim H_t^{pq}$ ne dépend pas de t pour $|t| < \varepsilon$ et que

$$H_o^{11} = H^{11}(M/F).$$

D'après le théorème 3.5 l'opérateur de Green K_t de D_t et le projecteur $P_t : \Omega^{11}(M_t/F_t) \to H_t^{11}$ dépendant différentiablement de t.

Pour terminer, on pose :

$$\tilde{\omega}_t = \frac{1}{2}(P_t\omega_t + \overline{P_t\omega_t})$$

La 2-forme basique ω_t vérifie les propriétés suivantes

i) $P_o\tilde{\omega}_o = \omega_o$; d'où $\tilde{\omega}_o = \omega_o$

ii) $d\tilde{\omega}_t = 0$

iii) $\tilde{\omega}_t$ est définie positive car P_t dépend différentiablement de t.

On en déduit que la métrique hermitienne $\tilde{h}_t$ associée à $\tilde{\omega}_t$ est de Kähler. Ce qui démontre le théorème.

REFERENCES

[1] BAILY, W.L. - The decomposition theorem for V-manifolds. Amer. J. of Math. 78 (1956) 862-888.

[2] EL KACIMI ALAOUI, A. - Opérateurs transversalement elliptiques sur les feuilletages riemanniens. Preprint 1986 - Lille.

[3] EL KACIMI ALAOUI, A. et HECTOR, G. - Décomposition de Hodge basique pour un feuilletage riemannien. Ann. Inst. Fourier t.36. Fasc 3 (1986) (à paraître)

[4] GIRBAU, J. and NICOLAU, M. - Pseudo-différential operators on V-manifolds and foliations. Part I, Collect. Math. 30 (1979) p.247-265 ; Part II Collect. Math 31 (1980) 63-95.

[5] GIRBAU, J. , HAEFLIGER, A. and SUNDARARAMAN - On deformations of transversely holomorphic foliations. J. Für die reine und Wangendte Mathematik ; Band 345 (1983) 122-147.

[6] KODAIRA, K. - Complex manifolds and deformation of complex structures Grundlehren der Mathematischen Wissenshaften 283 Springer-Verlag (1985).

[7] KODAIRA, K and SPENCER, D.C. - On deformations of complex analytic structures, III. Ann. of Math. Vol. 71, n°1, (1960) 43-76.

[8] REINHART, B.L. - Foliated manifolds with bundle-like metrics. Ann of Math., 69 (1959), 119-132.

[9] REINHART, B.L. - Harmonic integral on foliated manifolds. Amer. J. of Math. (1959), 529-586.

[10] RUMMLER, H. - Quelques notions simples en géométrie riemannienne et leurs applications aux feuilletages compacts. Comment. Math. Helvetici 54 (1979) 224-239.

[11] SATAKE, I. - On a generalization of the notion of manifold. Proc. Nat. Acad. Sci., USA 42 (1956) 359-363.

[12] WELLS, R.O. - Differential analysis on complex manifolds. G.T.M. n°65, Springer-Verlag (1979)

Université de Lille III
UFR de Mathématiques et
Sciences Economiques
59653 Villeneuve d'Ascq Cedex (France)

Université des Sciences et Techniques
de Lille Flandres-Artois
UA au CNRS n°751
UFR de Mathématiques Pures et Appliquées
59655 - Villeneuve d'Ascq Cedex (France)

CYCLIC RESULTANTS OF RECIPROCAL POLYNOMIALS

David Fried*

Suppose one is given a polynomial $p(x) = a_d x^d + \ldots + a_1 x + a_0$ over $\mathbf{R}$ with $a_d > 0$ that is reciprocal, i.e. $a_i = a_{d-i}$ for $0 \leq i \leq d$. Define the nth cyclic resultant r_n of $p(x)$ to be the resultant of $p(x)$ and $x^n - 1$. We will study the problem of reconstructing $p(x)$ from the sequence $b_n = |r_n|$, $n = 1, 2, \ldots$.

As motivation, consider the case when $p(x)$ is the Alexander polynomial of a knot K. Then the reciprocality of $p(x)$ is a consequence of Poincaré duality in the infinite cyclic cover of the knot complement $S^3 - K$[M] but was first proven by Seifert [S] by other means. A theorem of Fox says that b_n is the order of the first homology of the cyclic branched cover of S^3 with branching locus K and n sheets, with the proviso that this homology group is infinite $\Leftrightarrow b_n = 0$[F]. In this context the question whether b_n determined $p(x)$ was posed by Prof. Gonzalez-Acuña.

Our main result is that for $p(x)$ as above

Proposition. *If $b_n \neq 0$ for all $n > 0$ then b_n determines $p(x)$.*

The condition $b_n \neq 0$ means that $p(x)$ has no root at an nth root of unity. We will also give examples showing that this condition on $p(x)$ is necessary.

We recall an earlier result in this direction due to H. Stark [DG]. If γ is a closed orbit of a Lagrangian system then the linear Poincaré map A on the hypersurface of fixed energy is symplectic and so its characteristic polynomial $p(x) = \det(A - x)$ is monic and reciprocal. The work of Duistermaat and Guillemin on the distributional trace of wave operators showed that if no eigenvalue of A is a root of unity and if no other closed orbit of the given energy has period a multiple of the period of γ then the quantities $|\det(I - A^n)|$, $n > 0$, can be calculated from an appropriate distribution. They asked whether this sequence determines the spectrum of A. Stark showed that it determines the eigenvalues of modulus $\neq 1$, and also the spectrum of A^N for some unspecified $N > 0$. Since, with our notation, $b_n = |\det(I - A^n)|$ the proposition above settles this question of Duistermaat and Guillemin in the affirmative.

As a third geometric application, let A be an integral $d \times d$ symplectic matrix and $p(x) = \det(A - x)$. Then again $b_n = |\det(I - A^n)|$. The latter, however, is the number of points of period n for the map $f : T^d \to T^d$ induced by A, where $b_n \neq 0$ assures that this number is finite. Thus if a symplectic toral automorphism f has only finitely many points b_n of period n for each $n > 0$ then the sequence b_n determines the spectrum of the linear map $A : \mathbf{R}^d \to \mathbf{R}^d$ that lifts f.

* Partially supported by the National Science Foundation, the Sloan Foundation and the IHES.

To a map f with $N_n < \infty$ points of period n for each $n > 0$, Artin and Mazur associate the formal power series or generating function

$$\varsigma(t) = \exp \sum_{n>0} N_n \frac{t^n}{n}$$

suggested by Weil's work on equations over finite fields. In many cases they proved $\varsigma(t)$ is rational [AM] and this holds for various other dynamical systems as well [W], [Fr]. Motivated by our third application, we consider

$$B(t) = \exp \sum_{n>0} b_n \frac{t^n}{n}$$

and prove it is rational.

First we examine the sign of r_n, $n > 0$. We have

$$r_n = (a_0)^n \cdot \prod_\lambda (\lambda^n - 1).$$

where λ runs over the zeroes of $p(x)$ (with multiplicity). A complex conjugate pair of roots contributes a positive term to b_n. A real root λ contributes to the sign a factor $+1$ if $\lambda > 1$, -1 if $-1 < \lambda < -1$, and $(-1)^n$ if $\lambda < -1$. Thus

$$\begin{aligned} \text{sgn } (r_n) &= \epsilon \cdot \delta^n, \quad \epsilon = (-1)^E \ , \ \delta = (-1)^D \\ E &= \# \text{ roots of } p(x) \text{ in } (-1, 1) \\ D &= \# \text{ roots of } p(x) \text{ in } (-\infty, -1). \end{aligned}$$

Note that $b_2 \neq 0$ implies $+1$, -1 are not roots of b_n.

It follows that

$$b_n = \epsilon \cdot \delta^n \cdot (a_0)^n \cdot \prod_\lambda (\lambda^n - 1) = \Sigma \pm \mu^n$$

where the μ's are defined by expanding out the product over λ, i.e. each μ is the product of δa_0 and some subset of the λ's. But

$$\exp \Sigma - \mu^n \cdot \frac{t^n}{n} = 1 - \mu t$$

by the power series for $\log 1 - x$ and so

$$B(t) = \prod_\mu (1 - \mu t)^{\mp 1}$$

is rational, with divisor

$$\beta = \Sigma \mp [\mu^{-1}] \ \in \mathbf{Z}\ \mathbf{C}^*.$$

Here for a nonzero complex number ν we write $[\nu]$ for the corresponding divisor in the integral group ring $\mathbf{Z}\ \mathbf{C}^*$. Note that we can factor β

$$\beta = -\epsilon[(\delta a_0)^{\ \ 1}] \prod_\lambda ([\lambda^{\ \ 1}] - 1)$$

An explicit formula for $B(t)$ can be readily found. First use Sturm's Theorem to compute $\delta, \epsilon[U]$. Then take a matrix A with $\det(A - x) = p(x)$, say the companion matrix of p, and compute its exterior powers $\Lambda^k A$. Then

$$B(t) = Z_A(\delta a_0 t)^{\epsilon}$$

where

$$Z_A(t) = \prod_{k=0}^{d} \det(I - t\Lambda^k A)^{(-1)^{k+1}}.$$

This easily follows from the fact that d is even and the standard identities

$$\prod_{\lambda}(\lambda^n - 1) = \sum_{k=0}^{d}(-1)^k Tr(\Lambda^k A)^n$$

$$\exp \sum_{n>0}(Tr\ C^n)\frac{t^n}{n} = \det(I - tC)^{-1}$$

where the latter is applied to $C = \Lambda^k A,\ k = 0, \ldots, d$.

In a group ring $\mathbf{Z}G$ of an abelian group G, elements of the form $\pm g,\ g \in G$, are called trivial units. We will write $r \sim s$ if the ring elements r and s differ only by a factor of a trivial unit. We will show

Lemma 1. *For $g_0 \in G$ of infinite order, $g_0 - 1$ is a nondivisor of zero.*

Lemma 2. *If $\beta \in \mathbf{Z}G$ satisfies $\beta \sim \prod_{i=1}^{d} g_i - 1$, $g_i \in G$ of infinite order, $g_{d+1-i} = g_i^{-1}$, then β determines the factors $g_i - 1$.*

Of course, Lemma 1 is known but we will prove it in the course of proving Lemma 2. Both Lemmas clearly reduce to the case of finitely generated G. We can choose a homomorphism $\phi : G \to \mathbf{Z}$ so that $\phi(g_0) \neq 0$ or $\phi(g_i) \neq 0$, $i = 1, \ldots, d$, respectively. Then we imbed $\mathbf{Z}G = R$ in $R[t, t^{-1}] = S$ by

$$\psi(\Sigma c_g g) = \Sigma c_g g t^{\phi(g)}.$$

Then $\psi(g_0 - 1)$ is clearly a nondivisor of zero is S so $g_0 - 1$ was one in R, proving Lemma 1. For Lemma 2 we have

$$\psi(\beta) = U \cdot \prod_{i=1}^{d} t^{\phi(g_i)} g_i - 1$$

where U is a trivial unit. Note d is even (else $g_{d/2}$ has order ≤ 2) and we can reorder the g_i so that $\phi(g_i) > 0$ for $i \leq d/2$. Then the highest degree term in $\psi(\beta)$ is

$$T_1 = U \cdot g_1 \cdot \ldots \cdot g_{d/2} t^{\Sigma},\ \Sigma = \sum_{i=1}^{d/2} \phi(g_i)$$

and the term of second highest degree is

$$T_2 = -2T_1(\Sigma'_i g_i^{-1})t^{-q}$$

where q is the least value of $\phi(g_i)$, $i = 1, \ldots, d/2$, and the sum runs only over those i with $\phi(g_i) = q$. The 2 occurs because each solution g_i contributes twice, once by deleting g_i and taking a product of $(d/2) - 1$ terms, once by adding in g_{d+1-i} and taking a product of $(d/2) + 1$ terms. Thus β determines $\Sigma'_i g_i^{-1}$ and so, by Lemma 1, one can cancel out all the factors $g_i - 1$, $g_{d+1-i} - 1$ with $\phi(g_i) = q$. By induction on d we see that Lemma 2 holds.

Now we apply Lemma 2 with $G = \mathbf{C}^*$, $g_i = [\lambda^{-1}]$, λ running over the roots of $p(x)$. We find that the b_n's determine the λ's and also the trivial unit $U = -\epsilon[(\delta a_0)^{-1}]$, hence $a_0 = |(\delta a_0)^{-1}|^{-1}$. Thus the b_n's determine $p(x) = a_0 \cdot \prod_\lambda (x - \lambda)$, as desired. Since $-x(1-x^{-1}) = 1-x$ one sees easily that the reciprocality of $p(x)$ is needed for the proposition to hold.

Now we will produce examples of Alexander polynomials that cannot be distinguished by b_n's. Let $\Phi_m(x)$ be the mth cyclotomic polynomial. As $\prod_{\substack{m|n \\ m>1}} \Phi_m(1) = (t^n - 1)'(1) = n$, we find that

$$\Phi_m(1) = \begin{cases} 0, & m = 1 \\ p, & m = p^k > 1, \; p \text{ prime} \\ 1, & \text{other } m \end{cases}$$

Thus a polynomial $Q = \Pi \Phi_m^{e_m}$ is an Alexander polynomial if each m (with $e_m \neq 0$) has at least 2 prime factors, where we use Seifert's characterization of Alexander polynomials as reciprocal integral polynomials with value ± 1 at $x = 1$ [S]. As λ runs over roots of Q we have for $n > 0$

$$\begin{aligned} \prod_\lambda 1 - \lambda^n &= \prod_m \Big(\prod_\mu (1-\mu)^{<m,n>}\Big)^{e_m} \\ &= \prod_m (\Phi_{m/(m-n)}(1))^{<m,n>e_m} \end{aligned}$$

where μ runs over elements of $\mathbf{C}^*$ of order $m/(m,n)$, $\phi(m) = \deg \Phi(m)$ and

$$<m,n> = \phi(m)/\phi(m/(m,n)).$$

The latter product over m is nonnegative, so it equals b_n. We choose Q so $e_m = 0$ except for certain $m = p^a q^b$, $a \geq 1$, $b \geq 1$, where we have 2 fixed distinct primes p, q. Then we write e_{ab} for e_m.

If $e_{11} > 0$ then $b_n = 0 \Leftrightarrow pq|n$, $b_n = 1$ if neither $p|n$ nor $q|n$. Taking n divisible by p but not q gives $b_n = q^k$ where

$$k = \Sigma' < m,n > e_m$$

the sum running over m such that $m/(m,n)$ is a power of q. This k depends only on the number of times p divides n and it is a linear combination of the quantities $\sum_b e_{ab}$. Viewing

e_{ab} as a matrix entry, this is the sum of the ath row. Reversing the role of p and q, one sees that the row and column sums of (e_{ab}) determine the b_n's. This yields our examples, e.g. the polynomials $Q = \Phi_{pq}\Phi_{p^2q}\Phi_{pq^2}$ and $P = \Phi_{pq}^2\Phi_{p^2q^2}$ have the same b_n's.

We thank the mathematicians of UNAM for their hospitality and especially Prof. Gonzalez-Acuña for the interesting discussions that stimulated this paper. We also thank the referee for helpful comments.

BIBLIOGRAPHY

[AM] Artin, M. and B. Mazur, On periodic points, *Ann. of Math.* **81** (1965) 82-89.

[DG] Duistermaat, J.J. and V. Guillemin, The spectrum of positive elliptic operators and periodic bicharacteristics. *Inv. Math.* **25** (1975) 39-79.

[F] Fox, R.H. Free differential calculus III. Subgroups. *Ann. of Math.* **64** (1956) 407-419.

[Fr] Fried, D. Rationality for isolated expansive sets. *Advances in Math.*, to appear.

[M] Milnor, J. Infinite cyclic coverings. Topology of Manifolds (Michigan State Univ. 1967). PW& S, 1968.

[S] Seifert, H. Uber das Geschlecht von Knoten. *Math. Ann.* **110** (1934) 571-592.

[U] Uspensky, J.V. Theory of Equations. McGraw-Hill, 1948.

[W] Williams, R.F. The zeta function of an attractor. Topology of Manifolds (Michigan State Univ. 1967). PW& S, 1968.

Mathematics Department
Boston University
111 Cummington St.
Boston MA 02215
U.S.A.

PERSISTENT CYCLES FOR HOLOMORPHIC FOLIATIONS HAVING A MEROMORPHIC FIRST INTEGRAL*

Xavier Gómez-Mont and Jesús Muciño

The objective of this work is to extend Iliashenko's methods introduced in his paper "The origin of Limit Cycles under Perturbation of the equation $dw/dz = -R_z/R_w$, where $R(z,w)$ is a polynomial" [13]. In this seminal paper, Iliashenko introduces the notion of a persistent cycle under deformations and gives a way to compute the linear holonomy of a first order variation of a foliation in terms of an Abelian integral. We will give a general setting where these notions extend.

Let F be a codimension one holomorphic foliation with singularities in a complex manifold M and let δ be a closed loop contained in a leaf of F. If $\{F_t\}_{t\in T}$ denotes a family of foliations with $F_0 = F$, then it is possible to decide whether the loop δ may be deformed to other loops δ_t contained in leaves of F_t. This is possible since a closed loop may be detected as a fixed point of the holonomy map (or Poincaré return map) of F around δ, and the set of fixed points of the holonomy map of $\{F_t\}$ around δ will determine free homotopy classes of loops contained in leaves of F_t.

Let U be an open set in M where the foliation F may be described by a non-vanishing closed holomorphic 1-form ω. Choosing this 1-form chooses for us the following structure in $F|_U$. The integral $\int\omega$ gives rise to a multiple-valued holomorphic function on

---

* Supported by CONACYT.

U whose fibers $(\int\omega)^{-1}(c)$ are the leaves of F. Choosing a covering of U, the integral $\int\omega$ will induce a family of submersions to $\mathbb{C}$ whose fibers are plaques of the foliation F. The holonomy of $F|_U$ may be computed by composing with the changes of coordinates of the submersions, which may be seen to be translations, since ω is closed. Hence, $F|_U$ has a transverselly parallelisable structure. If the family of foliations $\{F_t\}$ may be described in U by $\Omega(z,t) = = w(z) + \eta(z)t + \cdots$ then the first order variation F^1 is determined by $\omega + \eta t$. If δ is a closed loop contained in a leaf of F, then the linear holonoly of F^1 around δ takes the form

$$(t,s) \to (t,s + (\textstyle\int_\delta \eta) \cdot t) \qquad (0.1)$$

If M is now a compact complex manifold, we denote by $Fol\ (M,L)$ the set of foliations of codimension 1 in M defined by maps $L \to T^*M$. In section 1 we show that $Fol\ (M,L)$ has the structure of a projective variety and that its tangent space at the point ω, $T_\omega Fol\ (M,L)$, may be identified with the infinitesimal deformations of F_ω. If U is an open set in M where ω may be defined by a closed holomorphic 1-form and δ is a loop contained in a leaf of F_ω in U, then the map

$$T_\omega Fol\ (M,L) \to \mathbb{C} \qquad \eta \to \textstyle\int_\delta \eta \qquad (0.2)$$

associates to an infinitesimal deformation of F_ω the linear holonomy (0.1) of the deformation around δ. The map (0.2) depends on the choises made for defining it, but only up to multiplication by a nonzero constant. Hence δ determines in a canonical form a point in $\mathrm{Proj}\ (T^*_\omega Fol\ (M,L))$, which we have called an Iliashenko point. Note that a hyperplane in $\mathrm{Proj}\ (T^*_\omega Fol\ (M,L))$ corresponds to a line in $T_\omega Fol\ (M,L)$, and hence to an infinitesimal direction of deformation. Also note that the transversally parallelizable structure of $F|_U$ implies that the loop δ may be deformed to nearby leaves of F_ω as closed loops, and hence the corresponding Iliashenko points give rise to a holomorphic curve, that we have called an Iliashenko curve.

The intersection of the Iliashenko curves with the hyperplanes in $\mathrm{Proj}\ (T^*_\omega Fol\ (M,L))$ determines which loops persist under the deformation specified by the hyperplane. We apply these results to foliations having a meromorphic first integral and we show that the Iliashenko curves associated to the indeterminacy locus are conics. We will extend these results in [8] and [14].

INDEX

1. THE SPACE OF CODIMENSION ONE HOLOMORPHIC FOLIATIONS.

In this section we introduce codimension one holomorphic foliations in a complex manifold, and show how to put a complex analytic structure in the set of all foliations that have isomorphic defining cotangent bundles. We exemplify through the case of projective spaces.

1.1. Codimension 1 Holomorphic Foliations.

A codimension 1 holomorphic foliation (with singularities) in the complex manifold M may be given by a family of integrable holomorphic 1-forms ω_α defined on an open cover $\{U_\alpha\}$ of M, $\omega_\alpha \wedge d\omega_\alpha = 0$ satisfying $\omega_\alpha = \xi_{\alpha\beta} \cdot \omega_\beta$, with $\xi_{\alpha\beta}$ holomorphic never vanishing functions. Viewing the 1-forms as inducing bundle maps on U_α

$$U_\alpha \times \mathbb{C} \to T^*M|_{U_\alpha} \qquad (p,t) \mapsto \omega_\alpha(p) \cdot t$$

where T^*M is the cotangent bundle of M and similarly

$$U_\beta \times \mathbb{C} \to T^*M|_{U_\beta} \qquad (p,s) \mapsto \omega_\beta(p) \cdot s$$

we obtain on $U_\alpha \cap U_\beta$ the equality

$$\omega_\beta(p)s = \xi_{\alpha\beta}(p)\omega_\beta(p)t$$

If we glue $U_\beta \times \mathbb{C}$ with $U_\alpha \times \mathbb{C}$ over $U_\alpha \cap U_\beta$ with the bundle isomorphism $(p,s) \mapsto (p,\xi_{\alpha\beta}^{-1} \cdot s) = (p,t)$, we will obtain a bundle map $\omega: L \to T^*M$, where L is the bundle on M formed with the cocycle $(\xi_{\alpha\beta}^{-1})$. We say that $\omega: L \to T^*M$ is equivalent to $\omega': L' \to T^*M$ if there is a holomorphic bundle isomorphism $\rho: L \to L'$ such that $\omega' \circ \rho = \omega$.

DEFINITION 1.1. *A codimension 1 holomorphic foliation (with singularities)* F in the complex manifold M is an equivalence class of holomorphic bundle maps $\omega: L \to T^*M$ from a line bundle L to the cotangent bundle of M such that ω does not vanish identically on any connected component of M and such that in local trivializing coordinates, ω is given by integrable 1-forms ω_α (i.e. $\omega_\alpha \wedge d\omega_\alpha = 0$). The *singular set* Sing F *of* F is the analytic subspace of M defined by $\omega = 0$.

If $\omega: L \to T^*M$ is a holomorphic bundle map on the connected manifold M, described in trivializing connected coordinates $\{U_\alpha\}$ by the 1-forms ω_α; then if one of them satisfies the integrability condition $\omega_0 \wedge d\omega_0 = 0$, then all of them satisfy the integrability condition. Namely, if $U_0 \cap U_1 \neq \phi$, then in $U_0 \cap U_1$ we have $\omega_1 \wedge d\omega_1 = \xi_{10}\omega_0 \wedge d(\xi_{10}\omega_0) = \xi_{10}^2\omega_0 \wedge d\omega_0 = 0$, and hence ω_1 satisfies the integrability condition in U_1, by analytic continuation.

In M-Sing F we may obtain by the theorem of Frobenins ([2] p. 89) a cover $\{U_\alpha\}$ and biholomorphisms $\varphi_\alpha: U_\alpha \to V_\alpha$, with V_α open balls in $\mathbb{C}^n$ such that $\varphi_\alpha^*(dz_n) = \omega_\alpha$ are 1-forms describing the foliation. In such a coordinate cover, the transition of coordinates $\varphi_{\alpha\beta} = \varphi_\alpha \circ \varphi_\beta^{-1}: \varphi_\beta(U_\alpha \cap U_\beta) \to \varphi_\alpha(U_\alpha \cap U_\beta)$ are sending the hyperplanes defined by $z_n = K$ to themselves, i.e. if $\varphi = (\varphi^1,\ldots,\varphi^n)$ then the coordinate function φ^n is only a function of z_n.

Introduce in $\mathbb{C}^n = \mathbb{C}^{n-1} \times \mathbb{C}$ the Euclidian topology in the first factor and the discrete topology in the second. With this topology $\mathbb{C}^n$ becomes an uncountable complex manifold of dimension $n-1$, and the maps $\varphi_{\alpha\beta}$ are biholomorphisms of this $(n-1)$-dimensional manifold. We may induce in M-Sing F a new topology, called the leaf topology, by means of $(\varphi_\alpha, \varphi_{\alpha\beta})$ as an uncountable $(n-1)$-dimensional complex manifold. A connected component of M-Sing F with this new topology will be called a *leaf of* F, and M-Sing F is a disjoint union of all its leaves, M-Sing $F = \coprod L_i$.

If L is a holomorphic line bundle on the compact connected complex manifold M, then the set of holomorphic bundle maps from L to T*M form a finite dimensional $\mathbb{C}$-vector space E(L), by the Cartan-Serre theorem of finiteness of cohomology groups applied to the sheaf of sections of the bundle Hom (L,T*M) ([9] p.152). Let $\omega_0, \ldots, \omega_N$ be a $\mathbb{C}$-basis of E(L), and choose a trivialization of L on an open set U of M so that ω_i is represented in U by the 1-forms $\tilde{\omega}_i$. Let $(a_0, \ldots, a_N)$ be coordinates of E(L) with respect to the basis ω_i. If $\omega = \sum_{i=0}^{N} a_i \omega_i$, then as previously observed, ω determines a holomorphic foliation if and only if $\tilde{\omega} = \sum_{i=0}^{N} a_i \tilde{\omega}_i \not\equiv 0$ satisfies the integrability conditions in U:

$$0 = \tilde{\omega} \wedge d\tilde{\omega} = (\Sigma\, a_i \tilde{\omega}_i) \wedge (\Sigma\, a_j d\tilde{\omega}_j) = \sum_{i,j=0}^{N} a_i a_j (\tilde{\omega}_i \wedge d\tilde{\omega}_j) \tag{1.1}$$

Hence we obtain that a finite number of quadratic equations (1.1) in E - {0} determine those maps ω: L → T*M that describe foliations.

Two maps ω, ω': L → T*M determine the same foliation if and only if there is an isomorphism ρ: L → L such that $\omega' \circ \rho = \omega$. Assuming M compact, we claim that all isomorphisms ρ: L → L consist of multiplication by a non-zero scalar. To see this, let $\{U_\alpha\}$ be an open cover of M where L is given by the cocycle $(\xi_{\alpha\beta})$, and ρ by a collection of holomorphic functions $\rho_\alpha : U_\alpha \to \mathbb{C}^*$. In $U_\alpha \cap U_\beta$ they satisfy the compatibility condition $\rho_\alpha \xi_{\alpha\beta} = \xi_{\alpha\beta} \rho_\beta$. Since all the terms are non-zero scalars we obtain $\rho_\alpha = \rho_\beta$. Hence ρ is given by a holomorphic function on M, which is a non-zero constant by the assumption that M is compact (by the maximum principle). Hence we have:

DEFINITION 1.2. The complex analytic subset *Fol* (M,L) of the projective space Proj E(L) defined by equations (1.1) will be called the *space of foliations of codimension* 1 *in* M *and with cotangent* *space* L. There is a one to one correspondence between the points in *Fol* (M,L) and equivalence classes of foliations defined by $\omega\colon L \to T^*M$, and the complex structure in *Fol* (M,L) may be seen to satisfy a universal property (see [4], [6]). Note that if the dimension of M is 2, then *Fol* (M,L) = Proj E(L), since the conditions (1.1) are automatically satisfied.

1.2. Projective Spaces.

We exemplify with M the projective space $\mathbb{C}P^n$. $\mathbb{C}P^n$ may be constructed as the quotient of the space $\mathbb{C}_0^{n+1} = \mathbb{C}^{n+1} - \{0\}$ by the action of multiplication by non-zero scalars, $\lambda \cdot (z_0, \ldots, z_n) = (\lambda z_0, \ldots, \lambda z_n)$. One obtains affine coordinates of $\mathbb{C}P^n$ by considering in $\mathbb{C}^{n+1}$ the hyperplanes H_j defined by $z_j = 1$ that project as homeomorphisms $H_j \to U_j \subset \mathbb{C}P^n$ onto its image. The open covering $\{U_j\}$ of $\mathbb{C}P^n$, with coordinates $(z_{j0}, \ldots, \hat{z}_{jj}, \ldots, z_{jn}) \in \mathbb{C}^n \cong U_j$ and transition coordinates

$$\varphi_{ij}(z_{j0}, \ldots, z_{jn}) = \left(\frac{z_{j0}}{z_{ji}}, \ldots, \frac{1}{z_{ji}}, \ldots, \frac{\hat{z}_{ji}}{z_{ji}}, \ldots, \frac{z_{jn}}{z_{ji}}\right) =$$

$$= (z_{i0}, \ldots, z_{ij}, \ldots, \hat{z}_{ii}, \ldots, z_{in})$$

defined on $z_{ji} \neq 0$ onto $z_{ij} \neq 0$ give a coordinate description of $\mathbb{C}P^n$.

Given an integer e, we define a bundle H(e) on $\mathbb{C}P^n$ as $U_j \times \mathbb{C}$ in the above coordinates, and gluing cocycle $\xi_{ij}\colon (U_i \cap U_j) \times \mathbb{C} \hookrightarrow U_j \times \mathbb{C} \to (U_i \cap U_j) \times \mathbb{C} \hookrightarrow U_i \times \mathbb{C}$ defined as $\xi_{ij}(z_{j0}, \ldots, z_{jn}) = z_{ji}^{-e}$. It may be shown (see [9] p. 144) that any holomorphic line bundle on $\mathbb{C}P^n$ is isomorphic to one and only one H(e), where e is the Chern class of H(e) in $H^2(\mathbb{C}P^2, Z)$. The bundle H(-1) is called the Hopf bundle, and it may also be obtained as the subline bundle of $H(-1) \hookrightarrow \mathbb{C}P^n \times \mathbb{C}^{n+1}$ such that $H(-1)_p$ is the line in $\mathbb{C}^{n+1}$ that p represents.

PROPOSITION 1.3. 1) There is a one to one correspondence between holomorphic maps $\omega\colon H(-e) \to T^*\mathbb{C}P^n$ and polynomial 1-forms

$\omega = \omega_{e-1} + \omega_{e-2} + \cdots + \omega_0$ in one of the canonical coordinate charts, say U_0, of $\mathbb{C}P^n$, where ω_j is a homogeneous 1-form of degree j in $(z_{01},\ldots,z_{0n})$ and the evaluation of ω_{e-1} on the radial vector field vanishes, $\omega_{e-1} \circ \left(\sum_{i=1}^{n} z_{0i} \frac{\partial}{\partial z_{0i}}\right) = 0$.

2) The set $E(H(-e))$ of maps from $H(-e)$ to $T^*\mathbb{C}P^n$ form a vector space of dimension

$$\binom{e+n-1}{e}(e-1)$$

3) For $\mathbb{C}P^2$, $\dim E(H(-e)) = \dim \mathcal{F}o\ell\,(\mathbb{C}P^2, H(-e)) + 1 = e^2 - 1$.

Proof. To simplify notation, let $(x_1,\ldots,x_n)$ and $(y_1,\ldots,y_n)$ be coordinates of $U_0 \cong \mathbb{C}^n$ and $U_1 \cong \mathbb{C}^n$ respectively, with transition coordinates

$$\varphi(x_1,\ldots,x_n) = \left(\frac{1}{x_1}, \frac{x_2}{x_1}, \ldots, \frac{x_n}{x_1}\right) = (y_1, y_2, \ldots, y_n) \tag{1.2}$$

Let ω be a polynomial 1-form on U_1 of degree less than of equal to d, and let $\omega = \sum_{i=1}^{n} \sum_{j=0}^{d} a_{ij} dy_i$, where a_{ij} are homogeneous polynomials of degree j in $y_1,\ldots,y_n$. ω is transformed to U_0 as

$$\varphi^*(\omega) =$$

$$= \sum_{j=0}^{d} \left[a_{ij}\left(\frac{1}{x_1},\ldots,\frac{x_n}{x_1}\right)\left(-x_1^{-2}dx_1\right) + \sum_{i=2}^{n} a_{ij}\left(\frac{1}{x_1},\ldots,\frac{x_n}{x_1}\right)x_1^{-2}(-x_i dx_1 + x_1 dx_i)\right]$$

$$= \sum_{j=0}^{d} \left[\left(-a_{1j}(1,\ldots,x_n) - \sum_{i=2}^{n} a_{ij}(1,\ldots,x_n)x_i\right)x_1^{-j-2}dx_1 + \right.$$

$$\left. + \sum_{i=2}^{n} a_{ij}(1,\ldots,x_n)x_1^{-j-1}dx_i\right]$$

Hence $\varphi^*(\omega)$ has a pole of order $d+2$ on $\mathbb{C}P^n - U_1$, unless the term

$$a_{1d}(1,x_2,\ldots,x_n) + \sum_{i=2}^{n} a_{id}(1,x_2,\ldots,x_n)x_i = 0 \tag{1.3}$$

in which case it will have a pole of smaller order. Dividing (1.3) by x_1^{d+1}, (1.3) transform to the y_i-coordinates as

$$0 = \sum_{i=1}^{n} a_{id}(y_1,\dots,y_n)y_i = \left(\sum_{i=1}^{d} a_{id}dy_i\right)\cdot\left(\sum_{i=1}^{n} y_i \frac{\partial}{\partial y_i}\right) \qquad (1.4)$$

Hence we obtain that the set of 1-forms on $\mathbb{C}P^n$, holomorphic in U_1 and having a pole of order less than or equal to e on $\mathbb{C}P^n - U_1$ has a representation on U_1 as a polynomial 1-form

$\omega = \omega_{e-1} + \omega_{e-2} + \cdots + \omega_0$ and $\omega_{e-1} \circ \sum_{i=1}^{n} y_i \frac{\partial}{\partial y_i} = 0$. Any holomorphic 1-form on U_1 extends to $\mathbb{C}P^n$ as a holomorphic map $H(-e) \to$ $\to T^*\mathbb{C}P^n$ if and only if it exteds as a meromorphic 1-form with a pole on $\mathbb{C}P^n - U_1$ of order less than or equal to e. This proves part 1.

The dimension of $\{\omega_{e-1} + \cdots + \omega_0\}$ is $n \cdot \dim$ {homogeneous polynomials of degree $(e-1)$ in $(n+1)$-variables} $= n\binom{e+n-1}{e-1}$.

The condition that $\omega_{e-1} \circ \left(\sum_{i=1}^{n} z_{0i} \frac{\partial}{\partial z_{0i}}\right) = 0$ imposes as many conditions as the dimension of {homogeneous polynomials of degree e in n-variables}; hence the dimension of $E(H(-e))$ is

$$n\binom{e+n-1}{e-1} - \binom{e+n-1}{e} = \binom{e+n-1}{e}\left[\frac{ne}{n} - 1\right] = \binom{e+n-1}{e}(e-1)$$

This proves part 2. Part 3 follows since the integrability conditions are automatically satisfied. ‖

1.3. Families of Foliations.

We will end this section by defining the notion of families and deformations of holomorphic foliations.

LEMMA 1.4. Let M be a compact complex manifold, L a holomorphic line bundle on M, $E(L)$ the $\mathbb{C}$-vector space of holomorphic bundle maps $L \to T^*M$, $\text{Proj}\, E(L)$ the projective space of lines through 0 in $E(L)$ and $H(-1)$ the Hopf bundle on $\text{Proj}\, E(L)$. Then, there is a bundle map on $\text{Proj}\, E(L) \times M$, $\mathcal{W}: \Pi_1^*H(-1) \otimes \Pi_2^*L \to \Pi_2^*T^*M$, such that for any $\omega \in \text{Proj}\, E(L)$ the restriction of $\mathcal{W}$ to $\omega \times M$ is the map represented by ω.

Proof. Let $\omega_0, \ldots, \omega_n$ be a basis of $E(L)$, $(a_0, \ldots, a_n)$ the coordinates of $E(L)$ associated with this basis, and $\widetilde{\Pi}_i$ the projections of $E(L) \times M$ to both factors, $i = 1,2$. We may form a tautological bundle map on $E(L) \times M$

$$\widetilde{W}: \widetilde{\Pi}_2^* L \to \widetilde{\Pi}_2^* T^*M \qquad \widetilde{W}(a_0, \ldots, a_n ; p) = \sum_{i=0}^{N} a_i \omega_i (p) : L_p \to T_p^* M \tag{1.5}$$

Note that $\widetilde{W}(\lambda a_0, \ldots, \lambda a_n ; p) = \lambda \widetilde{W}(a_0, \ldots, a_n ; p)$, so that $\widetilde{W}$ does not descend to a bundle map on $\operatorname{Proj} E(L) \times M$.

Let $\Pi: E(L) - \{0\} = E(L)_0 \to \operatorname{Proj} E(L)$ be the natural projection. There is a natural section σ of the line bundle $\Pi^* H(-1)$ on $E(L)_0$, such that $\sigma(\omega) = \omega \in \Pi^* H(-1)_\omega = \mathbb{C} \cdot \omega$. This section has the property that $\sigma(\lambda\omega) = \lambda\sigma(\omega)$. Hence the bundle map on $E(L)_0 \times M$, $\frac{1}{\sigma}\widetilde{W}: \Pi^* H(-1) \otimes \widetilde{\Pi}_2^* L \to \Pi_2^* T^* M$ is invariant under multiplication by $\lambda \in \mathbb{C}^*$, so induces the map W in the statement of the Lemma. ‖

DEFINITION 1.5. Let L be a holomorphic line bundle on the compact complex manifold M, $\mathcal{Fol}\,(M,L) \subset \operatorname{Proj} E(L)$ the space of foliations of codimension 1 in M with cotangent space L and $W: \Pi_1^* H(-1) \otimes \Pi_2^* L \to \Pi_2^* T^* M$ the restriction of W in Lemma 1.4 to $\mathcal{Fol}\,(M,L) \times M$. *A family of holomorphic foliations in* M *(with cotangent space* L*) parametrized by the complex analytic space* S may be specified by a holomorphic map $f: S \to \mathcal{Fol}\,(M,L)$, and the bundle map on $S \times M$

$$f^* W: f^* \Pi_1^* H(-1) \otimes \Pi_2^* L \to \Pi_2^* T^* M \tag{1.6}$$

defines the family of foliations. If F is a foliation represented by $\bar{\omega}_0 \in \mathcal{Fol}\,(M,L)$, then a *deformation of* F *parametrized by a germ of an analytic space* $(S,0)$ may be given by a germ of a holomorphic map $(S,0) \to (\mathcal{Fol}\,(M,L), \bar{\omega}_0)$.

Remark. It may be shown using Douady's thesis [4] that the space $\mathcal{Fol}\,(M,L)$ is universal for an intrinsic definition of holomorphic families.

Let $f: S \to \mathcal{Fol}\,(M,L)$ be a map specifying a family $\widetilde{F}$ of foliations in $S \times M$ where S is a complex manifold, as in (1.6). The singular set of the family $\widetilde{F}$, is the analytic subspace $\operatorname{Sing} \widetilde{F}$ of $S \times M$ defined by $f^* W = 0$. We may apply the theorem of Frobenius with parameters to obtain a covering $\{U_\alpha\}$ of $S \times M - \operatorname{Sing} \widetilde{F}$ and biholomorphisms over S, $\Phi_\alpha: U_\alpha \to S_\alpha \times V_\alpha \hookrightarrow S \times \mathbb{C}^n$, where S_α is an open set in S, V_α is open in $\mathbb{C}^n$, $\Pi_1 = \Pi_1 \circ \Phi_\alpha$ and such that the relative 1-form $\Phi_\alpha^*(dz_n)$ is a defining equation for $\widetilde{F}$ in U_α. The transition of coordinates $\Phi_{\alpha\beta} = \Phi_\alpha \circ \Phi_\beta^{-1}$ are biholomorphisms over S such that $\frac{\partial \Phi^n_{\alpha\beta}}{\partial z_k} = 0$, $k = 1,\dots,n-1$. Putting in $S \times \mathbb{C}^{n-1} \times \mathbb{C}$ the discrete topology in the first and third factors, and the Euclidean topology in the middle one, we obtain a structure in $S \times M\text{-Sing}\,\widetilde{F}$ of an uncountable $(n-1)$-dimensional manifold. A connected component L will be called a *leaf of* $\widetilde{F}$, and any such leaf is contained in a Π_1-fibre.

2. FIRST VARIATIONS OF HOLOMORPHIC FOLIATIONS.

In this section we will interpret a tangent vector to the space of codimension one holomorphic foliations as a family of foliations parametrized by the analytic space A associated to the dual numbers $\mathbb{C} \oplus t\mathbb{C}$, $t^2 = 0$. We introduce the holonomy pseudogroup and we show how to compute its first variation. We finish the section by introducing the notion of persistent cycles under deformations.

2.1. First Variations and $T\,\mathcal{F}o\ell\,(M,L)$.

In this subsection we will give a method to construct the tangent space to $\mathcal{F}o\ell\,(M,L)$ at a point in $\mathcal{F}o\ell\,(M,L)$ given by $\omega: L \to T^*M$.

We will begin by recalling the description of the tangent space to the analytic space S at a point p. Let $\mathcal{O}_{S,p}$ be the local algebra of germs of holomorphic functions on S at p, with maximal ideal m_p, then the tangent space to S at p is $T_pS = \mathcal{H}om_{\mathbb{C}}(m_p/m_p^2, \mathbb{C})$ (see [15] p. 75). Let A be the analytic space consisting of one point 0 and ring of holomorphic functions isomorphic to $\mathbb{C} \oplus \mathbb{C} \cdot t$, with $t^2 = 0$.

LEMMA 2.1. There is a one to one correspondence between points in T_pS and holomorphic maps $\phi: A \to S$ such that $\phi(0) = p$.

Proof. A holomorphic map $\phi: A \to S$, $\phi(0) = p$, is completely determined by the morphism of local $\mathbb{C}$-algebras $\phi^*: \mathcal{O}_{S,p} \to \mathbb{C} \oplus \mathbb{C}t$ (i.e. $\phi^{*-1}(\mathbb{C}t) = m_p$, since the morphism is local ([11], p. 73)). We then have maps $\phi^*: m_p \to \mathbb{C}t$ and $\phi^*(m_p^2) \subset \phi^*(m_p)^2 = 0$, so that ϕ^* induces a map $\phi^*: \frac{m_p}{m_p^2} \to \mathbb{C}t$ giving rise to an element in $\mathrm{Hom}_{\mathbb{C}}(m_p/m_p^2, \mathbb{C}) = T_pS$. Conversely, given a $\mathbb{C}$-linear map $\lambda: m_p/m_p^2 \to \mathbb{C}$ construct the local algebra morphism $\lambda: \mathcal{O}_{S,p} \to \mathbb{C} \oplus \mathbb{C}t$ as $\lambda(f) = f(0) + \lambda \circ \Pi(f - f(0))$, where $\Pi: m_p \to m_p/{}^2_p$ is the quotient map. ‖

Given a complex manifold, we will examine the analytic space $A \times M$, where A is the analytic space associated to the dual numbers $\mathbb{C} \oplus t\mathbb{C}$, $t^2 = 0$. We will view it as a family of complex manifolds parametrized by A. The analytic space $A \times M$ is set theoretically the same as M, it has also the same topology, but it has a different function theory. If U is an open set of M and $\mathcal{O}_U$ denotes the ring of holomorphic functions of M defined in U, then the ring of holomorphic functions on $A \times U$ is the ring $\mathcal{O}_U \oplus t\mathcal{O}_U$, $t^2 = 0$.

If L is a holomorphic line bundle on M, defined on an open cover $\{U_\alpha\}$ by the cocycle $(\xi_{\alpha\beta})$, then Π_2^*L is a holomorphic line bundle on $A \times M$ defined in the open cover $\{A \times U_\alpha\}$ by the cocycle $(\xi_{\alpha\beta} \cdot \mathrm{Id})$: Namely, on U_α it is the bundle $U_\alpha \times (\mathbb{C} \oplus \mathbb{C}t)$ and we glue with the cocycle $\xi_{\alpha\beta} \cdot \mathrm{Id}$, where Id is the 2×2 identity matrix. Hence we have that $\Pi_2^*L = L \oplus tL$. Similarly, the bundle $\Pi_2^*T^*M$ on $A \times M$, called the *relative cotangent bundle* of $A \times M \to M$, is isomorphic to $T^*M \oplus tT^*M$. If $f + tg$ is a function on $A \times U$, and $\omega + t\eta$ is a relative 1-form, then $(f + tg) \cdot (\omega + t\eta) = f\omega + t(g\omega + f\eta)$.

A bundle map $\Phi: L \oplus tL \to T^*M \oplus tT^*M$ on $A \times M$ is a vector bundle map on M which is $\mathcal{O}_U \oplus t\mathcal{O}_U$-linear over any open set U. Since $L \oplus tL$ is locally isomorphic to $\mathcal{O}_U \oplus t\mathcal{O}_U$, we may find a cover $\{U_\alpha\}$ of M where L is described by the cocycle $(\xi_{\alpha\beta})$, and in this description Φ is given in U_α by multiplication with a relative 1-form $\omega_\alpha + t\eta_\alpha$. From the cocycle condition obtained, we observe that (ω_α) and (η_α) glue to give bundle maps $\omega, \eta: L \to T^*M$, so that $\Phi = \omega + t\eta$.

LEMMA 2.2. Let M be a compact complex manifold, L a holomorphic line bundle on M and $E(L)$ the finite dimensional vector space of bundle maps from L to T^*M. Let $\widetilde{\mathcal{Fol}}\,(M,L)$ be the analytic subspace of $E(L) - \{0\}$ formed by those maps $\omega\colon L \to T^*M$ that satisfy the integrability conditons (1.1), then:

1. There is a one to one correspondence between the tangent space $T_\omega E$ to E at ω and bundle maps $\omega + t\eta\colon L \oplus tL \to T^*_M \oplus tT^*_M$ on $M \times A$.

2. If $\omega \in \widetilde{\mathcal{Fol}}\,(M,L)$, then $\omega + t\eta$ represents a vector tangent to $\widetilde{\mathcal{Fol}}\,(M,L)$ at ω if and only if for local 1-forms (ω_α), (ω_α) describing ω and η, we have

$$\omega_\alpha \wedge d\eta_\alpha + \eta_\alpha \wedge d\omega_\alpha = 0 \tag{2.1}$$

Proof. By Lemma 2.1 there is a one to one correspondence between $T_\omega E$ and holomorphic maps $\phi\colon (A,0) \to (E(L),\omega)$. By pulling back the family $\widetilde{\omega}$ in Lemma 1.4, we obtain for each such ϕ a bundle map over $M \times A$: $L \oplus tL \to T^*_M \oplus tT^*_M$ which as shown above is of the form $\omega + t\eta$, for $\eta \in E(L)$. Using the expression (1.5) of $\widetilde{\omega}$, we see that if ϕ represents $\eta \in T_\omega E(L)$ then $\phi^*(\widetilde{\omega}) = \omega + t\eta$. This proves part 1.

To prove 2, let $\omega + t\eta$ represent a tangent vector to $E(L)$ at ω giving rise to a map $\phi\colon (A,0) \to (E(L),\omega)$. ϕ is tangent to $\widetilde{\mathcal{Fol}}\,(M,L)$ if and only if the pull back to A of equation (1.1) is satisfied identically; namely

$$(\omega_\alpha + t\eta_\alpha) \wedge (d\omega_\alpha + td\eta_\alpha) = \omega_\alpha \wedge d\omega_\alpha + t(\omega_\alpha \wedge d\eta_\alpha + \eta_\alpha \wedge d\omega_\alpha) \pmod{t^2} \tag{2.2}$$

Since $\omega_\alpha \wedge d\omega_\alpha = 0$ by hypothesis, the vanishing of (2.2) is equivalent to vanishing of (2.1). ‖

PROPOSITION 2.3. With the same hypothesis as Lemma 2.2, let $\omega\colon L \to T^*M$ be a map defining a point $\bar{\omega} \in \operatorname{Proj} E(L)$, then:

1. There is a one to one correspondence between the tangent space $T_{\bar{\omega}}\operatorname{Proj} E(L)$ to $\operatorname{Proj} E(L)$ at $\bar{\omega}$ and points in $E(L)/\mathbb{C}\cdot\omega$; the correspondence is establishen by associating to a class $[\eta] \in E(L)/\mathbb{C}\cdot\omega$ the map $\omega + t\eta\colon L \oplus tL \to T^*M \oplus tT^*M$.

2. If $\bar{\omega} \in \mathcal{Fol}\,(M,L)$, then $T_{\bar{\omega}}\,\mathcal{Fol}(M,L)$ is the subspace of

$E/\mathbb{C}\cdot\omega$ formed of classes $[\eta]$ satisfying (2.1).

Proof. Let $\Pi\colon E(L)-\{0\}\to \mathrm{Proj}\ E$ be the map defining $\mathrm{Proj}\ E(L)$. The tangent spaces at ω and $\bar{\omega}$ are related by the exact sequence

$$0\to\mathbb{C}\cdot\omega\to T_{\omega}E(L)\xrightarrow{D\Pi}T_{\bar{\omega}}\mathrm{Proj}\ E(L)\to 0$$

The proposition follows from this sequence and Lemma 2.2. ∥

DEFINITION 2.4. Let F be a holomorphic foliation of codimension 1 represented by a point $\bar{\omega}\in \mathcal{F}o\ell\,(M,L)$. The family of foliations on M parametrized by the analytic space A associated to the dual numbers $\mathbb{C}\oplus\mathbb{C}t$ constructed from the tangent vector $\bar{\eta}\in T_{\bar{\omega}}\mathcal{F}o\ell\,(M,L)$ will be called an *infinitesimal deformation of* F (keeping M and L fixed), and will be denoted by F^1 or $F_{\bar{\eta}}^{1}$.

We will now see how we may choose coordinate covers of M adapted to a foliation F defined by a bundle map $\omega\colon L\to T^*M$. Let $M' = M\text{-Sing}\,F$ and let $\{U_\alpha\}$ be a coordinate cover of M' with coordinates $\phi_\alpha\colon U_\alpha\to V_\alpha\hookrightarrow\mathbb{C}^n$ with coordinates $(z_{\alpha 1},\ldots,z_{\alpha n})$ where V_α is an open ball in $\mathbb{C}^n$, with transition coordinates $\phi_{\alpha\beta}$: $\phi_\beta(U_\alpha\cap U_\beta)\to\phi_\alpha(U_\alpha\cap U_\beta)$, and assume that F is described by $dz_{\alpha n}$ in U_α. Note that any 1-form defining F in U_α is of the form $f_\alpha(z_{\alpha 1},\ldots,z_{\alpha n})dz_{\alpha n}$, and if we require that this 1-form is closed, then $\omega_\alpha = f_\alpha(z_{\alpha n})dz_{\alpha n}$. Assume that (ω_α) is a family of closed 1-forms defining F in $M\text{-Sing}\,F$. If $\xi_{\alpha\beta}\colon U_\alpha\cap U_\beta\to\mathbb{C}^*$ is the cocycle defined by $\omega_\alpha=\xi_{\alpha\beta}\omega_\beta$, then the cocycle $(\xi_{\alpha\beta})$ depends only on the variable $z_{\beta n}$; that is, if a foliation is defined by closed 1-forms, then the cocycle $(\xi_{\alpha\beta})$ is constant along the leaves.

LEMMA 2.5. Let $\{U_\alpha\}$, ω_α and $\xi_{\alpha\beta}$ be as above with ω_α closed 1-forms, and let $F_{\bar{\eta}}^{1}$ be an infinitesimal deformation of F associated to $\eta\colon L\to T^*M$. Then

1. In the above coordinates of M', η is described by a cocycle (η_α), $\eta_\alpha=\xi_{\alpha\beta}\eta_\beta$, such that η_α is F-closed; i.e. for any leaf L of F, η restricted to L as a 1-form in L is closed.

2. Let ω_1 be a non-vanishing closed 1-formed defined in the open subset U of M defining F, then the infinitesimal deformation $F_{\bar{\eta}}^{1}$ is defined in U by a 1-form η_1 that is F-closed.

Proof. The integrability condition (2.1) reduces to $\omega_\alpha\wedge d\eta_\alpha = 0$

if ω_α is a closed 1-form. Choosing local coordinates so that $\omega_\alpha = f(z_n)dz_n$ we have

$$\omega_\alpha \wedge d\eta_\alpha = (fdz_n) \wedge d\left(\sum_{i=1}^{n} \eta_i dz_i\right) = f\left[\sum_{j=1}^{n-1} \frac{\partial}{\partial z_j}\left(\sum_{i=1}^{n-1} \eta_j\right) dz_j \wedge dz_i\right] \wedge dz_n$$

Hence the term inside the bracket is zero, which means that the 1-form on the leaves of F is d_F-closed. This proves part 1. Part 2 also follows since ω_1 gives a trivialization of L restricted to U. ‖

The following Lemma is a Frobenius Theorem for infinitesimal deformations:

LEMMA 2.6. Let F be a holomorphic foliation in M, M' = M-Sing F and let $\{U_\alpha\}$ be an open cover of M' with biholomorphisms $\phi_\alpha : U_\alpha \to D_\alpha \times W_\alpha \subset \mathbb{C}^{n-1} \times \mathbb{C}$, where D_α are balls in $\mathbb{C}^{n-1}$ and W_α open sets in $\mathbb{C}$, $\phi_\alpha^* dz_n$ defines F in U_α and the transition coordinates of the submersions $\bar{\phi}_\alpha = \Pi_2 \circ \phi_\alpha : U_\alpha \to W_\alpha$ are $\bar{\phi}_{\alpha\beta} : \bar{\phi}_\beta(U_\alpha \cap U_\beta) \to \bar{\phi}_\alpha(U_\alpha \cap U_\beta)$. Then, for any infinitesimal deformation F^1 of F there are holomorphic maps $\bar{\phi}_\alpha + t\bar{\psi}_\alpha : A \times U_\alpha \to A \times W_\alpha$ with transition coordinates $\bar{\phi}_{\alpha\beta} + t\bar{\psi}_{\alpha\beta}$ such that $(\bar{\phi}_\alpha + t\bar{\psi}_\alpha)^* dz_{\alpha n}$ defines F^1 in U_α.

Proof. Let $\omega_\alpha = \phi_\alpha^* dz_{\alpha n}$ be the 1-forms defining F, and η_α the F-closed 1-forms describing F^1, as in Lemma 2.5, so that $\omega_\alpha + t\eta_\alpha$ describes F^1. Note that for every holomorphic function f_α in U_α, F^1 may also be defined by $(1 + tf_\alpha)(\omega_\alpha + t\eta_\alpha) = \omega_\alpha + t(\eta_\alpha + f_\alpha\omega_\alpha)$. Hence any 1-form $\tilde{\eta}_\alpha$ whose restriction to every leaf of F in U_α coincides with η_α, also serves to define F^1 in U_α.

Choose a section Σ_α to the foliation in U_α and define $\bar{\psi}_\alpha(p) = \int_\delta \eta_\alpha$, where δ is a path contained in a leaf of F going from a point in Σ_α to p. Clearly $d\bar{\psi}_\alpha$ coincides with η_α restricted to any leaf, hence $(\phi_\alpha + t\psi_\alpha)^* dz_{\alpha n} = \omega_\alpha + t\psi_\alpha^* dz_{\alpha n}$ defines F^1 in U_α. Define $\psi_\alpha : U_\alpha \to \mathbb{C}^n$ as $\psi_\alpha = (0,\ldots,0,\bar{\psi}_\alpha)$, then $(\phi_\alpha + t\psi_\alpha) : A \times U_\alpha \to A \times (D_\alpha \times W_\alpha)$ is an isomorphism of analytic spaces and $(\phi_\alpha + t\psi_\alpha)dz_n$ defines F^1 in U_α. If $\phi_{\alpha\beta} + t\psi_{\alpha\beta} = (\phi_\alpha + t\psi_\alpha)^{-1} \circ (\phi_\beta + t\psi_\beta)$ is the transition of coordinates, then $(\phi_{\alpha\beta} + t\psi_{\alpha\beta})^* dz_{\beta n} = \sum_{i=1}^{n} \left(\frac{\partial \phi^n_{\alpha\beta}}{\partial z_{\alpha i}} + t \frac{\partial \psi^n}{\partial z_{\alpha i}} \right) dz_{\alpha i}$ is of the form $(f + tg) \cdot dz_{\alpha n}$ since $(\phi_{\alpha\beta} + t\psi_{\alpha\beta})$ is preserving the foliations on $A \times \mathbb{C}^n$. Hence $\phi^n_{\alpha\beta}$ and $\psi^n_{\alpha\beta}$ are functions of $z_{\beta n}$ alone, which we denote by $\bar{\phi}^n_{\alpha\beta}$ and $\bar{\psi}^n_{\alpha\beta}$. The transition coordinates of the submersions $\bar{\phi}_\alpha + t\bar{\psi}_\alpha$ are then $\bar{\phi}^n_{\alpha\beta} + t\bar{\psi}^n_{\alpha\beta}$. This proves the lemma. ‖

2.2. First Variations of Holonomy Maps.

In this section we define the holonomy pseudogroup of a foliation, and show how to obtain from a holomorphic family of foliations, an induced family of holonomy pseudogroups. Then we will obtain from an infinitesimal deformation of a foliation an infinitesimal deformation of the holonomy pseudogroup, and we will show how to compute some of its elements using the preceeding subsection.

Let F be a holomorphic foliation of codimension 1 in M, defined by $\omega: L \to T^*M$. In $M' = M\text{-Sing } F$ we obtain a non-singular foliation of codimension 1. Let $\{U_\alpha\}$ be a countable open cover of M' and biholomorphisms $\phi_\alpha: U_\alpha \to V_\alpha$, with V_α open balls in $\mathbb{C}^n$, such that $\phi^*_\alpha(dz_n) = \omega_\alpha$ are 1-forms describing F, as in section 1.3. We will denote by $\bar{\phi}_\alpha: U_\alpha \to \bar{V}_\alpha \subset \mathbb{C}$ the maps obtained by applying ϕ_α and then composing with projection to the last factor, and we assume that $\bar{V}_\alpha \cap \bar{V}_\beta = \phi$ for $\alpha \neq \beta$ (obtained after suitable translations) and that $\bar{\phi}_\alpha$ has connected fibers. Denote by $\bar{\phi}_{\alpha\beta}: \bar{\phi}_\beta(U_\alpha \cap U_\beta) \to \bar{\phi}_\alpha(U_\alpha \cap U_\beta)$ the biholomorphism of open sets in $\mathbb{C}$ defined as $\bar{\phi}_\alpha \circ \bar{\phi}_\beta^{-1}$.

DEFINITION 2.7. The collection of biholomorphisms between open sets of $\mathbb{C}$ $\{\bar{\phi}_{\alpha_r\alpha_{r-1}} \circ \cdots \circ \bar{\phi}_{\alpha_1\alpha_0}\}$, with domain of definition the maximal open set where the composition makes sense, is called the *holonomy pseudogroup of the foliation* F (with respect to the submersions $\bar{\phi}_\alpha$). Many properties of the holonomy pseudogroup do not depend on the submersions $\{\bar{\phi}_\alpha : U_\alpha \to \bar{V}_\alpha \subset \mathbb{C}\}$; for a notion of equivalence of pseudogroups, see [10].

Given points p_0 and p_1 in a leaf L of F, let $p_0 \in U_0$ and $p_1 \in U_1$, and for every close path $\delta: [0,1] \to L$, $\delta(0) = p_0$, $\delta(1) = p_1$, we will show how to obtain an element of the holonomy pseudogroup, $h_\delta: (V_0, \phi_0(p_0)) \to (V_1, \phi_1(p_1))$ which tells us how the leaves of F near p_0 are distributed near p_1 following the path δ. To define h_δ, let $0 = t_0 < t_1 < \cdots < t_r = 1$ be a partition of $[0,1]$, and $U_{\alpha_0} = U_0, U_{\alpha_1}, \ldots, U_{\alpha_{r-1}} = U_1$ be elements of the cover $\{U_\alpha\}$ such that $\delta[t_i, t_{i+1}] \subset U_{\alpha_i}$, $i = 0,\ldots,r-1$; then $h_\delta = \bar{\phi}_{\alpha_{r-1}, \alpha_{r-2}} \circ \cdots \circ \bar{\phi}_{\alpha_1, \alpha_0}$. It is shown in [10] that h_δ does not depend on the covering or the partition used, and that it is also independent of the homotopy class of δ, with fixed end points. If $p_0 = p_1$ and $U_0 = U_1$, we obtain the *holonomy representation*

$$h: \Pi_1(L, p_0) \to \mathit{Bih}\,(\mathbb{C}, \phi_0(p_0))$$

of the fundamental group of L at p_0 into the germ of local biholomorphisms of $\mathbb{C}$ at $\phi_0(p_0)$. The *linear holonomy representation*

$$Dh: \Pi_1(L, p_0) \to \mathbb{C}^*$$

is obtained by taking the derivatives at $\phi_0(p_0)$ of the holonomy representation.

Let $\tilde{F}$ be a family of holomorphic foliations in M parametrized by the complex manifold S, or by the analytic space $S = A$ associated to the dual numbers $\mathbb{C} \oplus t\mathbb{C}$, and let $\{\tilde{U}_\alpha\}$ be an open cover of $S \times M\text{-Sing}\,\tilde{F}$ with biholomorphisms $\Phi_\alpha: \tilde{U}_\alpha \to S_\alpha \times V_\alpha \subset S \times \mathbb{C}^n$ over S such that the relative 1-forms $\Phi_\alpha^*(dz_n)$ define $\tilde{F}$ in U_α, as in section 1.3. Let $\bar{\Phi}_\alpha: \tilde{U}_\alpha \to S_\alpha \times \bar{V}_\alpha \subset S \times \mathbb{C}$ be the maps obtained by applying Φ_α and then composing with the projection $S \times \mathbb{C}^n \to S \times \mathbb{C}$ to the z_n-coordinate, and we assume that $\bar{V}_\alpha \cap \bar{V}_\beta = \phi$ for $\alpha \neq \beta$ and that $\bar{\Phi}_\alpha$ has connected fibers. Denote by $\bar{\Phi}_{\alpha\beta}: \bar{\Phi}_\beta(\tilde{U}_\alpha \cap \tilde{U}_\beta) \to \bar{\Phi}_\alpha(\tilde{U}_\alpha \cap \tilde{U}_\beta)$ the biholomorphism over S of open sets in $S \times \mathbb{C}$ defined as $\bar{\Phi}_\alpha \circ \bar{\Phi}_\beta^{-1}$.

DEFINITION 2.8. The collection of biholomorphisms between open sets of $S \times \mathbb{C}$ $\{\bar{\Phi}_{\alpha_r\alpha_{r-1}} \circ \cdots \circ \bar{\Phi}_{\alpha_1\alpha_0}\}$ with domain of definition the maximal open set where the composition makes sense is called the *holonomy pseudogroup of the family* $\tilde{F}$.

If $\tilde{F}$ is a deformation of $F = \tilde{F}_0$ parametrized by a germ of a complex manifold, and L is a leaf of F, then the *holonomy representation of the deformation along* L is the representation

$$h: \Pi_1(L,p_0) \to Bih\ (S \times \mathbb{C}, \Phi_0(p_0))$$

obtained for $\delta \in \Pi_2(L,p_0)$ as the composition

$\Phi_\delta = \bar{\Phi}_{\alpha r-1\alpha r-2} \circ \cdots \circ \bar{\Phi}_{\alpha 1 \alpha 0}$ associated to a partition

$0 < t_0 < \quad < t_r = 1$ and elements $\tilde{U}_{\alpha_0} = \tilde{U}_0, \tilde{U}_{\alpha_1}, \ldots, \tilde{U}_{\alpha_{r-1}} = U_0$

such that $\delta[t_i, t_{i+1}] \subset \tilde{U}_{\alpha_i}$. The *linear holonomy representation of the deformation along* L

$$Dh: \Pi_1(L,p_0) \to \left\{ \left(\begin{array}{c|c} I_n & \mathbb{C}^n \\ \hline & \mathbb{C}^* \end{array} \right) \right\} \subset GL(n+1, \mathbb{C})$$

is the representation obtained by taking the derivate at $\Phi_0(p_0)$ of the holonomy representation h; n is the dimension of the tangent space to S at 0, and I_n is the identity in $T_0 S$.

THEOREM 2.9. Let F be a codimension 1 holomorphic foliation in M defined by $\omega: L \to T^*M$, U and open set in M where F may be defined by a non-vanishing holomorphic closed 1-form ω_1 and δ a closed loop in U contained in a leaf of F, then:

1. For any infinitesimal deformation F^1 of F the linear holonomy of F^1 along δ takes the form

$$\begin{pmatrix} 1 & a(F^1) \\ 0 & 1 \end{pmatrix}, \; a(F^1) = \int_\delta \eta_1 \tag{2.3}$$

where $\omega_1 + t\eta_1$ describes F^1 in U.

2. If M is compact, the map $F^1 \to a(F^1)$ in (2.3) induces a linear map

$$\int_\delta : T_{\bar{\omega}} Fol\ (M,L) \to \mathbb{C} \tag{2.4}$$

This map depends only on F^1 up to multiplication by a non-zero scalar (i.e. For other elections of ω, U, ω_1, describing F, δ up to homology in a leaf, and the submersion where the holonomy is defined, we obtain a function of F^1 which is a scalar multiple of (2.4)).

Proof. Let $\{U_\alpha\}$, $\phi_\alpha : U_\alpha \to D_\alpha \times W_\alpha \hookrightarrow \mathbb{C}^{n-1} \times \mathbb{C}$ be an open cover by coordinate charts of U, as in Lemma 2.6, with $\omega_\alpha = \phi_\alpha^*(dz_{\alpha n})$ equal to ω_1 in U_α, and let $\bar{\phi}_\alpha : U_\alpha \to W_\alpha$ be the projection to the second factor and $\bar{\phi}_{\alpha\beta}(z_{\beta n}) = z_{\beta n} + c_{\beta n}$ be the transitions of coordinates. Since $\omega_\alpha = \omega_\beta$ in $U_\alpha \cap U_\beta$, the transition of coordinates for L in this cover are $\xi_{\alpha\beta} = 1$ and hence the infinitesimal deformation F^1 is described by a 1-form η in U with $\eta_\alpha = \eta \mid U_\alpha$ (see Lemma 2.6). As in the proof of Lemma 2.6, define

$$\bar{\psi}_\alpha(p) = \int_{p_\alpha}^{p} \eta_\alpha$$

where the integral is taken in a path from $p_\alpha \in \Sigma_\alpha$ to p in a leaf, then $\bar{\phi}_\alpha + t\bar{\psi}_\alpha$ are submersions defining the infinitesimal deformation in U_α. If $U_\alpha \cap U_\beta \neq \phi$, the transition of coordinates $(\mathrm{Id} + c_{\alpha\beta}) + t\bar{\psi}_{\alpha\beta}$ satisfies for $p \in U_\alpha \cap U_\beta$

$$\bar{\phi}_\alpha(p) + t\int_{p_\alpha}^{p} \eta_\alpha = \left((\mathrm{Id} + c_{\alpha\beta}) + t\bar{\psi}_{\alpha\beta}\right)\left(\bar{\phi}_\beta(p) + t\int_{p_\beta}^{p} \eta_\beta\right) =$$

$$= (\bar{\phi}_\beta(p) + c_{\alpha\beta}) + t\left(\int_{p_\beta}^{p} \eta_\beta + \bar{\psi}_{\alpha\beta}(\bar{\phi}_\beta(p))\right)$$

and since $\bar{\phi}_\alpha(p) = \bar{\phi}_\beta(p) + c_{\alpha\beta}$ and using that $\eta \mid U_\alpha = \eta_\alpha$ and $\eta \mid U_\beta = \eta_\beta$ we obtain

$$\bar{\psi}_{\alpha\beta}(\bar{\phi}_\beta(p)) = \int_{p_\alpha}^{p} \eta_\alpha - \int_{p_\beta}^{p} \eta_\beta = \int_{p_\alpha}^{p_\beta} \eta$$

The transition of coordinates is then

$$\left(\mathrm{Id} + t\int_{p_\alpha}^{p_\beta} \eta\right) + c_{\alpha\beta}$$

If $U_{\alpha_0}, \ldots, U_{\alpha_{r-1}} = U_{\alpha_0}$ is a covering of δ as in Definition 2.8, the holonomy map obtained by following δ has the form

$$z \mapsto z + t\int_{\delta_z} \eta \tag{2.5}$$

where δ_z is the closed loop near δ obtained by joining points on Σ_α over z and the constant is zero since δ is a closed loop. Hence we obtain (2.3).

From the expression (2.3), it follows that the map (2.4) is

$\mathbb{C}$-linear. Since the matrix (2.3) is the linear holonomy of F^1, it is uniquely defined up to conjugation and since

$$\begin{pmatrix}1 & b\\ 0 & c\end{pmatrix}^{-1}\begin{pmatrix}1 & a\\ 0 & 1\end{pmatrix}\begin{pmatrix}1 & b\\ 0 & c\end{pmatrix} = \begin{pmatrix}1 & ca\\ 0 & 1\end{pmatrix}$$

we obtain that fixing any election of ω, U, ω_1, δ and submersion, the map (2.4) obtained as a function of F^1 is a non-vanishing constant times (2.4). ‖

DEFINITION 2.10. With the assumptions in the theorem, there is a canonically defined point $\int_\delta \in \text{Proj } T^*_\omega \mathcal{F}o\ell(M,L)$, that we will call the *Iliashenko point of* δ.

2.3. Persistent Cycles under Deformations.

If a closed loop δ in a leaf of a foliation F has trivial holonomy, then there is a canonical way to associate to near leaves a homotopy class $[\delta_u]$ of loops near δ. If $\tilde{F}$ is a deformation of F, then some of the homotopy classes $[\delta_u]$ persist under the deformation, while others open up. In this section we will analyse this process, as well as the associated infinitesimal concept.

Let F be a holomorphic foliation of codimension 1 in M, L a leaf of F in M-Sing F and δ a curve in L representing a free homotopy class in L with trivial holonomy, $h_\delta = \text{id}$. Let $\{U_0, \ldots, U_r\}$ be a covering of δ by coordinate charts as in Lemma 2.6, such that there is a partition $0 = t_0 < t_1 < \cdots < t_{r-1} = 1$ with $\delta[t_i, t_{i+1}] \subset U_i$ for $i = 0,\ldots,r-1$. The element of the holonomy pseudogroup $\bar{\phi}_{r-1,r-2} \circ \cdots \circ \bar{\phi}_{1,0}$ is the holonmy of δ, which by hypothesis is the identity. Choose transversals Σ_i to F at $\delta(t_i)$, $i = 0,\ldots,r-1$, $\Sigma_{r-1} = \Sigma_0$. For any point near to $\delta(t_0)$ on Σ_0 there is a unique homotopy class of paths in U_i in a leaf going from a point in Σ_0 to Σ_1; and then in U_2 from Σ_1, to Σ_2, etc. In this way we construct homotopy classes of loops $\{\delta_u\}$ near to δ and on leaves near L.

DEFINITION 2.11. Let $\tilde{F}$ be a deformation of $F = \tilde{F}_0$ parametrized by a germ of a complex manifold S, L a leaf of F and δ a free homotopy class in $\Pi_1(L)$.

Let $\Phi_\delta(s,z) = (s,h(s,z)): (S \times \mathbb{C},(0,0)) \to (S \times \mathbb{C},(0,0))$ be the holonomy map associated to δ in definition 2.8. Let Z be the germ of a subset of $S \times \mathbb{C}$ determined by the equation $h(s,z) - z = 0$. Choosing transversals $\tilde{\Sigma}_i$ to the foliation $\tilde{F}$ in $S \times M\text{-Sing } F$ at δ, we may as before for every point z in Z sufficiently close to $(0,0)$ associate a free homotopy class δ_z in a leaf of $\tilde{F}$. We will say that δ_z *is obtained by following* δ *in* $\tilde{F}$. Note that Z is $S \times \mathbb{C}$ or has codimension 1 in $S \times \mathbb{C}$. If $Z_1 \cap (0 \times \mathbb{C})$ has 0 as an isolated zero, we will say that δ is a *limit cycle*, and the multiplicity will be called the *multiplicity of the limit cycle*. If δ is a limit cycle, then the projection to the first factor $Z \to S$ is a finite map, and the sum of the multiplicities of the cycles over $s \in S$ equals the multiplicity of δ (see [5]).

Assume now that S has dimension 1 and that the holonomy of δ in F is the identity. Then, we may write $h(s,z) - z = sg(s,z)$, where g is a germ of a holomorphic functions. If $g(0,0) = 0$ and $s \neq g(s,z)$, we will say that δ *is persistent thorugh the 1-dimensional deformation*; and otherwise we will say that δ *is not persistent.*

Let again δ be a free homotopy class in a leaf L of F with trivial holonomy, F^1 be an infinitesimal deformation of F, and $h_\delta = \mathrm{Id} + th': (A \times \mathbb{C},0) \to (A \times \mathbb{C},0)$ the holonomy representation of F^1. We will say that δ *is persistent for the infinitesimal deformation* if 0 is an isolated zero of h'.

PROPOSITION 2.12. Let δ be a free homotopy class in a leaf L of a foliation F with trivial holonomy, $\tilde{F}$ a deformation of F parametrized by $(\mathbb{C},0)$ and F^1 the induced infinitesimal deformation, then:

1) If δ is persistent for F^1, then it is persistent for $\tilde{F}$.

2) If ω_1 is a closed non-vanishing 1-form defining F in the open subset U containing δ and if $\{\delta_u\}$ are the free homotopy classes obtained from δ in the leaves $\{L_u\}$ of F near L, then F^1 may be defined by a 1-form $\omega + t\eta$ in U and the map which associates to δ_u its Iliasenko point in $\mathrm{Proj}\, T^*_\omega Fol(M,L)$ gives rise to a holomorphic map

$$u \mapsto \int_{\delta_u} \tag{2.6}$$

proof. Using the previously used notation, the holonomy of $\tilde{F}$ around δ is $h(s,z) = z + sg(s,z)$, and the holonomy of F^1 is $h(z) = = z + th'(z)$, where $h'(z) = \frac{\partial h}{\partial s}(0,z) = g(0,z)$. If δ is persistent for F^1, then 0 is an isolated zero of h', the implies $g(0,0) \neq 0$ and s does not divide g. Hence δ is persistent for F^1. This proves 1.

In (2.5) it was shown that by choosing special coordinates, the holonomy map around δ has the form $h_\delta = \mathrm{Id} + t\int_{\delta_u} \eta$, hence the linear holonomy at u, which is $\frac{\partial h}{\partial s}(u) = \int_{\delta_u} \eta$. If $\eta_0, \ldots, \eta_N$ is a basis of $T^*_\omega \mathcal{F}o\ell(M,L)$ then since the maps $\int_{\delta_u} \eta_i$ are holomorphic, we obtain that (2.6) is a holomorphic map. ‖

3. GOOD MEROMORPHIC FIRST INTEGRALS.

In this section we describe a class of meromorphic functions having simple singularities, and modelled in Lefschetz's pencils (see [1]). A holomorphic foliation of codimension 1 whose leaves coincide with the fibers of one of these functions will be said to have a good meromorphic first integral.

3.1. Lefschetz's Pencils.

Let $(z_0 : \cdots : z_N)$ be homogeneous coordinates of $\mathbb{C}P^N$. A hyperplane in $\mathbb{C}P^N$ is described by an equation $\sum_{i=0}^{N} a_i z_i = 0$, and the set of hyperplanes of $\mathbb{C}P^N$ is parametrized by $(\mathbb{C}P^N)^* = = \{(a_0 : \cdots : a_N)\}$.

Let $\ell = \Sigma a_i z_i$ and $\ell' = \Sigma b_i z_i$ be equations defining two distinct hyperplanes in $\mathbb{C}P^N$. The linear family of hyperplanes $\{\alpha\ell + \beta\ell' \mid (\alpha : \beta) \in \mathbb{C}P^1\} \subset (\mathbb{C}P^N)^*$ consist of those hyperplanes that contain the codimension 2 linear subvariety

$$K = \{z \in \mathbb{C}P^N \quad \ell(z) = \ell'(z) = 0\}$$

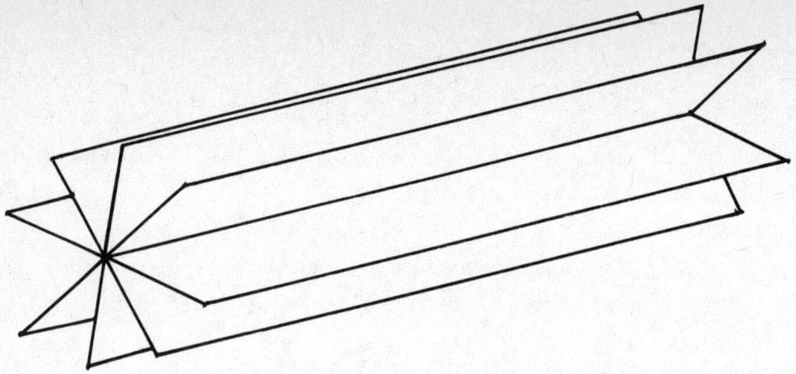

This family of hyperplanes may also be described as the closure of the fibers of the rational map $f = \frac{\ell}{\ell'}$ that has its indeterminacy locus at K. Using homogeneous coordinates for the range, $f = (\ell : \ell')$ is a rational map from $\mathbb{C}P^N$ to $\mathbb{C}P^1$ giving rise to a holomorphic map outside of K, $f: \mathbb{C}P^N - K \to \mathbb{C}P^1$.

The set of linear families of hyperplanes is parametrized by the Grassmanian of projective lines in $(\mathbb{C}P^N)^*$, Grass $(\mathbb{C}P^1, (\mathbb{C}P^N)^*)$.

LEMMA 3.1. To any linear family of hyperplanes $f = (\ell : \ell') : \mathbb{C}P^N - \{\ell = \ell' = 0\} \to \mathbb{C}P^1$ we may asociate a holomorphic foliation of codimension one $df: H(-2) \to T^*\mathbb{C}P^N$ whose leaves are the f-fibers. This family of foliations in Proj $\Gamma(\mathbb{C}P^N, H(-2)) \otimes \otimes T^*\mathbb{C}P^N)$ has dimension $2(N-1)$.

Proof. Given $f = (\ell : \ell')$, choose coordinates $(z_0, \ldots, z_N)$ of $\mathbb{C}^{N+1}$ such that $\ell = z_0$ and $\ell' = z_1$. In affine coordinates $(y_1, \ldots, y_N) = \left(\frac{z_1}{z_0}, \ldots, \frac{z_N}{z_0}\right)$ $f = y_1$. The 1-form $df = dy_1$ is tangent to the f-fibers and by Lemma 1.3 it has a pole of order 2 as a rational 1-form on $\mathbb{C}P^N$, so it gives rise to a holomorphic bundle map $df: H(-2) \to T^*\mathbb{C}P^N$. Since dim Grass $(\mathbb{C}P^1, (\mathbb{C}P^N)^*) = 2(N-1)$, the Lemma is proved. ‖

We will now blow up $\mathbb{C}P^N$ along K, and we will analyse the induced foliation. Let $f = (\ell : \ell') : \mathbb{C}P^N \to \mathbb{C}P^1$ be the rational map associated to a linear family of hyperplanes, with indeterminacy locus the linear subvariety $K = \{\ell = \ell' = 0\}$ of codimension 2.

Let $\widetilde{\mathbb{C}P}^N$ be the subvariety of $\mathbb{C}P^N \times \mathbb{C}P^1$ defined by the equation

$$\ell w_1 = \ell' w_0 \qquad (z_0 : \cdots : z_N) \in \mathbb{C}P^N;\ (w_0 : w_1) \in \mathbb{C}P^1$$

and let $\sigma\colon \widetilde{\mathbb{C}P}^N \to \mathbb{C}P^N$ and $\tilde{f}\colon \widetilde{\mathbb{C}P}^N \to \mathbb{C}P^1$ be the restriction to $\widetilde{\mathbb{C}P}^N$ of the projections to the factors of $\mathbb{C}P^N \times \mathbb{C}P^1$. σ is called the *blow up of* $\mathbb{C}P^N$ *along* K and $\sigma^{-1}(K) = K \times \mathbb{C}P^1$ is called the *exceptional divisor*. $\sigma\colon \widetilde{\mathbb{C}P}^N - \sigma^{-1}(K) \to \mathbb{C}P^N - K$ is a biholomorphism, $\tilde{f}$ is a $\mathbb{C}P^{N-1}$-fiber bundle and $\tilde{f} = f \circ \sigma$ (see [15] p. 98).

Any holomorphic line bundles on $\mathbb{C}P^N \times \mathbb{C}P^1$ is isomorphic to a bundle of the form $\Pi_1^*(H(n_1)) \otimes \Pi_2^*(H(n_2))$, and any holomorphic line bundle on $\widetilde{\mathbb{C}P}^N$ is isomorphic to one an only one of the restrictions of the above bundles to $\widetilde{\mathbb{C}P}^N$, that we will denote by $H(n_1, n_2)$. The exceptional divisor $K \times \mathbb{C}P^1$ is the zero set of a holomorphic section of the bundle $H(1,-1)$, which we will denote by e; it is defined uniquely up to multplication by a non-zero scalar.

LEMMA 3.2. Let $\omega\colon H(-2) \to T^*\mathbb{C}P^N$ be the holomorphic foliation associated to the linear family of hyperplanes $f = (\ell : \ell')$, having singular locus along K. Let $\sigma\colon \widetilde{\mathbb{C}P}^N \to \mathbb{C}P^N$ be the blow up of $\mathbb{C}P^N$ along K, and let $\tilde{\omega}\colon H(-2,0) = \sigma^*(H(-2)) \to T^*\widetilde{\mathbb{C}P}^N$ be the holomorphic bundle map induced from ω by means of the coderivative of σ: $\tilde{\omega} = {}^t(D\sigma)\omega$. Then $\tilde{\omega}$ vanishes of order 2 along the exceptional divisor $K \times \mathbb{C}P^1$, and $\frac{1}{e^2}\tilde{\omega}\colon H(0,-2) \to T^*\widetilde{\mathbb{C}P}^N$ is never vanishing and it describes the non-singular foliation whose leaves are the $\tilde{f} = f \circ \sigma$ fibers.

Proof. Take coordinate charts $(z_0 : \cdots : z_N)$ of $\mathbb{C}P^N$ such that $\ell = z_0$ and $\ell' = z_1$.

A typical coordinate chart of $\widetilde{\mathbb{C}P}^N \hookrightarrow \mathbb{C}P^N \times \mathbb{C}P^1$ is

$$\mathbb{C}^N \ni (z_1, \ldots, z_{N-1}, w_0) \to (z_1 w_0 : z_1 : \cdots : z_{N-1} : 1\,;\, w_0 : 1) \in \widetilde{\mathbb{C}P}^N$$

so that the blowing up receives an expression in affine charts

$$\sigma(z_1, \ldots, z_{N-1}, w_0) = (z_1 w_0, z_1, \ldots, z_{N-1})$$

the foliation in $\mathbb{C}P^N$ is defined in $\mathbb{C}^N$ as $z_1 dz_0 - z_0 dz_1$, hence

$$\sigma^*(z_1 dz_0 - z_0 dz_1) = z_1(z_1 dw_0 + w_0 dz_1) - (z_0 w_0) dz_1 = z_1^2 dw_0$$

Since the exceptional divisor is defined by $z_1 = 0$, we have that

$\sigma^*(\omega)$ vanishes of order 2 along $K \times \mathbb{C}P^1$, and hence $\frac{1}{e^2}\,{}^t(D\sigma)\omega =$ $= \omega$: $H(-2,0) \otimes H(2,-2) \to T^*\mathbb{C}P^N$ describes a non-singular foliation, whose leaves are the $\tilde{f}$-fibers. Note that $\tilde{\omega} = D\tilde{f}$: $\tilde{f}^*T^*\mathbb{C}P^1 \to T^*\mathbb{C}P^N$. ∥

Let now M be a compact complex manifold embedded in $\mathbb{C}P^N$ $i: M \hookrightarrow \mathbb{C}P^N$ and such that it is not contained in any hyperplane. Let $f = (\ell : \ell')$ be a linear family of hyperplanes such that $K = \{\ell = \ell' = 0\}$ intersects M in a subvariety $K' = K \cap M$ of codimension 2 in M. The restriction of f to M induces a rational function on M, having indeterminacy locus on K'; and the restriction of f to $M - K'$ induces a holomorphic map $M - K' \to \mathbb{C}P^1$. Let ω: $H(-2) \to T^*\mathbb{C}P^N$ be the codimension one holomorphic foliation associated to f, $\omega = df$, and L is the line bundle $i^*(H(-2))$ on M obtained by restriction, the coderivative induces on M a codimension 1-holomorphic foliation F' defined by $\omega' = (di)^*\omega$: $L \to T^*M$. The singular set of F' is K' plus the tangency points of M with the family of hyperplanes $\{\lambda\ell + \mu\ell'\}$. The leaves of F' in M-Sing F' are the intersection of the hyperplanes $\lambda\ell + \mu\ell' = 0$ with M. We will say that *the foliation* F' *is induced in* $M \hookrightarrow \mathbb{C}P^N$ *by the family of hyperplanes* $\{\lambda\ell + \mu\ell'\}$.

EXAMPLE 3.3. Let $d > 0$ be a positive integer, and let $z_0^d, z_0^{d-1}z_1^d, \dots, z_n^d$ be a basis of monomials of degree d in $\mathbb{C}^{n+1}$, and let ρ_d be the holomorphic map

$$\rho_d = (z_0^d : \cdots : z_n^d) : \mathbb{C}P^n \to \mathbb{C}P^N \qquad N = \binom{n+d}{n} - 1$$

ρ_d is an embedding of $\mathbb{C}P^n$, called the d-uple embedding (see [11] p. 13). A hyperplane H in $\mathbb{C}P^N$ with coordinates $\{(w_{i_0,\dots,i_n}) = (w_I) \mid |I| = i_0 + \cdots + i_n = d\}$ is given by an equation

$\sum_{|I|=d} a_I w_I = 0$; hence $\rho_d(\mathbb{C}P^n) \cap M$ is the subvariety of $\mathbb{C}P^n = \rho_d(\mathbb{C}P^n)$ defined by $\Sigma\, a_I z^I = \Sigma\, a_{i_0,\dots,i_n} z_0^{i_0},\dots,z_n^{i_n} = 0$, and hence there is a one to one correspondance between hypersurfaces of degree d in $\mathbb{C}P^n$ and hyperplanes in $\mathbb{C}P^N$. A linear family of hyperplanes $(\ell : \ell')$ in $\mathbb{C}P^N$ determines when intersecting with $\rho_d(\mathbb{C}P^n)$ a linear family of hypersurfaces of degree d; namely the family $\{\lambda(\ell \circ \rho_d) + \mu(\ell' \circ \rho_d)\}$ and hence $f \circ \rho_d = \frac{\ell \circ \rho_d}{\ell' \circ \rho_d}$ is a rational function on $\mathbb{C}P^n$ where numerator and denumerator are homogeneous polynomials of degree d. The condition that $K' = K \cap \rho_d(\mathbb{C}P^n)$ has codimension 2 in $\rho_d(\mathbb{C}P^n)$ is equivalent to asking that $\ell \circ \rho_d$ and $\ell' \circ \rho_d$ have no common factor.

DEFINITION 3.4. A holomorphic foliation of codimension 1 in M is a *Lefschetz pencil* if it is the foliation induced in M by a linear family of hyperplanes $\{\lambda\ell + \mu\ell'\}$ in $\mathbb{C}P^N$ via an embedding $i: M \hookrightarrow \mathbb{C}P^N$ and satisfying

1) $K = \{\ell = \ell' = 0\}$ intersects M transversely.

2) Each hyperplane $\lambda\ell + \mu\ell' = 0$ intersects M transversely, except possibly at one point, where it has a non-degenerate tangency (i.e. the holomorphic map $\frac{\ell}{\ell'}: M - K \to \mathbb{C}P^1$ has only Morse-type singularities, each with a distinct value).

The smooth submanifold K is called the *basis of the pencil*.

LEFSCHETZ THEOREM ([1]). Let M be a connectd complex submanifold of $\mathbb{C}P^N$ not contained in any hyperplane of $\mathbb{C}P^N$, then

1) There exists a Zariski dense open subset in Grass $(\mathbb{C}P^1, (\mathbb{C}P^N)^*)$ such that the holomorphic foliation they induce in M is a Lefschetz pencil.

2) If H is a hyperplane in $\mathbb{C}P^N$ intersecting transversely M, then the inclusion $M \cap H \hookrightarrow M$ induces isomorphisms between the homotopy groups $\Pi_k(M \cap H) \to \Pi_k(M)$, for $k = 0,\dots,\dim M - 2$.

3.2. Good Meromorphic First Integrals.

Based on the Lefschetz pencils, we introduce the following definition:

DEFINTION 3.6. Let $\omega: L \to T^*M$ be a holomorphic foliation of codimension 1 in M.

1) ω has a *meromorphic first integral* if there is a meromorphic function f on M and a Zarizki dense subset U of M where $f: U \to \mathbb{C}P^1$ is holomorphic and the leaves of ω in U coincide with the connected components of the fibers of $f|_U$.

2) ω has a *good meromorphic first integral* if the singular locus $\{\omega = 0\}$ of ω has codimension bigger than 1 and if there

is a meromorphic function f on M such that

a) The indeterminacy locus K of f is a submanifold of M of codimension two, and around each point p of K we may find coordinates $(z_1,\ldots,z_n)$ of an open set W in M, $p = 0$, such that the foliation in $W - K \cap W$ is described by $z_2^2 d(z_1 z_2^{-1}) = = z_2 dz_1 - z_1 dz_2$.

b) The critical points of the holomorphic map $f_1\colon M - K \to \mathbb{C}P^1$ obtained by restricting f to $M - K$ has codimension bigger than one (i.e. $\operatorname{cod}\{Df_1 = 0\} > 1$).

Remark. If ω has a good meromorphic first integral f and if W and $(z_1,\ldots,z_n)$ are as above, then f restricted to W receives an expression of the form $f(z_1,\ldots,z_n) = g(z_1 z_2^{-1})$, where $g\colon \mathbb{C}P^1 \to \mathbb{C}P^1$ is a rational map of $\mathbb{C}P^1$. Assumption b forces g to be a non-branched map, and so g has degree one. Hence, after a change of coordinates we actually have $f = z_1 z_2^{-1}$ in W.

LEMMA 3.6. Let $\omega\colon L \to T^*M$ be a holomorphic foliation of codimension 1 in M with a good meromorphic first integral f. If K is the indeterminacy locus of K, and if $\lambda_1,\ldots,\lambda_r$ are the critical values in $\mathbb{C}P^1$ of f on $M - K$, then the map $f_2\colon M - [\bigcup_i f^{-1}(\lambda_i) - K] \to \mathbb{C}P^1 - \bigcup_i\{\lambda_i\}$ obtained by restricting f has the structure of a C^∞-fibre bundle.

Proof. Let $\sigma\colon \widetilde{M} \to M$ be the blowing up morphism of M along K and let $\widetilde{\omega} = {}^t(D\sigma)\omega\colon \sigma^*L \to T^*\widetilde{M}$ be the holorphic bundle map induced from ω by means of the coderivative of σ. Similar as in Lemma 3.2 $\widetilde{\omega}$ vanishes of order two along the exceptional divisor $\sigma^{-1}(K)$, and if $\sigma^{-1}(K)$ is defined by $e = 0$, then $\frac{1}{e^2}\widetilde{\omega}$ describes a foliation $\widetilde{F}$ in M. The closure of the leaves of $\widetilde{F}$ coincide with the fibers of the holomorphic map $\widetilde{f} = f \circ \sigma\colon \widetilde{M} \to \mathbb{C}P^1$. By generic smoothness ([11] p. 272) there are only a finite number $\lambda_1,\ldots,\lambda_r$ of critical values, and $\widetilde{f}$ restricted to $\widetilde{M} - \widetilde{f}^{-1}\{\lambda_1,\ldots,\lambda_r\}$ has the structure of a C^∞-fibre bundle. The map $\widetilde{f}$ restricted to $\sigma^{-1}(K)$ is also a fibre bundle, with fiber isomorphic to K. $\widetilde{f}$ restricted to $\widetilde{M} - (f^{-1}\{\lambda_1,\ldots,\lambda_r\} \cup \sigma^{-1}(K))$ is also a C^∞-fibre bundle, and it is biholomorphic to f_2 via σ. This proves the Lemma. ‖

We will illustrate for projective spaces some of the preceeding definitions.

LEMMA 3.7. 1) The set of rational functions $\mathbb{C}P^n \to \mathbb{C}P^1$ such that the inverse image of a generic point is a hypersurface of degree d is a projective variety of dimension $2\binom{d+n}{n} - 4$ which is naturally embedded in $\operatorname{Proj} E(H(-2d))$, that has dimension $\binom{2d+n-1}{2d}(2d-1) - 1$.

2) For $n = 2$, the above set of rational function has dimension $d^2 + 3d - 2$, and $\operatorname{Proj} E(H(-2d)) = \mathcal{Fol}\,(\mathbb{C}P^2, H(-2d))$ has dimension $4d^2 - 2$.

Proof. The vector space of homogeneous polynomials of degree d in $n + 1$ variables has dimension $\binom{d+n}{d}$, so the Grassman manifold of 2-planes has dimension $2\left[\binom{d+n}{d} - 2\right]$. Part 1 follows then from Proposition 1.3, since $g^2 d\left(\frac{f}{g}\right) = gdf - fdg$ has degree $2d - 1$, and its terms of top degree annhilate the radial vector field. Part 2 follows from part 1. ‖

Remark. For every $d \geqslant 1$, we obtain a subvariety of $\mathcal{Fol}\,(\mathbb{C}P^n, H(-2d))$. We will show elsewhere ([8]) that for $n > 2$ this is an irreducible component of $\mathcal{Fol}\,(\mathbb{C}P^n, H(-2d))$; for $n = 2$ it has codimension $3d^2 - 3d$. There are other foliations having meromorphic first integrals, due to cancellations in branching sets, i.e. $d(f^m/g^{m'})$.

LEMMA 3.10. 1) A foliation F in $\mathcal{Fol}\,(\mathbb{C}P^2, H(-e))$ with isolated singularities of multiplicity 1 has $e^2 - 3e + 3$ singular points.

2) If F is a Lefschetz pencil in $\mathcal{Fol}\,(\mathbb{C}P^2, H(-2d))$, then the indeterminacy locus has d^2 points, and there are $3d^2 - 6d + 3$ Morse type singular points.

Proof. The number of singular points is computed as the second Chern class of $T^*\mathbb{C}P^2 \otimes H(e)$ (see [3]), which using the Euler Sequence ([11] p. 176) tensored with $H(e)$.

$$0 \to \Omega^1_{\mathbb{C}P^n} \to H(-1)^{\oplus (n+1)} \to H(0) \to 0$$

and the properties of Chern classes ([12])

$$(1+dt)[(1+c_1(\Omega^1(d))t+c_2(\Omega^1_d))t^2] = (1+(d-1)t)^3$$

$$1+(d+c_1(\Omega^1(d))t+(c_2(\Omega^1(d))+dc_1(\Omega^1(d)) = 1+3(d-1)+3(d-1)^2t^2$$

Hence $c_1(\Omega^1(d)) = 2d-3$ and $c_2(\Omega^1(d)) = d^2-3d+3$.

To prove 2, observe that the indeterminacy locus is obtained by intersecting two elements of the pencil, which by Bezout's theorem is d^2. Since the foliation of the pencil is in $\mathcal{Fol}\ (\mathbb{C}P^2, H(-2d))$, by part 1 there are $4d^2 - 6d + 3$ singular points, so there are $3d^2 - 6d + 3$ Morse type singularities. ‖

4. THE ILIASHENKO CURVES.

In this section we define the Iliashenko curves of a foliation having a good meromorphic first integral, and show that some of these curves are conics.

4.1. The Iliashenko Curves.

Let $\omega\colon L \to T^*M$ be a holomorphic foliation of codimension 1 in the compact manifold M, having a good meromorphic first integral $f\colon M \to \mathbb{C}P^1$ and indeterminacy locus K. The fibers of f restricted to $M - K$ will be denoted by $F_\lambda = f^{-1}(\lambda)$, and we will denote by $\Lambda = \{\lambda_1, \ldots, \lambda_r\}$ the set of critical values of $f|_{M-K}$.

Let $\tau\colon W \to \mathbb{C}P^1 - \Lambda$ be the universal covering space of $\mathbb{C}P^1 - \Lambda$ viewed as homotopy classes of paths with fixed end points, starting in λ_0. By Lemma 3.8, the restriction of f, $f_2\colon M - [K \cup f^{-1}(\Lambda)] \to \mathbb{C}P^1 - \Lambda$, has the structure of a C^∞-fibre bundle, and pulling it back to W, we obtain a fibre bundle that is C^∞-bundle isomorphic to $F_{\lambda_0} \times W$ if W is biholomorphic to $\mathbb{C}$ or to the unit disc (i.e. if $r \geqslant 1$. If there are no critical values we still have $f_{2*}H_1(F_\lambda, Z) = H_1(F_\lambda, Z) \times \mathbb{C}P^1$). Hence, given $\delta \in \Pi_1(F_{\lambda_0})$ for any element $\lambda \in W$ there is a well defined free homotopy class δ_λ obtained from δ by deforming it along the path λ to a free homotopy class in $\Pi_1(F_{\tau(\lambda)})$. Composing f_2 in the range with a Moebius transformation, we may obtain $f_2(\delta_\lambda) \neq \infty$, and we may use df_2 in $M - (K \cup f_2^{-1}(\infty))$ to apply theorem 2.9 to δ_λ, and hence the Iliashenko point of δ_λ is well defined for $\lambda \in W$.

DEFINITION 4.1. Let F be a holomorphic foliation of codimension 1 having a good meromorphic first integral $f\colon M \to \mathbb{C}P^1$ and $\delta \in \Pi_1(F_{\lambda_0})$, then the map

$$I_\delta\colon W \to \text{Proj } T^*_F \, \mathcal{F}o\ell\,(M,L) \tag{4.1}$$

obtained by associating to $\lambda \in W$ the Iliashenko point of δ_λ (see definition 2.10) will be called the *Iliashenko curve* of δ.

THEOREM 4.2. Let F be a holomorphic foliation of codimension 1 in the compact manifold M having a good meromorphic first integral $f\colon M \to \mathbb{C}P^1$, then:

1) The Iliashenko curve I_δ is holomorphic, and depends only on the homology class of δ in $H_1(F_{\lambda_0}, Z)$.

2) If δ, $\delta' \in H_1(F_{\lambda_0}, Z)$, then $I_{\delta+\delta'} = I_\delta + I_{\delta'}$.

3) There is a one to one correspondance between infinitesimal directions of deformations (i.e. given by $\omega + t(a\eta)$, $a \in \mathbb{C}^*$) and hyperplanes H_η in Proj $T^*_F \, \mathcal{F}o\ell\,(M,L)$. If $I_\delta(W) \not\subset H_\eta$ then the set of points $I_\delta(W) \cap H_\eta$ corresponds to those homology classes δ_λ which are persistent for the infinitesimal direction η of deformation.

Proof. Part 1 follows from Proposition 2.12, and the expression of the map as an integral, which is zero for commutator paths.

Part 2 follows from the additive properties of the integral (2.6).

The equations defining a hyperplane in Proj $T^*_F \, \mathcal{F}o\ell\,(M,L)$ belong to $\text{Hom}\,(T^*_F \, \mathcal{F}o\ell\,(M,L), \mathbb{C}) = T_F\,(\mathcal{F}o\ell\,(M,L))$, and so hyperplanes correspond to points in Proj $T_F \, \mathcal{F}o\ell\,(M,L)$. A point λ lies in $I_\delta(W) \cap H_\eta$ if and only

$$\int_{\delta_\lambda} \eta = 0$$

Hence by Theorem 2.9 δ_λ is persistent for the infinitesimal deformation F_η. ‖

4.2. The Conics Associated to the Indeterminacy locus.

Let F be a codimension 1 holomorphic foliation having a good meromorphic first integral $f: M \to \mathbb{CP}^1$ with indeterminacy locus K; and let $(z_1, \ldots, z_n)$ be coordinates of an open set W in M, with $0 \in K \cap W$ where the foliation is described by the 1-form $z_2dz_1 - z_1dz_2$. Let W' be the two dimensional manifold transversal to K defined by $z_3 = \cdots = z_n = 0$ with coordinates (z_1, z_2). The foliation F' in W' induced from F consist of lines through the origin in W'. If S'_r denotes a small sphere $|z_1|^2 + |z_2|^2 = r$, then S'_r will intersect each leaf $L_{(a:b)} = \{t(a,b) \in W' \mid t \in \mathbb{C}^*\}$ in a closed loop $\delta_{(a:b)}$. Since we may take f to be $z_1z_2^{-1}$ in W', then it is clear that the family of loops $\delta_{(a:b)}$ are obtained from the other by deformation inside the leaves of F. Since the critical points of f are far from W', we also see that there is no monodromy around the critical values for these loops, and that they may be extended to loops over the critical values. This implies that the Iliashenko curve is defined as a map with domain $\mathbb{CP}^1 - \Lambda$, since there is no monodromy around the critical values for $\delta_{(a:b)}$, and by continuity of the formula (2.6) it extends also to the points of Λ. Hence the Iliashenko curve is described by a holomorphic map

$$I_\delta: \mathbb{CP}^1 \to \text{Proj } T^*_F \, Fol\,(M,L) \tag{4.2}$$

We will now give an expression for this map. To do this, let $\omega = z_2dz_1 - z_1dz_2$ define F in W and let $\eta_1, \ldots, \eta_N$ be the holomorphic 1-forms in W obtained from a basis of $T_F \, Fol\,(M,L)$ using the trivialization of L given by ω. Since

$$d\left(\frac{z_1}{az_1 + bz_2}\right) = \frac{b}{(az_1 + bz_2)^2}(z_2dz_1 - z_1dz_2) = \frac{b\omega}{(az_1 + bz_2)^2}$$

it follows that dividing ω by $(az_1 + bz_2)^2$ we make ω a closed 1-form. Choose coordinates so that $a = 0$, $b = 1$ (to simplify notation), and in the trivialization of L on $W_1 - \{z_2 = 0\}$ induced by the closed 1-form $z_2^{-2}\omega$, a basis of $T_F \, Fol\,(M,L)$ is $z_2^{-2}\eta_1, \ldots, z_2^{-2}\eta_N$. Expand η_k in power series

$$\eta_k = \sum_{i,j \geq 0} a_{ij} z_1^i z_2^j dz_1 + \sum_{i,j \geq 0} b_{ij} z_1^i z_2^j dz_2 \tag{4.3}$$

The form $z_2^{-2}\eta_k$ restricted to the line $t \mapsto t(a,b)$ has the expression in t

$$b^{-2}t^{-2}\left(\sum_{i,j\geqslant 0} (a_{ij}a^{i+1}b^j + b_{ij}a^i b^{j+1})t^{i+j}dt\right)$$

and hence the integral of $z_2^{-2}\eta_k$ along the loop $\delta_{(a:b)}$ is equal by the Residue Theorem to

$$\int_{\delta_{(a:b)}} \eta_k = 2\Pi i b^{-2} \sum_{i+j=1} (a_{ij}a^{i+1}b^j + b_{ij}a^i b^{j+1}) =$$

$$= \frac{2\Pi i}{b^2}[a_{10}a^2 + (a_{01} + b_{10})ab + b_{01}b^2] \qquad (4.4)$$

Repeating a similar calculation for the other forms η_k, the map (4.2) has an expression

$$(a:b) \mapsto (c_1a^2 + c_1'ab + c_1''b^2 : \cdots : c_Na^2 + c_N'ab + c_N''b^2)$$

We have proved:

PROPOSITION 4.3. Let δ be a loop going once around one of the connected components of the indeterminacy locus, then either:

1) The Iliashenko curve of δ is not defined (namely, the integrals in (2.3) vanish for any $\delta_{(a:b)}$).

2) The Iliashenko curve degenerates to a point (namely, the linear maps in (2.4) for any $\delta_{(a:b)}$ are constant multiples one of the others).

3) The Iliashenko curve of δ is a smooth conic in Proj $T_{F}^{*}\, \mathcal{Fol}\,(M,L)$.

We may interpret the conic Q corresponding to the above loop δ as follows. Let H_η be a hyperplane in Proj $T_F^*\, \mathcal{Fol}\,(M,L)$ defined by the infinitesimal deformation η in $T_F\, \mathcal{Fol}\,(M,L)$. If $Q \cap H_\eta$ consists of two points, this is selecting for us two fibers of f, such that when deforming in the direction of η the corresponding two loops are persisting. The singularity of F at K is an example of a Kupka phenomena (see [8]). The Kupka phenomena is persistent in the sense that if $\{F_t\}_{t\in\mathbb{C}}$ is a 1-parameter deformation of F there is a smooth submanifold K_s of codimension 2 near K formed of singular points of F_t, $F_t \cap W' = F_t'$ is a foliation by curves in the complex surface W' with an isolated singularity at

$K_t \cap W' = \{p_t\}$ and the foliation $\mathcal{F}_t$ in a neighbourhood of K_t is locally modelled on the foliation $\mathcal{F}_t'$ in W' near p_t product a disc in $\mathbb{C}^{n-2}$. Hence we may restrict to W' and analyse the family $\mathcal{F}_t'$. Assume that $\mathcal{F}_t'$ is defined by

$$(z_2 dz_1 - z_1 dz_2) + t(\Sigma a_{ij} z_1^i z_2^j dz_1 + \Sigma b_{ij} z_1^i z_2^j dz_2) + t^2(\cdots) \qquad (4.5)$$

with first order part (4.3). The vector field

$$X_t = \begin{pmatrix} 1 & 0 \\ 0 & 1 \end{pmatrix}\begin{pmatrix} z_1 \\ z_2 \end{pmatrix} + t\left\{\begin{pmatrix} -b_{00} \\ a_{00} \end{pmatrix} + \begin{pmatrix} -b_{10} & -b_{01} \\ a_{10} & a_{01} \end{pmatrix}\begin{pmatrix} z_1 \\ z_2 \end{pmatrix} + \cdots\right\} + t^2(\cdots)$$

is also determining $\mathcal{F}_t$. Assume that the linear part of the first order variation has distinct eigenvalues. Then by a linear change of coordinates in $W' = \mathbb{C}^2$ we may put this linear part in diagonal form, so that $a_{10} = b_{01} = 0$ and $-b_{10} \neq a_{01}$. The linear part of X_t at p_t is

$$\begin{pmatrix} 1 - b_{10}t & 0 \\ 0 & 1 + a_{01}t \end{pmatrix} + t^2(\quad)$$

and hence the eigenspaces of X_t at p_t for $t \neq 0$ small are approximating the z_1 and z_2 axis. Note that in this coordinates the vanishing of the integral in (4.4) is $ab = 0$, hence it is also detecting the two axes.

Formulating the conclusion in intrinsec terms: Assume that the deformation $\mathcal{F}_t$ is such that the linear part of the first order variation of the transversal model (4.5) of the Kupka phenomena along K_t of $\mathcal{F}_t$ has distinct eigenvalues, then the eigendirections of the linear parts of X_t tend to the directions at p specified by the solutions of (4.4).

If one further assumes that the quotient of the above eigenvalues is a non-real complex number, then by Pincaré's Linearization Theorem we may conclude that there are two separatrix manifolds passing through K_t, for $t \neq 0$, and tending t to 0 they are approaching the fibers $F_{(a:b)}$ of f, with $(a:b)$ the points of intersection $Q \cap H_\eta$.

We will now see that if the indeterminacy locus K of a good meromorphic first integral $f: M \to \mathbb{CP}^1$ has at least two irreducible components, then we have an infinite number of Iliashenko curves that are conics. This is the case if M has dimension 2, where

K consists of a finite number of points. By Lemma 3.10, the indeterminacy locus of a Lefschetz pencil in $\mathcal{Fol}(\mathbb{C}P^2, H(-2d))$ consists of d^2 points.

Let δ and δ' be loops around distinct connected components of K and let $\delta'' = m\delta + n\delta'$ with $m,n \in Z$. Since the monodromy action around the critical values of δ and δ' is trivial, we obtain that it acts trivially also on δ'' (as a 1-cycle in homology). In $M - (K \cup F_\infty)$ the foliation is defined by the closed holomorphic 1-form df, and we obtain as in Theorem 2.9 maps

$$I, I', I'' : \mathbb{C}P^1 - \{\infty\} \to T_{\mathcal{F}}\, \mathcal{Fol}\,(M,L)$$

whose projectivisation are the Iliashenko curves associated to δ, δ' and δ''. By the defining formula (2.3), we have $I'' = mI + nI'$. From (4.4) we see that I'' has a quadratic expression in a/b; hence the Iliashenko curve $Q_{m,n}$ is also a conic (if non-degenerate, as in Proposition 4.3). Since $\frac{m}{n} I + \frac{m'}{n'} I' = \frac{1}{mn'} [mn'I + m'nI']$ observe that if the conics $Q_{1,0}$ and $Q_{0,1}$ are non degenerate and distinct, then the closure of $\{Q_{m,n} \mid m,n \in Z\}$ is the image of $Q_{1,0} \times \mathbb{R}P^1$. Similar constructions holds for any 1-homology class in the subgroup of $H_1(F_0, Z)$ generated by the loops around the indeterminacy locus of f.

We finish by observing that the formalism presented here allows one to extend the results in Iliashenko's work [13] to rational functions. A polynomial function is not a good first integral since it is branching on the line at infinity. Using the techniques developed in [7], one may extend the formalism of this paper allowing the complex structure of M and L to vary. This line of approach will be continued in [14].

REFERENCES

[1] A. Andreotti, Th. Frankel: The Second Lefschetz Theorem of Hyperplane Sections, Global Analysis papers in honor of K. Kodaira. University of Tokyo (1969), 1 - 20.

[2] C. Chevalley: Theory of Lie Groups, Princeton University Press, 1946.

[3] S. Chern: Meromorphic vector fields and characteristic numbers, Scripta Math. XXIX, (1973), 243 - 251.

[4] A. Douady: Le probleme des modules pour les sous-espaces analytiques compacts d'un espace analytique donnè, Ann. Inst. Fourier, 16 (1966), 1-95.

[5] A. Douady: Flatness and privilege. Enseignement Math. 14 (1968), 47-74.

[6] X. Gómez-Mont: Universal families of foliations by curves, in Proc. Conf. Dyn. Syst. Dijon, 1985. ed by Cerveau, R. Moussu, Astérisque 1987.

[7] X. Gómez-Mont: The transverse dynamics of a holomorphic flow, To appear in Ann. Math.

[8] X. Gómez-Mont, A. Lins: Structural stability for holomorphic foliations having a meromorphic first integral, preprint.

[9] Ph. Griffiths, J. Harris: Principles of Algebraic Geometry. Wiley Intersc., 1978.

[10] A. Haefliger: Grupoides d'holonomie et classifiants. Astérisque 116 (1984), 70-97.

[11] R. Hartshorne: Algebraic Geometry. Springer-Verlag, 1977.

[12] F. Hirzebruch: Topological Methods in Algebraic Geometry. Springer-Verlag, 1966.

[13] J. Iliashenko: The origin of limit cycles under perturbations of the equation dwldz = -Rz/Rw, where R(z,w) is a polynomial, Math. USSR, Sbornik 7, (1969) 353-364.

[14] J. Muciño: Ph. D. Thesis. México. To appear.

[15] I. Shafarevich: Basic Algebraic Geometry, Springer-Verlag, 1965.

Instituto de Matemáticas
Universidad Nacional Autónoma de México
Cd. Universitaria
México 04510, D. F.
MEXICO

WEAKLY MIXING BILLIARDS

by

E. Gutkin* and A. Katok**

1. Introduction.

In this paper we make a modest new contribution to the study of dynamical properties of polygonal billiards using categorial approach.

In very general terms, the approach is based on Baire category theorem and on an approximation principle which says that if a Baire space B has a dense set of elements satisfying an approximate version of a certain property then it contains a dense G_δ set of elements which possess that property exactly. Without trying to discuss here what properties can be studied that way, we refer to [K1] where categorial approach is developed systematically for various spaces of dynamical systems. In a number of cases categorial approach or its modification provide the only known way to establish the existence of dynamical systems with a particular property. Existence of ergodic billiards [KMS], [K2], discussion below, is a good example of such a situation.

Let P be a connected polygon in Euclidean plane $\mathbb{R}^2$. The billiard flow B_P^t is defined on the space Y_P of all unit tangent vectors to $\mathbb{R}^2$ with footpoints in P. It can be described as follows. A vector $v \in Y_P$ with the footpoint $p \in P$ moves with the unit speed along the straight line $p + vt$, $t \in \mathbb{R}$ until it reaches the boundary of P, then it instantly changes its direction according to the rule "the angle of incidence is equal to the angle of reflection" and continues until the next collision with the boundary and so on. If a vector hits a vertex of P, the flow is not defined after the collision. The billiard flow thus defined preserves the Liouville measure on Y_P which is the product of Lebesgue measure on P and the angular measure on the circle of directions. The set of vectors which eventually hit a vertex of P has Liouville measure zero so that from the point of view of ergodic theory the billiard flow is well defined.

---

*Department of Mathematics, University of Southern California, Los Angeles, CA 90089; partially supported by NSF Grant DMS84-03238.

**Department of Mathematics, California Institute of Technology, Pasadena, CA 91125; partially supported by NSF Grant DMS85-14630.

The phase space Y_P of the billiard flow is three-dimensional and in general very little is known about ergodic properties of that flow. About the only general statement of that kind is that the entropy of B_P^t is equal to zero. This is true not only for the Liouville measure but for any Borel invariant measure as well [K3]. However for certain classes of polygones more information is known. A polygon P is called *rational* if all of its angles are commensurate with π. For any rational polygon P the space Y_P splits into a one-parameter family of two-dimensional subsets $Y_{P,\theta}$, $0 \le \theta < \frac{\pi}{N(P)}$ invariant with respect to the billiard flow [ZK], [G]. Here N(P) is the least common multiple of the denominators of the numbers $\frac{\alpha}{\pi}$ where α runs over the set of angles of P. By appropriate identification, the set $Y_{P,\theta}$ is made into a compact surface. Let us denote the restriction of B_P^t to $Y_{P,\theta}$ by $B_{P,\theta}^t$ and call it the *directional billiard flow*. The number of ergodic invariant measures for such a flow, which are not supported by periodic orbits, is bounded [S]. The flow $B_{P,\theta}^t$ is not mixing [K4]. A recent fundamental result [KMS] says that for almost every θ the flow $B_{P,\theta}^t$ is uniquely ergodic. When the number N(P) becomes large, the surfaces $Y_{P,\theta}$ become more and more uniformly distributed in Y_P. This sets the stage for the application of the categorial approach [K2], [KMS] which allows in particular to establish the existence of billiard flows ergodic in the whole space Y_P. This argument mimics an earlier similar argument [ZK] related to topological transitivity. It is still not known whether for a generic rational polygon P for most θ the flows $B_{P,\theta}^t$ are weakly mixing. "Most" may mean either a set of full measure or a dense G_δ. In this paper we solve this question in the sense of category for certain classes of rational polygons. Namely, we consider polygons for which the number N(P) is equal to 2, 3, 4 or 6. Each of these classes contains a dense subset of so-called almost integrable polygons (see Definition 3 below) which do have non-constant eigenfunctions [G]. Within our classes the almost integrable polygons are characterized by rational values of some natural parameters. When denominators of those parameters go to infinity, the non-constant eigenfunctions become more and more oscillating and eventually disappear for polygons with irrational but very well approximable values of the parameters.

An interesting open problem is the existence and genericity of billiards which are weakly mixing in whole phase space Y_P. Let us fix the topology of the billiard table P, i.e., the number of connected components of the boundary of P and the number of vertices on each boundary component. Let n be the total number of vertices. Let $\tilde{P}$ be the space of all such billiard tables with topology given by parametization by the coordinates of vertices. $\tilde{P}$ is a non-compact manifold of dimension 2n.

<u>Theorem</u> [KMS], [K2]. The set of all polygons $P \in \tilde{P}$ such that the billiard flow B_P^t is ergodic is a dense G_δ subset of $\tilde{P}$.

It is not difficult to see that the set $\tilde{P}_{mix}$ of all $P \in \tilde{P}$ for which B_P^t is weakly mixing is a G_δ.

<u>Conjecture</u>. The set $\tilde{P}_{mix}$ is a dense G_δ subset of $\tilde{P}$.

2. <u>Preliminaries.</u> <u>Statement of Results</u>.

For any polygon P we denote by U_P^t the one-parameter group of unitary operators on $L_2(Y,\mu)$ corresponding to the billiard flow B_P^t. Here μ is the (unnormalized) Liouville measure on Y_P. We assume $\mu(Y_P) = |P|$, which is the area of P.

The group G generated by Euclidean motions and dilations of the plane acts naturally on $\tilde{P}$ and the quotient $\tilde{P}/G$ can be identified with the submanifold P of $\tilde{P}$ consisting of polygons P with a distinguished vertex at the origin of $\mathbb{R}^2$, the first side on the positive x-axis and $|P| = 1$. Clearly, $\dim P = 2n - 4$ and, because the action of G is compatible with the flows B_P^t, it suffices to study those flows for $P \in P$.

We identify the set of directions θ on the plane with the circle $S^1 = \{0 \le \theta < 2\pi\}$ where $\theta = 0$ corresponds to the direction of the positive x-axis.

<u>Definition 1</u>. A polygon is called *integrable* if it tiles the plane under reflections.

It is well known that the only integrable polygons are rectangles, the equilateral triangles, the $\pi/2$, $\pi/4$, $\pi/4$-triangles, and the $\pi/2$, $\pi/3$, $\pi/6$-triangles.

We fix an integrable polygon Δ and denote by Γ the lattice obtained by tiling the plane by reflections of Δ. For instance, if Δ is the unit square, Γ is the square lattice.

Definition 2. A polygon $P \in \mathcal{P}$ is of *Δ-class* if the sides of P are parallel to the lines of Γ.

For instance, if Δ is a rectangle, P of Δ-class means that the sides of P are either horizontal or vertical. In what follows we denote by $\mathcal{P}$ the set of polygons of Δ-class (Δ is fixed) satisfying the previous assumptions.

Polygons $P \in \mathcal{P}$ are rational, i.e., their angles are rational multiples of π, hence, as we mentioned before, the flow B_P^t decomposes into the one-parameter family of directional billiard flows $B_{P,\theta}^t$ [Z-K], $0 \le \theta \le \pi/N(\Delta)$, where $N(\Delta) = 2, 3, 4$ or 6 depending on the type of Δ (see above). The flows $B_{P,\theta}^t$ for $0 < \theta < \pi/N(\Delta)$ live on the surface S_P, which is tiled by $2N(\Delta)$ copies of P, and preserve the Lebesgue measure μ on S_P.

Let e,f be a pair of generators of Γ. A direction θ is called *irrational* (resp. *rational*) if for a vector pe + qf in direction θ the ratio p/q is irrational (resp. rational). The definition does not depend on the choices involved.

Definition 3 [G]. A polygon $P \in \mathcal{P}$ is called *almost integrable* if it is homothetical to a polygon drawn on the lattice Γ.

The set $\mathcal{P}_I$ of almost integrable polygons is dense in $\mathcal{P}$. For an almost integrable polygon P the flow $B_{P,\theta}^t$ is ergodic if θ is irrational and periodic if θ is rational [G].

By *combinatorics* of a connected polygon P we will mean the following: The number of connected components of the boundary of P, the number of vertices and the angle at each vertex. Let n be the total number of vertices for polygons in $\mathcal{P}$.

Now we can formulate the first main result of this paper.

Theorem 1. Let Δ be an integrable polygon and let $\mathcal{P}$ be the manifold of polygons of Δ-class with fixed combinatorics. For any direction θ

denote by $P_{mix}(\theta) \subset P$ the set of polygons P such that the flow $B^t_{P,\theta}$ is weakly mixing. Then

i) Let Δ be a rectangle and $n > 4$. For any $\theta \neq 0, \pi/2$ the set $P_{mix}(\theta)$ is a dense G_δ in P.

ii) Let Δ be a triangle and $n > 3$. For any irrational direction θ the set $P_{mix}(\theta)$ is a dense G_δ.

Definition 4. A polygon M with 2n sides $a_1, b_1, \ldots, a_n, b_n$ is called *matched* if there are n parallel translations $g_1, \ldots, g_n$ such that $b_j = g_j a_j$, $j = 1, \ldots, n$.

For every direction θ we define the *linear flow* $L^t_{M,\theta}$ in direction θ on M as follows. A point in M flows in direction θ with the unit speed until it reaches the boundary of M. If this happens on a_j (resp. b_j), the point gets transferred to the side b_j (resp. a_j) by the translation g_j (resp. g_j^{-1}) and continues to move in the same direction. The Lebesgue measure on M is preserved by the flows $L^t_{M,\theta}$.

In what follows we normalize our matched polygons M by requiring that the sides a_1, b_1 be horizontal.

Definition 5. A matched polygon M is called *elementary of type* α, $0 < \alpha \leq \pi/2$, if M has only horizontal sides and sides making angle α with the x-axis.

We fix α and denote by M the set of elementary matched polygons of type α with a fixed number 4n of sides and a fixed combinatorics. The set M endowed with its natural topology is a manifold. The following theorem is a close counterpart of Theorem 1.

Theorem 2. Let $n > 1$. For any direction θ the set $M_{mix}(\theta)$ of polygons $M \in M$ such that the flow $L^t_{M,\theta}$ is weakly mixing is a dense G_δ.

The reader should keep in mind that an elementary matched polygon M with 4n sides can have less than 4n geometric vertices. A gnomon, for instance, has 8 sides and 6 geometric vertices. In other words, some of the angles of M may be equal to π.

Theorems 1 and 2 are derived via categorial approach from a result which describes the discrete spectrum of linear flows in almost integrable polygons. We need more definitions to state the corresponding theorem.

Definition 6. We say that an elementary matched polygon M of type α *is modelled on the parallelogramm* A if M is tiled by translated copies of A and A is a maximal parallelogramm to tile M. In what follows we simply say that *M is modelled on* A.

Definition 7. Let A be a parallelogramm spanned by e and f. A direction θ is called *irrational* (resp. *rational*) *with respect to* A if for a vector ae + bf in direction θ the number a/b is irrational (resp. rational).

Let M be modelled on A and let A_i, $i \in I$, be the copies of A tiling M. For any $i \in I$ we identify functions on A_i and A.

Definition 8. Let notation be as above. A function f on M is called *A-periodic* if the restrictions of f on A_i, $i \in I$, are all equal.

We denote by $L_2^{(d)}(M)$ the Hilbert space of square integrable (with respect to the Lebesgue measure) A-periodic functions on M. By a natural isomorphism, $L_2^{(d)}(M) = L_2(A)$.

Theorem 3 (cf. [G], Theorem 3). Fix a parallelogram A and let M be a polygon modelled on A. Then

i) The flow $L^t_{M,\theta}$ is uniquely ergodic if θ is irrational and periodic otherwise.

ii) For any irrational direction θ the discrete spectrum component of $L_2(M)$ for the flow $L^t_{M,\theta}$ is the space $L_2^d(M)$ of A-periodic functions. The identification $L_2^d(M) = L_2(A)$ induces a natural isomorphism of $L^t_{M,\theta}$ restricted to $L_2^d(M)$ with $L^t_{A,\theta}$.

Consider the space $\mathcal{Q} = \mathcal{P} \times S^1$ of pairs (P,θ) where $\mathcal{P}$ is the space of polygons of Δ-class. We want to show that for a typical pair (P,θ) the flow $B^t_{P,\theta}$ is weakly mixing.

Theorem 4. Let $\mathcal{P}$ be the space of polygons of Δ-class (Δ is fixed) and let $\mathcal{Q}_{mix}$ be the set of pairs $Q = (P,\theta)$ such that $B^t_{P,\theta}$ is weakly mixing. Then $\mathcal{Q}_{mix}$ is a dense G_δ in $\mathcal{Q}$.

3. Proofs.

Proof of Theorem 3. Let M be any matched polygon with the pairs a_i, b_i, $i = 1, \ldots, n$, of parallel sides. Identifying a_i with b_i for all i we obtain a closed surface S_M and the flows $L^t_{M,\theta}$ live on S_M.

The surface S_A corresponding to a parallelogramm A is a torus and the flow $L^t_{A,\theta}$ is the linear flow in direction θ on the torus S_A.

The tiling of M by copies of A defines the projection $p:S_M \to S_A$ which commutes with the flows $L^t_{M,\theta}$ and $L^t_{A,\theta}$ for all θ. Now we are in the setting of Theorem 3 of [G] and we refer the reader to the proof of that theorem.

Proof of Theorem 2. We consider polygons $M \in \mathcal{M}$ of area one such that (0,0) is a vertex of M and obviously it suffices to prove the assertion for the manifold (denoted again by $\mathcal{M}$) of polygons satisfying these conditions.

Fix a direction θ and choose a parallelogramm A_θ, $|A_\theta| = 1$, with angle α such that θ is irrational with respect to A_θ. Denote by (x,y) the linear coordinates defined by A_θ so that $A_\theta = \{(x,y): 0 \leq X \leq 1,\ 0 \leq y \leq 1\}$. Let e_0 and f_0 be the vectors spanning A_θ. Denote by $\mathcal{A}$ the set of parallelogramms A spanned by $e = re_0$ and $f = sf_0$ where r and s are rational and let $\mathcal{M}_I \subset \mathcal{M}$ be the subset of polygons modelled on A, $A \in \mathcal{A}$. Let a_i, b_i and c_i, d_i, $i = 1, \ldots, n$, be respectively the pairs of horizontal sides and the sides forming angle α with horizontal direction. Then $M \in \mathcal{M}_I$ if and only if the numbers $|a_i|/|e_0|$, $|c_i|/|f_0|$ are rational for $i = 1, \ldots, n$. Here the absolute value sign denotes the length of a vector. From now until the end of the proof we delete θ from notation.

We denote by 1_M the indicator function M. We have the natural embedding $L_2(M) \to L_2(\mathbb{R}^2)$ and the projection $L_2(\mathbb{R}^2)$ on $L_2(M)$ given by $f \to 1_M f$. Using this we extend the flows L^t_M and the unitary groups U^t_M to $\mathbb{R}^2$ and $L_2(\mathbb{R}^2)$ respectively by identity on $\mathbb{R}^2 \setminus M$. We use the same symbols for the extended L^t_M and U^t_M and denote by $\langle f,g\rangle$ the scalar product in $L_2(\mathbb{R}^2)$.

The flow L^t_M is weakly mixing if for any $f \in L_2(\mathbb{R}^2)$ the function $t \to \langle U^t_M f, 1_M f\rangle$ strongly converges in the sense of Cesaro (see [H] or [W]) to $|\langle f, 1_M\rangle|^2$ as $|t| \to \infty$. We will need the following.

Lemma 1. For any $f,g \in L_2(\mathbb{R}^2)$, any t and any $\varepsilon > 0$ the set of polygons $M \in \mathcal{M}$ such that

$$|\langle U_M^t f, 1_M g\rangle - \langle f, 1_M\rangle\langle 1_M, g\rangle| < \varepsilon \qquad (1)$$

is open in $\mathcal{M}$.

<u>Proof</u>. For any t, f and g the functions $M \to \langle f, 1_M\rangle$ and $M \to \langle U_M^t f, 1_M g\rangle$ are continuous on $\mathcal{M}$. □

We choose a dense in $L_2(\mathbb{R}^2)$ sequence f_i, i = 1, 2 ... and for any t and $N \geq 1$ denote by $\mathcal{M}_{t,N} \subset \mathcal{M}$ the set of polygons $\mathcal{M}$ such that for i = 1, ..., N

$$|\langle U_M^t f_i, 1_M f_i\rangle - |\langle f_i, 1_M\rangle|^2| < 1/N. \qquad (2)$$

In view of Lemma 1, $\mathcal{M}_{t,N}$ is open in $\mathcal{M}$ for any t and N and we set $\mathcal{M}_N = \bigcup_t \mathcal{M}_{t,N}$. Thus, $\mathcal{M}_N$ is open and $\bigcap_{N=1}^{\infty} \mathcal{M}_N$ is a G_δ. We claim that $\mathcal{M}_{mix} = \bigcap_N \mathcal{M}_N$.

If $M \in \mathcal{M}_{mix}$ then (cf. [H] or [W]) for any $f, g \in L_2(\mathbb{R}^2)$ there is a set $T_{f,g}$ of density one in $\mathbb{R}$ such that $\langle U_M^t f, 1_M g\rangle$ converges to $\langle f, 1_M\rangle\langle 1_M, g\rangle$ when $|t| \to \infty$ in $T_{f,g}$. Intersection of a finite number of sets of density one has density one, hence nonempty, therefore $\mathcal{M}_{mix} \subset \bigcap_N \mathcal{M}_N$.

Assume that the opposite inclusion fails, i.e., that there exists $M \in (\bigcap_N \mathcal{M}_N)\setminus\mathcal{M}_{mix}$. Then there is an eigenfunction $f \in L_2(M)$ of U_M^t such that $\langle f, 1_M\rangle = 0$ and $\|f\| = 1$. Let $U_M^t f = \exp(\sqrt{-1}\,at)f$. Fix $\varepsilon > 0$ and let f_i be such that $\|f-f_i\| < \varepsilon$. For any $t \in \mathbb{R}$

$$U_M^t f_i = U_M^t(f_i-f) + U_M^t f = U_M^t(f_i-f) + \exp(\sqrt{-1}\,at)f.$$

Therefore

$$\langle U_M^t f_i, 1_M f_i\rangle = \langle U_M^t(f_i-f) + \exp(\sqrt{-1}\,at)f,\ 1_M(f_i-f) + f\rangle$$

$$= \langle U_M^t(f_i-f), 1_M(f_i-f)\rangle + 2\exp(\sqrt{-1}\,at)\mathrm{Re}\,\langle (f_i-f), f\rangle + \exp(\sqrt{-1}\,at)$$

which implies the estimate

$$|\langle U_M^t f_i, 1_M f_i\rangle - \exp(\sqrt{-1}\, at)| < 2\varepsilon + \varepsilon^2. \qquad (3)$$

Since $|\langle f_i, 1_M\rangle| = |\langle f_i - f, 1_M\rangle| < \varepsilon$, we have for any t

$$|\langle U_M^t f_i, 1_M f_i\rangle - |\langle f_i, 1_M\rangle|^2| > |\langle U_M^t f_i, 1_M f_i\rangle| - \qquad (4)$$

$$|\langle f_i, 1_M\rangle|^2 > 1 - 2\varepsilon - 2\varepsilon^2.$$

Taking ε small enough in (4) we find an index i such that

$$|\langle U_M^t f_i, 1_M f_i\rangle - |\langle f_i, 1_M\rangle|^2| > \tfrac{1}{2} \qquad (5)$$

for all t. Hence, for $N = i + 1$, $M \notin \mathcal{M}_N$ in contradiction to the assumption.

We have shown that $\mathcal{M}_{mix} = \bigcap_n \mathcal{M}_n$ is a G_δ.

It remains to show that $\mathcal{M}_{mix}$ is dense. For $M \in \mathcal{M}_I$ denote by p(M) and q(M) the least common denominators of $|a_i|/|e_0|$, $i = 1, \ldots, n$ and $|c_i|/|f_0|$, $i = 1, \ldots, n$ respectively. If $p(M) = p$ and $q(M) = q$, M is tiled by copies of the parallelogramm $A_{p,q}$ spanned by $e = e_0/p$ and $f = f_0/q$. Denote the parallelogramm $\{xe_0+yf_0 : |x|, |y| \le N\}$ by B_N.

For $M \in \mathcal{M}_I$ denote by P_M^d (resp P_M^c) the projection on the nontrivial discrete spectrum, i.e., on the discrete spectrum inside the space $L_2(M)$ (resp. continuous spectrum) of U_M^t. Let $p(M) = p$, $q(M) = q$ and let $M \subset B_N$. By Theorem 3 $(P_M^d f)(x,y) =$

$$(pq)^{-1}\, 1_M(x,y) \sum_{i=-pN}^{pN} \sum_{j=-qN}^{qN} (1_M f)(x+i/p, y+j/q). \qquad (6)$$

Denote (x,y) by z and (i/p, j/q) by e_{ij}. We rewrite (6) as

$$(P_M^d f)(z) = (pq)^{-1} \sum_{i,j} (1_M f)(z+e_{ij}). \qquad (7)$$

Denote $P_M^d f$ by g. Since g is $A_{p,q}$-periodic, for any $z \in M$ there exists $u \in A_{p,q}$ such that $g(z) = g(u)$. Thus, for any $z, z' \in M$ there are $u, u' \in A_{p,q}$ so that

$$g(z) - g(z') = g(u) - g(u') = (pq)^{-1} \sum_{i,j} f(u+e_{ij}) - f(u'+e_{ij}) \qquad (8)$$

where the summation is over such pairs (i,j) that $e_{ij} \in M$.

Let f be a continuous function supported on B_N. For any $\varepsilon > 0$ there is $\delta(\varepsilon) > 0$ such that $|f(z)-f(z')| < \varepsilon$ if $|z-z'| < \delta(\varepsilon)$. Fix $\varepsilon > 0$ and assume that the diameter of $A_{p,q}$ is less than $\delta(\varepsilon)$. Then $|(u+e_{ij})-(u'+e_{ij})| < \varepsilon$ and, by (8), for any $z,z' \in M$

$$|g(z)-g(z')| < \varepsilon. \tag{9}$$

Integrating (9) over M we obtain that for any $z \in M$

$$|g(z) - \int_M g(\zeta)d\mu(\zeta)| < \varepsilon. \tag{10}$$

Since g is obtained from $1_M f$ by averaging

$$\int_M g(z)d\mu(z) = \int_M f(z)d\mu(z) = \langle f,1_M\rangle. \tag{11}$$

Recalling that $g = P_M^d f$ we have, by (10) and (11)

$$\|P_M^d f - \langle f,1_M\rangle 1_M\|_u < \varepsilon \tag{12}$$

where $\|\varphi\|_u = \max|\varphi(z)|$ over $z \in B_N$. Denote by $\|\varphi\|$ the L_2-norm. If φ is supported on M, $\|\varphi\| \le \|\varphi\|_u$ and, by (12)

$$\|P_M^d f - \langle f,1_M\rangle 1_M\| < \varepsilon. \tag{13}$$

We choose a dense in $L_2(\mathbb{R}^2)$ sequence of continuous functions f_i, $i = 1, 2, \ldots$ such that $\operatorname{supp} f_i \subset B_i$ and let $\mathcal{M}_N$, $N = 1, \ldots$ be the corresponding sequence of open sets in $\mathcal{M}$ where $\mathcal{M}_{mix} = \bigcap_N \mathcal{M}_N$. For any N we can find $\delta_N > 0$ such that $|z-z^1| < \delta_N$ implies $|f_i(z)-f_i(z')| < (2N \max_{i<N}\|f_i\|)^{-1}$ for $i = 1, \ldots, N$.

Let $M \in \mathcal{M}_I$ be contained in B_N and assume that diam $A_{p(M),q(M)} < \delta_N$. Then for any f

$$\langle U_M^t f, 1_M f\rangle - |\langle f, 1_M\rangle|^2 =$$

$$\langle U_M^t[(P_M^d f - \langle f, 1_M\rangle 1_M) + \langle f, 1_M\rangle 1_M + P_M^c f], 1_M f\rangle - |\langle f, 1_M\rangle|^2 = \tag{14}$$

$$\langle U_M^t(P_M^d f - \langle f, 1_M\rangle 1_M), 1_M f\rangle + \langle U_M^t P_M^c, P_M^c f\rangle.$$

For $i = 1, \ldots, N$ there exists a set $T_i \subset \mathbb{R}$ of density one such that $|\langle U_M^t P_M^c f_i, P_M^c f_i\rangle| < 1/2N$ if $t \in T_i$. Hence for $t \in \bigcap_{i=1}^{N} T_i$ which is nonempty, $|\langle U_M^t P_M^c f_i, P_M^c f_i\rangle| < 1/2N$ for all $i \le N$. By (13), for any $i \le N$ and any t

$$|\langle U_M^t(P_M^d f_i - \langle f_i, 1_M\rangle 1_M), 1_M f_i\rangle| < (2N \max_{i \le N} \|f_i\|)^{-1} \|1_M f_i\| \le 1/2N. \tag{15}$$

Hence, (14) implies that for $i \le N$ and $t \in \bigcap_{i=1}^{N} T_i$

$$|\langle U_M^t f_i, 1_M f_i\rangle - |\langle f_i, 1_M\rangle|^2| < 2(2N)^{-1} = 1/N \tag{16}$$

thus, $M \in \mathcal{M}_N$. Polygons

$$\{M \in \mathcal{M}_I : M \subset B_N \text{ and } \operatorname{diam} A_{p(M),q(M)} < \delta_N\}$$

are dense in the set $X_N = \{M \in \mathcal{M} : M \subset B_N\}$, thus the closure of $\mathcal{M}_N$ contains X_N. Since any polygon belongs to some X_N, $\bigcap_N \mathcal{M}_N$ is dense in $\mathcal{M}$. □

<u>Remark</u> <u>1</u>. If $n = 1$, $\mathcal{M}$ consists of parallelograms and Theorem 3 applies.

Let M be a matched polygon and let S_M be the corresponding surface (see proof of Theorem 3). The surface S_M is closed and orientable and its genus $g(S_M)$ is determined by the combinatorics of M. The conformal structure on S_M induced from M is singular at the vertices if $g(S_M) > 1$. The singularities can be resolved and S_M becomes a surface of constant negative curvature [G], but we are interested in the imposed on S_M flat conformal structure (with singularities if $g > 1$). We call such surfaces *almost flat* and denote their set by S.

Everything we said so far about matched polygons M extends to the case when M has selfoverlappings by regarding M as a union of polygons belonging to different copies of $\mathbb{R}^2$ and making natural identifications. From now on we allow M to have such selfoverlappings and denote the manifold of these polygons normalized as before by M. The mapping $M \to S$ is, by definition, onto and is locally one-to-one and thus supplies S with a structure of a manifold. If $S \in S$ we denote by $L^t_{S,\theta}$ the family of linear flows on S. The following assertion is immediate from Theorem 2.

Corollary 1. Let S be the manifold of almost flat surfaces obtained from the set M of elementary matched polygons with a fixed number $4n > 4$ of sides and a fixed combinatorics. For any θ the set $S_{mix}(\theta)$ of surfaces S such that the flow $L^t_{S,\theta}$ is weakly mixing is a dense G_δ.

Proof of Theorem 1. Let P be a rational polygon and let N be the least common multiple of the denominators of the angles $\pi m_i/n_i$ of P. Reflecting P in its sides $2N - 1$ times we obtain a matched polygon M [G]. The billiard flows $B^t_{P,\theta}$ unfold into the linear flows $L^t_{M,\theta}$ on M. Although M is not uniquely determined by P, the surface S_M does not depend on any particular way of unfolding P and $S_M = S_P$, the canonical surface defined by P [G]. Thus, we obtained a continuous mapping $s: P \to S$. Denote by D_N the dihedral group of order 2N. The image of s consists of surfaces S with an action of D_N and we have $P = s^{-1}(S) = S/D_N$.

Now we apply this to the polygons of Δ-class and notice that $N = N(\Delta)$ is equal to 2, 3, 4 and 6 if Δ is a rectangle, equilateral triangle, $\pi/4$ and $\pi/6$ triangle respectively. By fixing a way of unfolding P into M we obtain a continuous injective mapping $u: P \to M$ where M consists of elementary matched polygons of type $\alpha = \pi/2, \pi/6, \pi/2$ and $\pi/6$ when $N(\Delta) = 2, 3, 4$ and 6 respectively. We fix a direction θ and delete θ from our notation. Let M_N be the sequence of open sets introduced in the proof of Theorem 2. Since u is continuous and commutes with the flows B_P^t and L_P^t on P and M respectively, $P_N = u^{-1}(M_N)$ are open and $P_{mix} = \bigcap_N P_N$. It remains to show that P_{mix} is dense in P. We consider two cases in the theorem separately.

i) If $\theta \neq 0, \pi/2$ we can choose a rectangle Δ, $|\Delta| = 1$ such that θ is irrational with respect to Δ. Let e and f be the horizontal and the vertical vectors of Δ respectively. For any $r,s > 0$ denote by $\Delta_{r,s}$ the rectangle spanned by re and sf and let $P_I \subset P$ be the set of polygons which can be tiled by $\Delta_{r,s}$ under reflections where r and s are rational. Clearly, P_I is a countable dense subset of P. The rest of the proof is analogous to the second part of the proof of Theorem 2. For $P \in P_I$ we define the integers p(P) and q(P) and show that for any N the polygon P belongs to P_N if p(P) and q(P) are big enough. Thus, P_N is dense in P, therefore $P_{mix} = \bigcap_N P_N$ is a dense G_δ.

ii) We can no longer vary Δ but if θ is irrational (with rspect to Δ) we can repeat the argument of i) with obvious modifications. We spare the details.

<u>Proof of Theorem 4</u>. Let $\mathbb{Q}(\theta) = P \times \{\theta\}$. Choose a countable dense in $L_2(\mathbb{R}^2 \times S^1)$ sequence $f_i(x,y;\theta)$ such that f_i continuously depend on θ and for any fixed θ the functions $f_i(x,y;\theta)$ make a dense in $L_2(\mathbb{R}^2)$ sequence. The open sets $P_N(\theta)$ defined similar to the sets M_N (cf. (2)), continuously depend on θ and $\mathbb{Q}_{mix} \cap \mathbb{Q}(\theta) = P_{mix}(\theta) = \bigcap_N P_N(\theta)$. Set $\mathbb{Q}_N(\theta) = P_N(\theta) \times \{\theta\}$ and $\mathbb{Q}_N = \bigcup_\theta \mathbb{Q}_N(\theta)$.

Since $\mathbb{Q}_N(\theta)$ is open in $\mathbb{Q}(\theta)$ for any θ and depends continuously on θ, the set $\mathbb{Q}_N$ is open. The intersection $\mathbb{Q}(\theta) \cap (\bigcup_N \mathbb{Q}_N) = \bigcap_N \mathbb{Q}_N(\theta) = P_{mix}(\theta) \times \{\theta\} = \mathbb{Q}_{mix} \cap \mathbb{Q}(\theta)$, hence, $\bigcap_N \mathbb{Q}_N = \mathbb{Q}_{mix}$ is a G_δ. Since $\mathbb{Q}_{mix} \cap \mathbb{Q}(\theta)$ is dense in $\mathbb{Q}(\theta)$ at least for irrational θ which are dense in S^1, $\mathbb{Q}_{mix}$ is dense in $\mathbb{Q}$.

□

<u>References</u>

[G] E. Gutkin, Billiards on almost integrable polyhedral surfaces, Erg. Th. and Dyn. Syst., <u>4</u>, N4(1984), 569-584.

[H] P. R. Halmos, Lectures on Ergodic Theory, Tokyo, 1956.

[K1] A. Katok, Constructions in Ergodic Theory, Part 1, preprint, to appear in Birkhauser, Progress in Math.

[K2] A. Katok, Ergodicity of generic irrational billiards, Abstracts from workshop on 2-manifolds and Geometry, Oct. 1984, MSRI, Jan. 1986.

[K3] A. Katok, The growth rate for the number of singular and periodic orbits for a polygonal billiard, preprint, 1984.

[K4] A. Katok, Interval exchange transformations and some special flows are not mixing, Israel J. of Math. 35 (1980), 301-310.

[KMS] S. Kerckhoff, H. Masur, J. Smillie, Ergodicity of billiard lows and quadratic differentials, to appear in Ann. of Math.

[S] E. A. Sataev, On the number of invariant measures for flows on orientable surfaces, Math. of the USSR, Izrestija 9(1975), 813-830.

[W] P. Walters, An introduction to ergodic theory, Springer-Verlag, 1982.

[ZK] A Zemlyakov, A. Katok, Topological transitivity of billiards in polygons, Math. Notes of the USSR Acad. Sci. 18 N2(1975), 760-764.

BLOW UP TECHNIQUES IN THE KEPLER PROBLEM

Ernesto A. Lacomba*
Departamento de Matemáticas
Universidad Autónoma Metropolitana, Iztapalapa.
Apdo. Postal 55-534, 09340 México, D.F.

Guillermo Sienra
Instituto de Matemáticas, UNAM and
Instituto de Ciencias, Universidad Autónoma
de Puebla, México.

The purpose of this paper is to make a detailed comparative analysis of the different blow up treatments which have been applied to the Kepler problem, and in general to classical mechanical systems. We compare different sorts of blow up transformations motivated from celestial mechanics. The Kepler problem is the simplest setting for doing that, since it is an integrable problem of celestial mechanics. Projective blow ups turn out to be the best ones for a global description, but of course not any problem admits them.

In section 1 we describe the problem and the two singularities present: a collision and a escape motions singularity.

Section 2 is restricted to the collision singularity. We consider classical transformation like Mc Gehee, Sundman and Levi Civita, together with the projective blow up and their relationships. Since the projective blow up has a definite geometrical meaning by considering the canonical divisor, we interpreted Mc Gehee and Sundman transformations in this setting. For instance, iterating twice the Mc Gehee transformation gives Sundman transformation. The last one is on the other hand, the inverse of the projective transformation.

Section 3 is more algebraic than geometric and we study all transformations changing our energy relation (a cubic equation) into a quadratic one. Among all these, we find that in some sense the simplest one is precisely Mc Gehee transformation.

In section 4 we compactify our vector field to a foliation in $\mathbb{R}P^2$ with two singularities, proving that the escape singularity desingularizes after one projective blow up. The collision singularity is desingularized after five blow ups, and its divisors diagram is shown in Fig. 8. Section 5 is devoted to some conclusions.

---

* Member of CIFMA (México). Research partially supported by PRONAES (México), grant C86-010260.

1. Statement of the problem and description of singularities.

The Kepler problem in one dimension is given by the following system of differential equations in the phase space $\mathbb{R}^+ \times \mathbb{R}$:

$$(1) \qquad \begin{aligned} \dot{x} &= y \\ \dot{y} &= -x^{-2}. \end{aligned}$$

This is a Hamiltonian system of differential equations with Hamiltonian function

$$H(x,y) = y^2/2 - x^{-1}.$$

As it is well known, this is a constant of motion. Hence the level curves

$$(2) \qquad y^2/2 - x^{-1} = h$$

for fixed $h \in \mathbb{R}$, are invariant under the flow (they coincide with the flow up to orientation, since their dimension is one). In figure 1 we describe the phase portrait. Equation (2) is called the energy relation of system (1). A general reference for Hamiltonian flows is (|1|).

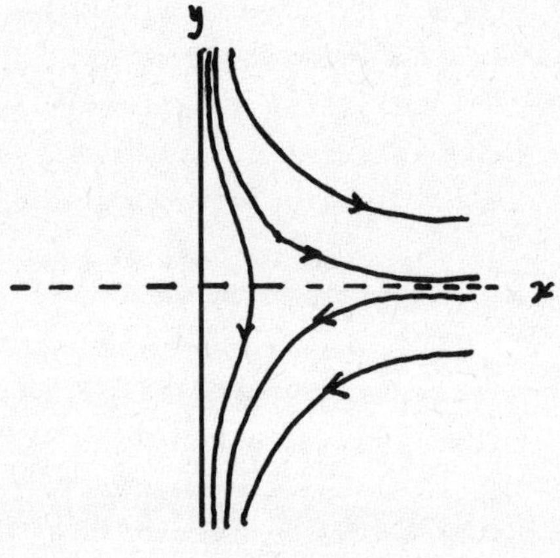

Fig. 1.

The coordinate x in system (1) is interpreted as the position of a particle, while the coordinate y is its velocity. The singularities of the system are $x \to 0$ while $y \to \pm\infty$ and $x \to \pm\infty$ while $y \to \pm\sqrt{2h}$, provided that $h \geq 0$. The first type of singularity is interpreted as collision of the particle with the attracting center $x = 0$, and the second one corresponds to <u>escape</u> motions of the particle. Notice that all singularities appear in the unbounded part of the plane.

2. Geometrical description of transformations.

In this section we are going to describe transformations which permit to study the collision singularity $x \to 0$, $y \to \pm\infty$. The treatment of the escape singularities will be deferred to section 4. We will begin by describing the classical transformations denoted by M, L, ζ, followed by the more geometrical ones π, M, T_x, T_y, R. We will also describe important relationships among them.

M) <u>Mc Gehee transformation</u> .- It corresponds to the coordinate transformation

$$u = x$$
$$v = x^{1/2}y$$

of the half-plane $x > 0$, with time rescaling given by $dt/d\tau = x^{3/2}$, for details see |7,8|. The energy relation (2) in these coordinates is

$$(2\text{-}M) \qquad v^2/2 - 1 = hu.$$

The flow (1) in these new coordinates can be easily computed, and the corresponding phase portrait is given in Figure 2. Observe that this is no longer a Hamiltonian flow

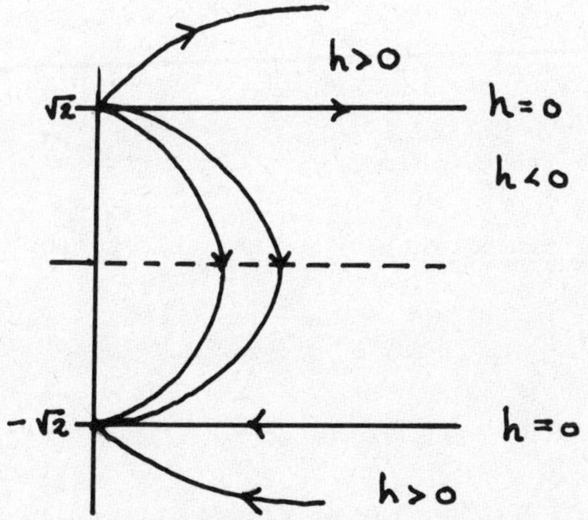

Fig. 2.

ζ) <u>Sundman</u> <u>transformation</u>.- In this case we apply the following coordinate transformation

$$u = x$$
$$v = xy,$$

with a time rescaling given by $dt/d\tau = x$. The energy relation (2) becomes now

$$(2\text{-}\zeta) \quad \frac{1}{2} v^2 - u = h\, u^2,$$

with a corresponding phase portrait given in figure 3.

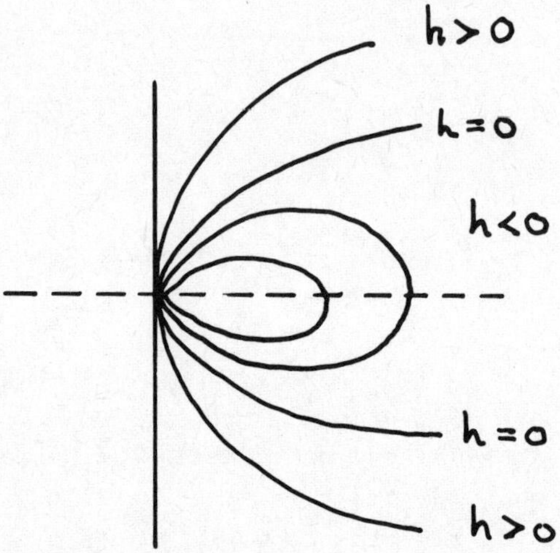

Fig. 3.

As in the above case, this is not a Hamiltonian flow.

ι) <u>Levi</u> <u>Civita</u> <u>transformation</u>.- Here we consider the coordinate transformations

$$u = x^{1/2}$$
$$v = x^{1/2} y$$

with a time rescaling $dt/d\tau = x$. The energy relation (2) becomes

$$(2-L) \qquad \frac{1}{2} v^2 - 1 = hu^2,$$

with a corresponding phase portrait as in Figure 4.

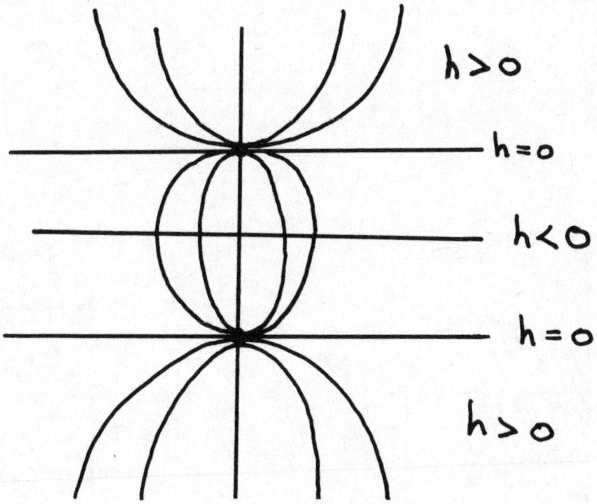

Fig. 4

This system is Hamiltonian on each energy level.

All of the above transformations have sense in celestial mechanics problems with more degrees of freedom. Mc Gehee transformation |8| can be applied to study the total collapse in any n-body problem, while Levi Civita and Sundman transformations (known as regularizations) can be applied to study any collisions of exactly 2 bodies in an n-body problem. See |3,7|.

π) Projective blow up.- This transformation is the standard blow up used in algebraic geometry and in a local coordinate neighborhood of the origin can be written as

$$u = x$$
$$v = y/x.$$

In this case the v axis is known as the canonical divisor of the blow up. Hence, its importance and geometrical meaning must be clear. See | 5, 6 |. Notice that this transformation takes the energy relation into a quintic equation $u^3v^2 - 2 = 2\,hu$.

M) This is defined as a transformation on the half-plane $x > 0$ Whose square under iteration is π, i.e.,

$$u = x$$
$$v = y/\sqrt{x}.$$

Notice that M sends a "parabolic cone" of the origin defined as the set $\{(x,y): y^2 \leq x,\ x > 0\}$ into the set $\{(u,v): |v| \leq 1,\ u > 0\}$, similarly to the way π behaves, where the set $x = 0$ is also a divisor. See Figure 5. It

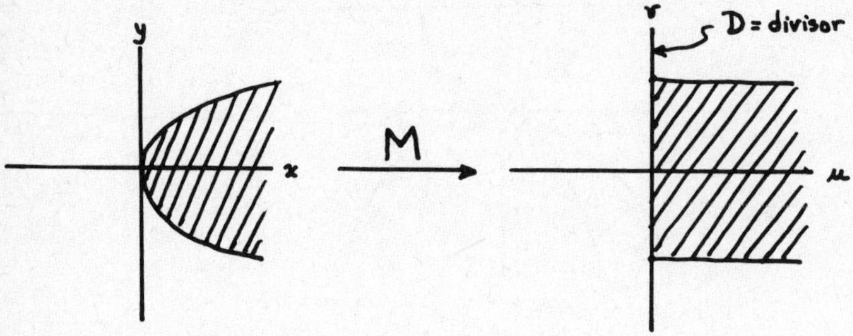

Fig. 5

sends any curve $y = \pm k\sqrt{x}$ into the line $(x,\pm k)$ for a given constant $k \geq 0$.

T_x, T_y) These transformations are the inversions in x (resp. y) coordinate, sending zero to infinity and viceversa:

$$T_x \begin{cases} u = 1/x \\ v = y \end{cases} \quad \text{and} \quad T_y \begin{cases} u = x \\ v = 1/y \end{cases}$$

They have a very clear geometrical meaning.

R) The next geometric transformation on the half-plane $x > 0$ is the one taking square root to the first coordinate:

$$R \begin{cases} u = x^{1/2} \\ v = y \end{cases}$$

From the above definitions, we obtain the following relationships:

(a) $\zeta = T_y \circ \pi \circ T_y = \pi^{-1}$

(b) $\mathcal{M} = T_y \circ M \circ T_y = M^{-1}$

(c) $\mathcal{M}^2 = \zeta$, or equivalently $M^2 = \pi$

(d) $\mathcal{L} = \zeta \circ R$.

These equations relate the classical transformations to the projective ones.

It is important to analyze the geometrical process for obtaining ζ by means of equation a). The inversion T_y sends the singularity $y \to \infty$ to the origin, and the level curves (2) into the cubics $x/2 - y^2 = h\,xy^2$, all of them passing through the origin. When π is applied, the cubics become $1/2 - uv^2 = hu^2v^2$. Notice that the axis $u = 0$ is the divisor of the transformation. Finally, applying T_v implies an exchange of the origin of the divisor with its infinity, and the energy relation becomes $v^2/2 - u = hu^2$. Hence, the vertical axis in Figure 3 can be interpreted as a divisor.

We remark that the standard blow up at ∞ in algebraic geometry is $\pi \circ T_y$. However, the reason for applying an extra T_v in ζ is that the neighborhood we choose to apply π does not contain the non zero part of the vertical axis, loosing information at infinity after π is applied.

The transformation $\mathcal{M}$ has a similar geometrical meaning. In place of the geometrical blow up π, it uses its square root M, as can be seen from c). As before, the vertical axis in Figure 2 can be interpreted as a divisor.

3. Mc Gehee transformation as a generator of blow ups.

As we see from the equations (2-ζ), (2-$\mathcal{L}$), (2-$\mathcal{M}$) in last section, the classical transformations of Sundman, Levi Civita and Mc Gehee take the energy relation (2), essentially a cubic polynomial, into a quadratic equation. This is a significant simplification, since the singularities of quadratic equations are simpler, even more if the equation has a special form. This remark motivates, the following proposition, which characterizes an important class of transformations whose energy relation is quadratic.

Let $p_1(u,v)$, $p_2(u,v)$, $p_3(u,v)$ be monic polynomials of degree at most two in 2 variables, such that

1) $p_1 \neq p_j$ for $i \neq j$

2) at most one of them is a constant

3) p_1/p_3 and p_2/p_3 are algebraically independent.

Let us define the set

$\tau = \{$words in S, R_y, $M = M^{-1}$, T_y, Q_y under composition, with M appearing at most twice$\}$, where $Q_y = R_y^{-1}$ and $S(u,v) = (v,u)$.

Proposition.- The energy relation $y^2 - 1/x = h$ takes the form $p_1 - p_2 = hp_3$ with p_1, p_2, p_3 as above, by applying a transformation in ζ.

Proof: The transformation $T(x,y) = (u,v)$ sending the equation $y^2 - \frac{1}{x} = h$ into the equation $p_1 - p_2 = hp_3$ has the property that $y^2 = p_1/p_3$ and $x = p_3/p_2$. Hence the possible pairs of the form $(p_1/p_3, p_2/p_3)$ with p_1, p_2, p_3 satisfying the above conditions are (u,v), (u,v^2), (u^2,v), (u^2,v^2), (u^2,uv), (uv,v), (uv,u^2), $(u/v,1/v)$, $(u/v,v)$, $(u/v,u)$, $(u^2/v,1/v)$, $(u^2/v,v)$, $(u^2/v,u)$, $(u,1/v)$, $(u,u/v)$, (u,u^2v), $(u/v^2,1/v^2)$, $(u/v^2,1/v)$, $(u^2/v^2,1/v^2)$, $(u^2/v^2,1/v)$, $(uv,1/v^2)$, $(1/v,1/uv)$, $(1/v,1/u)$, $(u/v,1/u)$, $(u^2/v,v)$ and $(u/v^2,u/v)$. All of them except the last two have an easy descomposition into a word in τ, which the reader can easily check. Regarding the last two, we have the following corresponding descompositions, respectively:

$$T_y \circ S \circ T_y \circ S \circ M \circ R_y \circ S \circ M$$

and $$Q_y \circ S \circ R_y \circ M \circ R_y \circ S \circ M.$$ This concludes the proof.

Q.E.D.

On the other hand, not any word in τ takes the energy relation to a quadratic form.

From this point of view, think of S, R_y, T_y, Q_y as trivial transformations. Then the simplest non trivial word in $\mathcal{I}$ is $M = M^{-1}$, the Mc Gehee transformation, which can be thought as a building block transformation. The next ones in "order of complexity" are $\zeta = M^2$ and $L = M^2 \circ R_x$, since none of the length two words except M^2 transform (2) into a quadratic equation.

4. Projective blow up of the two singularities

In this section we are going to study projective blow up transformations (as transformation π in Section 3), in order to desingularize the foliation described by Equation (1) extended to the projective space $\mathbb{R}P^2$, which is a compact space.

Let us consider the projective compactification $\mathbb{R}P^2$ of $\mathbb{R}^2$ and complete the foliation as follows:

If $x = 0$, add the leaf consisting of the y axis. If $x < 0$, add the leaves with equation

$$\frac{1}{2} y^2 + \frac{1}{|x|} = h$$

(this is interpreted as repulsive electrostatic problem). In figure 6 we show the foliation in $\mathbb{R}^2$.

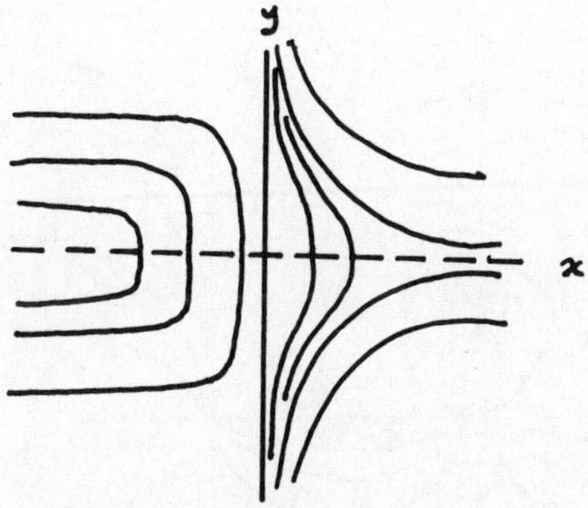

Fig. 6.

Let us consider the transformation $T: \mathbb{R}P^2 \to \mathbb{R}P^2$ which in cartesian coordinates is written as $(X,Y) = \big((x+y)/(-x+y),\ \sqrt{2}/(-x+y)\big)$. This transformation sends the y-axis direction to the line $X = 1$, and the x-axis direction to the line $X = -1$. The energy relation (2) in these coordinates becomes

$$(3) \qquad (X+1)^2(X-1) - 4\sqrt{2}\, Y^3 - 4hY^2(X-1) = 0$$

Substitution of the equation $x = (y^2/2 - h)^{-1}$ obtained from (2) into the image of T, gives

$$\left(\frac{-2 + (2h - y^2)y}{2 + (2h - y^2)y} \,,\quad \frac{\sqrt{2}\,(2h - y^2)}{2 + (2h - y^2)y} \right).$$

As $y \to \infty$, this expression goes to $(1,0)$, and as $x \to \infty$ (i.e. $y \to \pm\sqrt{2h}$) the expression goes to $(-1,0)$.

The line at infinity in the original coordinates, maps by T into the line $Y = 0$. It can be thought as the limiting leaf when $h \to +\infty$, since it is contained in the $h > 0$. The line $X = +1$ is not strictly any limiting level curve, since it bounds regions with $h \to +\infty$ and $h \to -\infty$. Nevertheless, we add it as another leaf of the foliation.

This two new leaves can be read out from (3) if we divide by h and make $h \to \pm\infty$. Thus, the original foliation is actually completed to a foliation with two singularities in $\mathbb{R}P^2$, which we will call F. This foliation and its two singularities is shown in figure 7 by using the X, Y coordinates

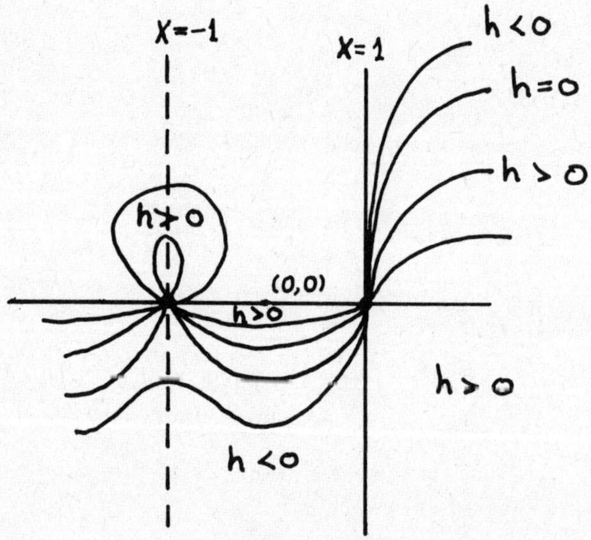

Fig. 7.

Besides the foliation, we are dealing with a vector field which in our new coordinates X, Y can easily be computed to be

$$(4)\qquad \begin{aligned} \dot{X} &= (X+1)^2/2 + \sqrt{2}\, Y^3/(X-1) \\ \dot{Y} &= Y(X+1)/2 + \sqrt{2}\, Y^4/(X-1)^2 \end{aligned}$$

This rational vector field is zero at the point (-1,0). Rescaling the time by a factor $(X-1)^2$, we see that it also becomes zero at the point (1,0).

Our following theorems deal with the desingularization of this vector field at the two singularities. This is done by a successive application of a projective blow up transformation π, as defined in Section 3. In fact, we prove that after a finite number of steps both singularities are desingularized.

Theorem 1. The singularity (-1,0) of the foliation F is desingularized after one blow up. The underlying vector field (4) also desingularizes in a neighborhood of the divisor.

Proof. In order to simplify the computations, let us choose more convenient coordinates

$$X' = (x+1)/(-x+1), \quad Y' = \sqrt{2}y/(-x+1).$$

they still bring the escape singularity to the point (-1,0), but the x-axis goes into the X'-axis. The other singularity remains in the unbounded part of the plane, and the foliation near the singularity in these coordinates looks as near (-1,0) in Figure 7, after a clockwise rotation of 90°.

Again from (2) we have $x = (y^2/2 - h)^{-1}$. Hence by substitution we get

$$X' = \frac{+2 + y^2 - 2h}{-2 + y^2 - 2h}, \quad Y' = \frac{\sqrt{2}\, y(y^2 - 2h)}{-2 + y^2 - 2h}$$

In order to apply π we first translate (-1,0) to the origin, getting new coordinates which we denote again by X', Y':

$$X' = \frac{-2 + 2h - y^2}{2 + 2h - y^2} + 1, \quad Y' = \frac{(2h - y^2)\sqrt{2}\, y}{2 + 2h - y^2}$$

Since π is defined by $\pi(X', Y') = (X', Y'/X')$, we obtain $\pi(X', Y') = \left(2(2h - y^2)(2 + 2h - y^2)^{-1}, y/\sqrt{2}\right)$. Taking the limit as $y \to \pm\sqrt{2h}$, $\pi(X',Y') \to (0, \pm\sqrt{h})$, which proves that the foliation desingularizes the horizontal cone neighborhood $\{(X',Y'): |Y'| \le |X'|\}$. Similary, with the transformation π' defined by $\pi'(X',Y') = (X'/Y',Y')$, we obtain that $\pi'(X',Y') \to (\pm 1/\sqrt{h},0)$ in the limit as $y \to \pm\sqrt{2h}$. This shows that the foliation desingularizes partially in the complementary cone neighborhood $\{(X',Y'): |Y'| \ge |X'|\}$. However, it is not well behaved in the zero level $h = 0$, which is pushed out to infinity. Hence, π is the good blow up to cover $h = 0$, while π' is good to cover $h = \pm\infty$.

After applying π, the system (1) in a neighborhood of the divisor in $\bar{X} = X'$, $\bar{Y} = Y'/X'$ coordinates can easily checked to be

$$\dot{\bar{X}} = \bar{Y}\bar{X}^2/\sqrt{2}$$

$$\dot{\bar{Y}} = -\bar{X}^2/\sqrt{2}(\bar{X} - 2)^2.$$

Evaluated at the point $(0, \pm\sqrt{h})$ gives a zero value, but rescaling by a $1/\bar{X}^2$ factor and evaluating at the same point, we get

$$\dot{\bar{X}} = \sqrt{h/2}\ , \quad \dot{\bar{Y}} = -\sqrt{2}/8.$$

Similary in the other cone neighborhood defined by π'.

Q.E.D.

Theorem 2.- The singularity (1,0) of the foliation F is desingularized after 5 blow ups. The sequence of divisors is as in Figure 8.

Proof.- In this case we use the more convenient coordinates

$$X' = x\sqrt{2}/(-y + 1) + 1, \quad Y' = 2/(-y + 1).$$

They still bring the collision singularity to the point (1,0) with the vertical axis going into the line $X' = 1$. Again the other singularity remains in the unbounded part of the plane.

Using again the formula $x = (y^2/2 - h)^{-1}$ and translating (1,0) to the origin, we get

$$X' = 2\sqrt{2}\ (y^2 - 2h)^{-1}(-y+1)^{-1}, \quad Y' = 2/(-y+1).$$

Since the foliation is tangent to the vertical axis, we have to begin applying the blow up π' in the vertical cone neighborhood:

1) $\quad \pi'(X',Y') = \left(\sqrt{2}/(y^2 - 2h),\ 2/(-y+1)\right).$

This tends to (0,0) in the limit as $y \to \pm\infty$, so the singularity persists.

Now we will write the results of applying successive blow ups on the left side of the following table, while on the right side appear the limit points when y tends to the singularity

2) $\left(\dfrac{-y+1}{\sqrt{2}(y^2 - 2h)}, \dfrac{2}{-y+1}\right) \qquad (0,0)$

3) $\left(\dfrac{(-y+1)^2}{2\sqrt{2}(y^2 - 2h)}, \dfrac{2}{-y+1}\right) \qquad \left(\dfrac{1}{2\sqrt{2}}, 0\right)$

4) $\left(\dfrac{1 + 2h - 2y}{2\sqrt{2}(y^2 - 2h)}, \dfrac{4\sqrt{2}(y^2 - 2h)}{(-y+1)(1+2h-2y)}\right) \qquad (0, 2\sqrt{2})$

5) $\left(\dfrac{(-y+1)(1+2h-2y)}{8[(3+2h)y - (1+6h)](y^2 - 2h)}, \dfrac{2\sqrt{2}[(3+2h)y - (1+6h)]}{(-y+1)(1+2h-2y)}\right)$

$$\left(-\frac{1}{2(3+2h)}, 0\right)$$

After each blow up, we translated the corresponding singular point to the origin.

Notice that since the degree of the denominator of the X' coordinate is decreasing by 1 in each step when π' is applied, we had to switch to π in step 4.

From the above list, it follows that the sequence of divisors of successive blow ups is as in Figure 8. This completes the proof.

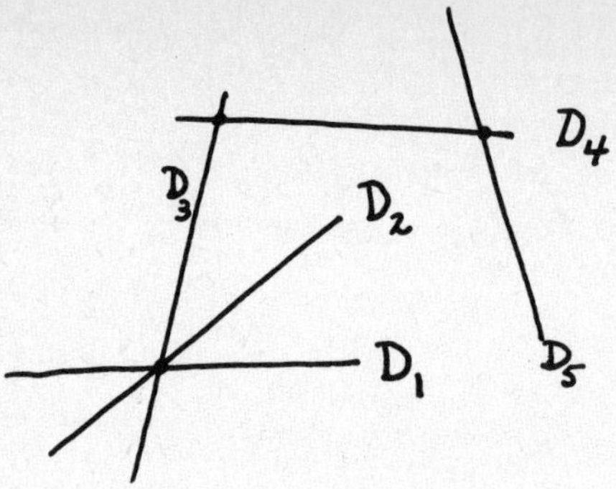

Fig. 8.

Q.E.D.

After a time rescaling, one checks that the final blow up vector field along the divisor D_5 (horizontal axis) is indeed non zero.

Topologically, for each blow up, the resulting manifold is in this case a connected sum of $\mathbb{R}P^2$ spaces. Hence, after 6 blow ups we obtain the non orientable compact manifold of 7 connected sums:

$$\mathbb{R}P^2 \# \mathbb{R}P^2 \# \ldots \# \mathbb{R}P^2,$$

and the divisors $D_1, \ldots, D_5$ are all topologically S^1.

5. Conclusions.- An important conclusion of section 3 is the minimality of Mc Gehee transformation M as a building block.

On the other hand, observe that classical blow ups do not separate energy levels, while as we have seen in Section 4, projective transformations permit to separate them eventually, giving a non singular global foliation in a certain manifold. But of course, the corresponding global vector field is singular for topological reasons.

An information about the singularity that we obtain from the blowing up process is the degree of tangency with which leaves aproach each other in a neighborhood of the singularity . In our case we can see from Section 4 that as $y \to \pm\infty$ in (2), the leaves aproach the y-axis with degree of tangency 5.

Finally, we can say that there is not a complete theory for studying blow ups of vector fields in dimension bigger than 2, see |2|, |4|.

Acknowledgement.- We would like to thank fruitful discussions with Felipe Cano (Spain).

Bibliography.

1. R. Abraham and J. Marsden, Foundations of Mechanics, Benjamin/ Cummings, Reading Mass., 1978.

2. C. Camacho, A. Lins Neto and P. Sad, Topological invariance and desingularization for holomorphic vector fields, J. Differential Geometry 20 (1984) 143-174.

3. R. Devaney, Singularities in Classical Mechanical Systems, in Ergodic Theory and Dynamical Systems I, (A. Katok, Ed.), Birkhauser, Basel p. 211, 1981.

4. F. Dumortier, R. Roussarie and J. Sotomayor, Generic 3-parameter families of vector fields in the plane, unifolding a singularity with nilpotent linear part, preprint 1986.

5. W. Fulton, Algebraic curves, an Introduction to Algebraic Geometry, W.A. Benjamin, New York 1969.

6. P. Griffiths and J. Harris, Principles of Algebraic Geometry, John Wiley, New York, 1978.

7. E. Lacomba, Blow up on energy levels in celestial mechanics, Publicacions Secció de Matematiques, Univ. Auton. de Barcelona, 28 (1984) 97-117.

8. Mc Gehee, Triple collisions in the colinear three body problem, Inventiones Math 27 (1974) 191-227.

Algebraic solutions of polynomial differential equations and foliations in dimension two

by

Alcides Lins Neto

§1. Introduction

In this paper we will consider holomorphic singular foliations by curves in $\mathbb{C}P(2)$, which in affine coordinates are defined by differential equations of the form

$$(1) \qquad \frac{dx}{dT} = P(x,y), \qquad \frac{dy}{dT} = Q(x,y)$$

where P and Q are polynomials in two complex variables. Another way of writing equation (1) is,

$$(1') \qquad P(x,y)dy - Q(x,y)dx = 0.$$

We say that $(x_o,y_o) \in \mathbb{C}^2$ is a singularity of (1) or (1') if $P(x_o,y_o) = Q(x_o,y_o) = 0$. We will denote by $\mathrm{sing}(\mathcal{F})$ the set of all singularities of the foliation $\mathcal{F}$ in $\mathbb{C}P(2)$, induced by (1) or (1'). Since we are interested in the algebraic solutions of $\mathcal{F}$, but not in their parametrization by the complex time, we will consider the associated equation always written as in (1'). Observe that if P and Q have a common factor, say $P = f.P_1$ and $Q = f.Q_1$, then the algebraic curve $\{(x,y);\ f(x,y) = 0\}$ is a solution of (1'). We will not consider this type of solution, so that we can cancel out the factor f in (1'), obtaining a <u>simplified</u>

equation $Q_1 dx - P_1 dy = 0$, where P_1 and Q_1 have no common factor. We will consider only this kind of equation.

1.1 - Algebraic solutions of a foliation. Let $\mathfrak{F}$ be a singular foliation on $\mathbb{C}P(2)$ and $S \subset \mathbb{C}P(2)$ be an irreducible algebraic curve. We say that S is a solution of $\mathfrak{F}$, if $S\text{-sing}(\mathfrak{F})$ is a leaf of $\mathfrak{F}$. This condition can be written in an affine coordinate system, as follows: suppose that $\mathfrak{F}$ is represented in this coordinate system by the equation $P(x,y)dy - Q(x,y)dx = 0$ and S by $\{(x,y); f(x,y) = 0\}$, where P, Q and f are polynomias, f irreducible. It is not difficult to see that $S\text{-sing}(\mathfrak{F})$ is a leaf of $\mathfrak{F}$, if and only if

$$(2) \qquad \frac{\partial f}{\partial x} \cdot P + \frac{\partial f}{\partial y} \cdot Q = g \cdot f$$

where g is some polynomial. In this definition we are not excluding the case where S has some singularities. It follows, from the definition, that all singularities of S are also singularities of $\mathfrak{F}$. However, it can happens that some singularities of $\mathfrak{F}$ are smooth points of S. In the example $axdy-bydx = 0$, $a \neq 0 \neq b$, the point $(0,0)$ is a singularity of $\mathfrak{F}$, but is not a singularity of the algebraic solution $S = \{(x,y); y = 0\}$.

1.2 - Statement of the results. Let S be an algebraic solution of a singular foliation $\mathfrak{F}$ on $\mathbb{C}P(2)$. Let $p \in S$ be a singularity of $\mathfrak{F}$. Suppose that S is defined in some coordinate system $\varphi = (x,y)\colon U \to \mathbb{C}^2$ by $f(x,y) = 0$, where $p \in U$ and $x(p) = y(p) = 0$. If p is a singularity of S and U is small enough, then we can decompose f as a product of locally irreducible analytic functions $f = f_1 \dots f_k$ (cf. [1]).

We call the set $B_j = \{q \in U;\ f_j(q) = 0\}$ a <u>local branch of</u> S <u>at</u> p and f_j the <u>equation of the branch</u> B_j. When $\frac{\partial f_j}{\partial x}(0,0) \neq 0$ or $\frac{\partial f_j}{\partial y}(0,0) \neq 0$, we say that B_j is a <u>smooth branch</u>. When p is not a singularity of S then there is an unique branch of S at p, which is smooth.

In the next section we will associate to each singularity p of $\mathfrak{F}$, such that $p \in S$, and each local branch B of S at p a complex number $i(B,\mathfrak{F})$, the <u>index of</u> B <u>with respect to</u> $\mathfrak{F}$. This number will be defined in terms of some residue. We will use the notation $C(S,\mathfrak{F})$ for the sum $\sum_{B \in \mathcal{B}} i(B,\mathfrak{F})$, where $\mathcal{B}$ denotes the set of all local branches of S through singularities of $\mathfrak{F}$. The main result of this paper is the following:

<u>Theorem A</u>. <u>Let</u> $\mathfrak{F}$ <u>be a singular foliation of</u> $\mathbb{C}P(2)$ <u>and</u> S <u>be an algebraic irreducible solution of</u> $\mathfrak{F}$. <u>Then</u>

$$C(S,\mathfrak{F}) = 3.dg(S) - \chi(S) + \sum_{B \in \mathcal{B}} \mu(B). \qquad (3)$$

In (3), $dg(S)$ denotes the degree of S in $\mathbb{C}P(2)$, $\mu(B)$ the <u>Milnor number of the local branch</u> B and $\chi(S)$ the <u>intrinsic Euler characteristic of</u> S. The Milnor number of the branch B is defined as follows: let $g\colon U \to \mathbb{C}$ be an equation of B, where $p \in B \subset U$ is the singularity of $\mathfrak{F}$. Then $\mu(B)$ is, by definition the intersection number of $\partial g/\partial x$ and $\partial g/\partial y$ at p. It can be proved that $\mu(B) = \dim_{\mathbb{C}} \mathcal{O}(2,p)/(\partial g)$, where $\mathcal{O}(2,p)$ is the ring of germs of holomorphic functions at p and (∂g) is the ideal generated by $\partial g/\partial x$, $\partial g/\partial y$ (cf. [2]). The intrinsic Euler characteristic of S is defined as follows: it is possible to prove that there exist a compact Riemann surface S' and a holomorphic map $h: S' \to S$ such that h is a homeomorphism outside a finite set $F \subset S'$, where $h(F) \subset \mathrm{sing}(S)$, the singular set of S. This fact can be proved by applying the blowing-up method at each singularity of S (cf. [7]). The intrinsic Euler characteristic of S is, by definition, the Euler characteristic of S'.

Remarks

1. When S has no singular points, Theorem A follows from a result of Camacho-Sad and from the genus formula (cf. [3] and [1]). As we will see in the next section, the index of singularity of $\mathfrak{F}$ in a smooth point of S, coincides with the index defined by Camacho and Sad in [3]. In [3] they prove the following result:

Theorem [3]. *Let* M *be a complex manifold of dimension* 2, $S \subset M$ *be a smooth embedded compact Riemann surface and* $\mathfrak{F}$ *be a singular foliation of dimension* 1, *defined in a neighborhood of* S. *Suppose that* S *is invariant by* $\mathfrak{F}$. *Then:*

$$\sum_{p \in S \cap \mathrm{sing}(\mathfrak{F})} i(p,\mathfrak{F},S) = c(S),$$

where $c(S)$ *is the Chern class of the normal bundle of* S *and for each* $p \in S \cap \mathrm{sing}(\mathfrak{F})$, $i(p,\mathfrak{F},S)$ *is the index of* $\mathfrak{F}$ *and* S *at* p.

When $S \subset \mathbb{C}P(2)$ is smooth, it is well known that $c(S) = d^2$, where $d = dg(S)$. In this case, the last term in the right side of (3) is zero, so that we have

$$c(S) = d^2 = 3d - \chi(S) \Leftrightarrow g(S) = \frac{1}{2}(2-\chi(S)) =$$
$$= \frac{1}{2}(d^2-3d+2) = \frac{1}{2}(d-1)(d-2)$$

which is the genus formula.

We would like to remark also that Camacho-Sad Theorem will be used in the proof of Theorem A.

2. Since the right side of (3) does not depend of $\mathcal{F}$, it follows that $C(S,\mathcal{F})$ depends only on S and in the way in which S is immersed in $\mathbb{C}P(2)$.

3. The number $C(S,\mathcal{F})$ is a positive integer for all S and $\mathcal{F}$ for which S is an algebraic solution. Moreover $C(S,\mathcal{F}) = 1$ if, and only if, S is a line in $\mathbb{C}P(2)$. This implies, in particular, that an algebraic solution of $\mathcal{F}$, contains at least one singularity of $\mathcal{F}$.

Proof: We have $dg(S) \geq 1$, $\chi(S) \leq 2$ and $\mu(B) \geq 0$ for all branchs B, through singularities of $\mathcal{F}$. Therefore

$$C(S,\mathcal{F}) = 3\, dg(S) - \chi(S) + \sum_{B\in\mathcal{B}} \mu(B) \geq 3-2 = 1.$$

Since the three terms in the right side of (3) are integers, $C(S,\mathcal{F})$ must be an integer. Moreover, $C(S,\mathcal{F}) = 1$, implies that $\sum_{B\in\mathcal{B}} \mu(B) = 0$ and $3dg(S) = 1 + \chi(S) \leq 3$. Hence $dg(S) = 1$, which implies that S is a line in $\mathbb{C}P(2)$.

Two interesting problems arise from Theorem A:

1st - Generalize Theorem A for highter dimensions.

2nd - Generalize Theorem A for foliations by curves in a compact 2-dimensional complex manifold.

In §1.4 we will define the degree of a foliation in $\mathbb{C}P(2)$ and the space $\mathfrak{X}_n$ of foliations of degree n.

Theorem A will be used to prove the following result:

Theorem B. For all $n \geq 2$, there exists an open and dense set $\mathcal{U}_n \subset \mathfrak{X}_n$, such that if $\mathcal{F}\in\mathcal{U}$, then $\mathcal{F}$ has no algebraic solution.

Remark 4. The fact that for $n \geq 2$, $\mathfrak{X}_n$ contains a generic set of foliations without algebraic solutions, was already known (cf.[4]). As far as I know, the fact that it contains an <u>open set</u> of foliations without algebraic solutions was not known. I would like to observe that the definition of degree of a foliation in [4], does not coincide with our definition. In fact, if $\hat{dg}(\mathfrak{F})$ is the degree defined in [4], we have $\hat{dg}(\mathfrak{F}) = dg(\mathfrak{F})+1$.

<u>Remark 5</u>. In §3.4 we will sketch the proof of an analogous of Theorem B, in the real case.

1.3 - <u>The index of a separatrix with respect to a foliation</u>.

Let $\mathfrak{F}$ be a singular foliation by curves on a complex 2-dimensional manifold M. Let $p \in M$ be a singularity of $\mathfrak{F}$. Since M has dimension 2, p is necessarily an isolated singularity. It happens that $\mathfrak{F}$ can be represented in a small neighborhood of p, say U, by an equation $\omega = 0$, where ω is some holomorphic 1-form in U such that $\omega(q) = 0$ if and only if $q = p$ and the leaves of $\mathfrak{F}$ in $U-\{p\}$ are integrals of the equation (cf. [5]). If we consider an analytic coordinate system in U, say $\varphi = (x,y): U \to \mathbb{C}^2$, where $\varphi(p) = (0,0)$, then equation $\omega = 0$ can be written as $P(x,y)dy-Q(x,y)dx = 0$, where P and Q are analytic functions in $V = \varphi(U) \subset \mathbb{C}^2$, such that $P(0,0) = Q(0,0) = 0$. In [3], Camacho and Sad prove that the equation $\omega = 0$ has always an analytic solution through p, if U is small enough. We call this type of solution, a <u>separatrix of $\mathfrak{F}$ through</u> p. A separatrix has an equation f=0, where f is irreducible and f(p) = 0. Observe that {f=0}, where f is irreducible, is an analytic solution of ω=0, if and only if

(4) $$df \wedge \omega = (\frac{\partial f}{\partial x}\cdot P + \frac{\partial f}{\partial y}\cdot Q) = dx \wedge dy = f.g\, dx \wedge dy,$$

where $g: U \to \mathbb{C}$ is analytic. This condition is analogous to (2) in §1.1. It is equivalent to the following: there exist germs at $(0,0)$, h, k and α, where h, k are analytic functions and α an analytic 1-form, such that k and f are relatively primes in the local ring $\mathcal{O}(2,p)$, $df \wedge \alpha \not\equiv 0$ or $\alpha \equiv 0$, and

(5) $\qquad k.\omega = hdf + f\alpha.$

In fact, if (5) is true then

$$kdf\wedge\omega = k\left(\frac{\partial f}{\partial x} P + \frac{\partial f}{\partial y} Q\right)dx\wedge dy = fdf\wedge\alpha = f.h_1\ dx\wedge dy,\ h_1\not\equiv 0, \text{ if } \alpha\not\equiv 0.$$

This implies that k divides $f.h_1$. Since k and f are relatively primes we must have $h_1 = k.g$, which implies (4).

Let us suppose now that (4) is true. Suppose also that $\{f=0\} \neq \{x=0\}$. In this case we can put $k = \partial f/\partial y$, $h = P$ and $\alpha = -gdx$ and it is not difficult to see that (4) $\Rightarrow$ (5).

I would like to observe that the decomposition in (5) is not unique, as the reader can easily verify.

Now, since f is irreducible, the analytic set $B = \{f=0\}$, has a Puiseaux's parametrization (cf. [6]) $\psi: D_1 \to B$, where $D_1 = \{T \in \mathbb{C};\ |T| < 1\}$ and ψ is a homeomorphism onto $\psi(D_1) \subset B$. This parametrization has an expression of the form $\psi(T) = (x(T),y(T))$, $x(T) = T^m \varphi_1(T)$, $y(T) = T^n \varphi_2(T)$, where $\varphi_1(0) \neq 0 \neq \varphi_2(0)$ and either $m=n$, or m and n are relatively primes. Let $\gamma(f) = \gamma(B)$, be the homology class in $H_1(B-\{0\},Z)$ of the curve $\theta \mapsto \psi(re^{i\theta})$, $0 \leq \theta \leq 2\pi$, where $0 < r < 1$.

Definition. The index of $\mathfrak{F}$ with respect to the local branch B of S at p, is by definition the residue

$$(6) \qquad i(B,\mathfrak{F}) = \frac{-1}{2\pi i}\int_{\gamma(B)} \frac{\alpha}{h}$$

where α and h are as in (5).

Remark. Observe that (6) is well defined because f does not divide h. In fact we have the following:

Lemma 1. The number defined in (6), depends only on B and $\mathfrak{F}$.

Proof: Let us prove first that (6) does not depend on the decomposition in (5). Suppose that $k_1.\omega = h_1 df + f\alpha_1$ is another decomposition. We must have

$$\frac{df}{f} = \frac{k}{f.h}\,\omega - \frac{\alpha}{h} = \frac{k_1}{f.h_1}.\omega - \frac{\alpha_1}{h_1} \Rightarrow (kh_1-hk_1)\omega = f(h_1\alpha - h\alpha_1).$$

Since f is irreducible and does not divide both components of ω, this implies that $kh_1-hk_1 = h_2.f$. Therefore

$$\frac{\alpha}{h} - \frac{\alpha_1}{h_1} = \frac{h_2}{h.h_1}\,\omega.$$

Now, since $\omega|_B \equiv 0$, this last equation implies that

$$\int_{\gamma(B)} \frac{\alpha}{h} = \int_{\gamma(B)} \frac{\alpha_1}{h_1}.$$

Let us prove that (6) does not depend on the equation which represents $\mathfrak{F}$ near p. Suppose $\omega_1 = 0$ is another equation representing $\mathfrak{F}$ near p. This implies that $\omega_1 = \lambda.\omega$ where $\lambda(0) \neq 0$. Therefore, if $kw = hdf + f\alpha$ is a decomposition for ω, then $k_1\omega_1 = h_1 df + f\alpha_1$ is a decomposition for ω_1, where $k_1 = \lambda k$, $h_1 = \lambda h$ and $\alpha_1 = \lambda\alpha$. This implies that $\alpha_1/h_1 = \alpha/h$, and so (6) does not depend on the equation of $\mathfrak{F}$ near p.

Finally, let us prove that (6) does not depend of the equation of B. Let $f_1 = 0$ be another irreducible equation of B. We must have $f = \lambda . f_1$, where $\lambda(0) \neq 0$. If $k\omega = hdf + f\alpha$ is a decomposition for ω (in terms of f), then $k\omega = \lambda hdf_1 + f_1(\lambda\alpha + hd\lambda)$ is decomposition for ω (in terms of f_1). Now,

$$\frac{1}{\lambda h}(\lambda\alpha + hd\lambda) = \frac{\alpha}{h} + \frac{d\lambda}{\lambda}.$$

Since $\lambda(0) \neq 0$, we must have $\int_{\gamma(B)} \frac{d\lambda}{\lambda} = 0$, which implies the lemma. ∎

<u>Particular cases</u>:

1. Let us suppose that $\mathcal{F}$ has a smooth invariant branch, say B, through the singularity $(0,0) \in U \subset \mathbb{C}^2$. In this case we can suppose that $B = \{(x,y) \in U;\ y=0\}$, and $\mathcal{F}$ can be represented by an equation of the form $\omega = Pdy - yqdx = 0$ in U, where $P,q\colon U \to \mathbb{C}$ are analytic. Therefore a decomposition of ω relative to B, can be given as $\omega = hdy + y\alpha$, where $h = P$ and $\alpha = -qdx$. From this, we get

$$i(B,\mathcal{F}) = -\frac{1}{2\pi i}\int_{\gamma(B)} \frac{\alpha}{h} = \frac{1}{2\pi i}\int_{\gamma(B)} \frac{q(x,0)}{P(x,0)}\,dx =$$

$$= \operatorname{Res}\left(\frac{q(x,0)}{P(x,0)},\ x=0\right) = \operatorname{Res}\left(\frac{\partial}{\partial y}\left(\frac{yq}{P}\right)_{(x,0)},\ x=0\right).$$

This coincides with the definition of Camacho and Sad in [3], for the smooth case. Since a non singular point $p \in S$ has just one branch of S through it, they use the notation $i(B,\mathcal{F}) = i_p(\mathcal{F},S)$.

In particular, if we represent $\mathfrak{F}$ by the differential equation $\dot{x} = P(x,y)$, $\dot{y} = Q(x,y)$, and the Jacobian of (P,Q) at $(0,0)$ has eigenvalues $\lambda_1 \neq 0$ and λ_2, where the eigen-space relative to λ_1 coincides with the tangent space of B at $(0,0)$, then $i(B,\mathfrak{F}) = \lambda_2/\lambda_1$ (cf. [3]).

2. Let us suppose that $\mathfrak{F}$, in a neighborhood U of $(0,0) \in \in \mathbb{C}^2$, is the foliation whose leaves are the level surfaces of some analytic function $f: U \to \mathbb{C}$, where $f(0,0) = 0$ and $f \not\equiv 0$. Let us suppose that $f^{-1}(0)$ has branches $B_1,\dots,B_k$ through $(0,0)$, where $B_j = f_j^{-1}(0)$, $1 \leq j \leq k$, and $f = f_1 \dots f_k$. If $k > 1$, let us denote by $[B_i,B_j]_0 = [f_i,f_j]_0$, the intersection number of f_i and f_j at $(0,0)$.

<u>Assertion</u>. <u>If</u> $k=1$, <u>then</u> $i(B_1,\mathfrak{F}) = 0$. <u>If</u> $k > 1$, <u>then</u>

$$i(B_j,\mathfrak{F}) = - \sum_{\substack{i=1 \\ i\neq j}}^{k} [B_i,B_j]_0 .$$

If $k > 1$, then $df = \sum_{i=1}^{k} f_1 \dots \hat{f}_i \dots f_k \, df_i$ and a decomposition for the branch $B_j = f_j^{-1}(0)$ is of course $df = h df_j + f_j \alpha$, where $h = f_1 \dots \hat{f}_j \dots f_k$ and

$$f_j \alpha = \sum_{i\neq j} f_1 \dots \hat{f}_i \dots f_k \, df_i = f_j [\sum_{i\neq j} f_1 \dots \hat{f}_i \dots \hat{f}_j \dots f_k \, df_i].$$

Therefore,

$$-\frac{\alpha}{h} = - \sum_{i\neq j} \frac{df_i}{f_i} \Rightarrow i(B_j,\mathfrak{F}) = - \sum_{i\neq j} \frac{1}{2\pi i} \int_{\gamma(B_j)} \frac{df_i}{f_i} .$$

Now, if $\varphi: D_1 \to B_j$ is a Puiseaux's parametrization of B_j, we have

$$[B_i,B_j]_o = \mathrm{ord}(f_i \circ \varphi(T),\ T=0) = \frac{1}{2\pi i}\int_\gamma \frac{d(f_i \circ \varphi)}{f_i \circ \varphi} = \frac{1}{2\pi i}\int_{\gamma(B_j)} \frac{df_i}{f_i}$$

for any $i \neq j$. This proves the assertion. ■

1.4 - Degree of a foliation.

Let $\mathfrak{F}$ be a singular foliation on $\mathbb{C}P(2)$ and $L \subset \mathbb{C}P(2)$ be a projective line, which is not an algebraic solution of $\mathfrak{F}$. We say that $p \in L$ is a tangency point of $\mathfrak{F}$ with L, if either $p \in \mathrm{sing}(\mathfrak{F})$ or $p \notin \mathrm{sing}(\mathfrak{F})$ and the tangent spaces of L and of the leaf of $\mathfrak{F}$ through p, at p, coincide. This can be expressed as follows: let $((x,y),\mathbb{C}^2)$ be an affine coordinate system in $\mathbb{C}P(2)$ such that $p \in \mathbb{C}^2$. Suppose that $\mathfrak{F}$ can be expressed by the differential equation $P(x,y)dy - Q(x,y)dx = 0$ and $L \cap \mathbb{C}^2$ can be parametrized by $\varphi(t) = (x_o+at,\ y_o+bt)$, in this coordinate system, where $p = \varphi(t_1)$. It is not difficult to see that p is a tangency point of $\mathfrak{F}$ with L, if and only if, t_1 is a root of the polynomial

$$h_L(t) = bP(\varphi(t)) - aQ(\varphi(t)).$$

The multiplicity of tangency of $\mathfrak{F}$ with L at p is, by definition, the multiplicity of t_1 as a root of h. We denote this number by $\#(\mathfrak{F},L;p)$. It is not difficult to prove that $\#(\mathfrak{F},L;p)$ does not depend on the parametrization of $L \cap \mathbb{C}^2$ and is invariant under analytic change of variables. From this fact, we can define the total number of tangencies of $\mathfrak{F}$ with L as

$$\#(\mathcal{F},L) = \sum_{p\in L} \#(\mathcal{F},L;P)$$

where $\#(\mathcal{F},L;p) = 0$ if $p \in L$ is not a tangency point.

<u>Lemma 2</u>. <u>Let</u> $\mathcal{F}$ <u>be as above</u>. <u>Then</u>:

(a) <u>If</u> L_1 <u>and</u> L_2 <u>are projective lines, which are not algebraic solutions of</u> $\mathcal{F}$, <u>then</u> $\#(\mathcal{F},L_1) = \#(\mathcal{F},L_2)$.
From this assertion we can define the <u>degree of</u> $\mathcal{F}$, $dg(\mathcal{F})$, as $\#(\mathcal{F},L)$, where L is a projective line which is not an algebraic solution of $\mathcal{F}$.

(b) <u>Suppose that</u> $\mathcal{F}$ <u>is expressed in the affine coordinate system</u> $((x,y),\mathbb{C}^2)$ <u>by the differential equation</u> $Pdy-Qdx = 0$, <u>where</u> $d = \max\{dg(P),dg(Q)\}$. <u>Then</u> $dg(\mathcal{F}) = d$ <u>or</u> $dg(\mathcal{F}) = d-1$. <u>Moreover, if</u> P_d <u>and</u> Q_d <u>are the homogeneous parts of degree</u> d <u>of</u> P <u>and</u> Q <u>respectively, then the following assertions are equivalent</u>:

(i) $dg(\mathcal{F}) = d$.

(ii) $yP_d(x,y) - xQ_d(x,y) \not\equiv 0$.

(iii) <u>The line at infinity</u> $L_\infty = \mathbb{C}P(2) - \mathbb{C}^2$, <u>is an algebraic solution of</u> $\mathcal{F}$.

<u>Proof</u>: Consider first the change of variables $x = 1/w$, $y = z/w$, where $\{w=0\}$ represents L_∞ in the affine coordinate system $((z,w),\mathbb{C}^2)$. This change of variables transforms equation $Pdy-Qdx = 0$ into

$$(*) \qquad w^{-d-2}[w\tilde{P}dz - (z\tilde{P}-\tilde{Q})\,dw] = 0,$$

where $\tilde{P}$ and $\tilde{Q}$ are the polynomials defined by $P(1/w,z/w) = \tilde{P}(z,w)/w^d$ and $Q(1/w,z/w) = \tilde{Q}(z,w)/w^d$. Multiplying (*) by

w^{d+2} in order to cancel the pole, we obtain

(**) $\qquad w\tilde{P}dz - (z\tilde{P}-\tilde{Q})dw = 0.$

If we write $z\tilde{P}-\tilde{Q} = R_o(z) + R_1(z)w + \ldots + R_d w^d$, then $R_o(z) = zP_d(1,z) - Q_d(1,z)$. Set $f(x,y) = yP_d(x,y) - xQ_d(x,y)$. If $f \equiv 0$, then we can write $z\tilde{P}-\tilde{Q} = wR(z,w)$, where R is a polynomial. Hence (**) can be written as $w(\tilde{P}dz-Rdw) = 0$, or dividing by w, as $\tilde{P}dz - Rdw = 0$. Observe now, that w is not a factor of $\tilde{P}$, because $zP_d(1,z) \equiv Q_d(1,z) \not\equiv 0$, and $d = \max\{dg(P),dg(Q)\}$. This implies that $\{w=0\}$ is not an algebraic solution of $\mathfrak{F}$. On the other hand, if $f \not\equiv 0$, then clearly w is not a factor of $z\tilde{P}-\tilde{Q}$ and it follows from (**), that $\{w=0\}$ is an algebraic solution of $\mathfrak{F}$. This proves (ii) $\Leftrightarrow$ (iii).

Let us prove (a). Let L be a projective line, which is not an algebraic solution of $\mathfrak{F}$. After a linear change of variables, we can suppose that $L \cap \mathbb{C}^2 = \{y=0\}$, in the affine coordinate system $((x,y),\mathbb{C}^2)$. In this case $h_L(t) = -Q(t,0)$, hence $\#(\mathfrak{F},L) = dg(Q(t,0)) + \#(\mathfrak{F},L;L \cap L_\infty)$. On the other hand, if we take the change of coordinates $x = 1/w$, $y = z/w$, then a parametrization of L in the new system is $z=0$, $w = s = 1/t$ and $L \cap L_\infty = \{z=w=0\}$. We have two cases: 1st. $f \not\equiv 0$ - Here $\#(\mathfrak{F},L;L \cap L_\infty)$ is the multiplicity of $s=0$ as a root of $\tilde{h}_L(s) = \tilde{Q}(0,s) = s^d Q(1/s,0)$, which is equal to $d-dg(Q)$. Hence $\#(\mathfrak{F},L) = d$, in this case. 2nd. $f \equiv 0$ - Here $\#(\mathfrak{F},L;L \cap L_\infty)$ is the multiplicity of $s=0$ as a root of $R(0,s) = -s^{-1}\tilde{Q}(0,s) = -s^{d-1}Q(1/s,0)$, which is equal to $d-1-dg(Q)$. In this case $\#(\mathfrak{F},L) = d-1$.

Since L is arbitrary, this proves (a) and (i) $\Leftrightarrow$ (ii) of (b). ■

Corollary. A foliation of degree n in $\mathbb{C}P(2)$ can be expressed in an affine coordinate system by a differential equation of the form $(P+xg)dy - (Q+yg)dx = 0$, where P, Q and g are polynomials such that:

(a) $P+xg$ and $Q+yg$ are relatively primes.

(b) g is homogeneous of degree n.

(c) $\max\{dg(P), dg(Q)\} \le n$.

(d) $\max\{dg(P), dg(Q)\} = n$ if $g \equiv 0$.

We leave the proof for the reader.

Example. Let $\mathfrak{F}$ be the foliation, defined in an affine coordinate system by the equation $axdy-bydx = 0$, $a \neq 0 \neq b$. According to our definition, $dg(\mathfrak{F}) = 0$ if $a = b$, and $dg(\mathfrak{F}) = 1$ if $a \neq b$.

Remark. Let A_{n+1} be space of polynomials of degree $\le n+1$ in two variables. Let $V \subset A_{n+1} \times A_{n+1}$ be the subspace of pairs of polynomials of the form $(P+xg, Q+yg)$, where P, Q and g are as in (b) and (c) of the corollary. Cleary V is a vector subspace of $A_{n+1} \times A_{n+1}$. Let PV be the projective space of lines through $0 \in V$. Since the differential equations $(P+sg)dy - (Q+yg)dx = 0$ and $\lambda(P+xg)dy - \lambda(Q+yg)dx = 0$, define the same foliation in $\mathbb{C}^2$, we can identify the set of all foliations of degree n in $\mathbb{C}P(2)$ with a subset $\mathfrak{X}_n \subset PV$. We consider $\mathfrak{X}_n$ with the topology induced by the topology of PV. We call $\mathfrak{X}_n$ the space of foliations of degree n in $\mathbb{C}P(2)$.

We observe, without proving, that $\mathfrak{X}_n$ is an open, dense and connected subspace of PV.

§2. Proof of Theorem A.

2.1 - A generalization of Camacho-Sad index Theorem

We will consider the following situation: let M be a complex 2-dimensional manifold and $S \subset M$ be a compact dimension one irreducible subvariety (singular or not). Let $\mathfrak{F}$ be a singular foliation by curves, such that S is a solution of $\mathfrak{F}$, that is, S is the union of a leaf with a finite number of singularities of $\mathfrak{F}$. This means that there is a covering of M by open sets, say $M = \bigcup_{i \in I} U_i$, and collections of holomorphic 1-forms $(\omega_i)_{i \in I}$ and non vanishing holomorphic functions $(g_{ij}: U_i \cap U_j \to \mathbb{C}^*)_{U_i \cap U_j \neq \emptyset}$, such that if $U_i \cap U_j \neq \emptyset$ then $\omega_i = g_{ij}\omega_j$. We suppose also that the singular set $\text{sing}(\omega_i) = \{g \in U_i; \omega_i(g) = 0\}$ is discrete. Since $\omega_i = g_{ij}\omega_j$ in $U_i \cap U_j$, the differential equations $\omega_i = 0$, $i \in I$, define a foliation by curves on $M\text{-sing}(\mathfrak{F})$, where $\text{sing}(\mathfrak{F}) = \bigcup_{i \in I} \text{sing}(\omega_i)$. On the other hand S can be defined in each U_i, $i \in I$, by an equation $f_i = 0$, where $f_i: U_i \to \mathbb{C}$ is analytic and $f_i = h_{ij}f_j$ in $U_i \cap U_j \to \mathbb{C}^*$. We suppose that for all $i \in I$, f_i is reduced, that is has no factors which are powers. The singular subvariety S will be a solution of $\mathfrak{F}$ if, and only if, for all $i \in U_i$, we have $df_i \wedge \omega_i = f_i\mu_i$, where μ_i is an holomorphic 2-form on U_i (see (2) in §1.1).

Under the above conditions, given a singularity of $\mathfrak{F}$, $p \in S$, such that S has branches $B_1, \ldots, B_k$ through p, it is possible to define the index $i(B_j, \mathfrak{F})$, of B_j with respect to $\mathcal{F}$, as in §1.3. We define $c(S,\mathcal{F})$ as the sum of all indexes with respect to $\mathcal{F}$, of all branches of S through all singularities of $\mathcal{F}$.

<u>Theorem C</u>. <u>Let</u> $S \subset M$ <u>be as above</u>. <u>Then</u>

(a) <u>If</u> S <u>is a solution of a singular foliation</u> $\mathcal{F}$, <u>then</u> $c(S,\mathcal{F})$ <u>is an integer</u>.

(b) <u>If</u> S <u>is a solution of two singular foliations</u> $\mathcal{F}_1$ <u>and</u> $\mathcal{F}_2$, <u>then</u> $c(S,\mathcal{F}_1) = c(S,\mathcal{F}_2)$.

(c) <u>If</u> S <u>is smooth, then</u> $c(S,\mathcal{F})$ <u>is the Chern class of the normal bundle of</u> S <u>in</u> M.

<u>Proof</u>: If S is smooth, then (a), (b) and (c) follow from Camacho-Sad Theorem (see Remark 1 in §1.2 and [3]). Let us suppose that S has some singularities, say $\text{sing}(S) = \{p_1,\dots,p_k\} \subset M$. In this case, by using the blowing-up method several, times, it is possible to construct a manifold $\tilde{M}$ and a proper analytic map $\pi: \tilde{M} \to M$ such that:

(i) For each $i \in \{1,\dots,k\}$, $\pi^{-1}(p_i) = D_i$ is a subvariety of $\tilde{M}$, which is a finite union of projective lines such that if two of them have non-empty intersection, then they intersect transversely in just one point. We will call D_i the <u>divisor associated to</u> p_i. The points of D_i in the intersection of two different projective lines in D_i, will be called <u>corners of</u> D_i.

(ii) If $D = \bigcup_{i=1}^{k} D_i$, then $\pi|(\tilde{M}-D): \tilde{M}-D \to M-\text{sing}(S)$ is a biholomorphism.

(iii) There exists a compact Riemann surface $\tilde{S} \subset \tilde{M}$, without singularities, such that:

1. $\tilde{S}$ does not contain corners of D and $\tilde{S}$ intersects D transversely.
2. $\pi(\tilde{S}) = S$ and $\pi|(\tilde{S}-D)\colon \tilde{S}-D \to S-\mathrm{sing}(S)$ is a biholomorphism.

For the proof see [7].

Now let $\mathfrak{F}$ be a singular foliation on M. Since we are using the blowing-up method to define $\pi\colon \tilde{M} \to M$, there is an unique singular foliation $\tilde{\mathfrak{F}}$ on $\tilde{M}$ such that:

3. $\mathrm{sing}(\tilde{\mathfrak{F}})$ is discrete, and $\mathrm{sing}(\mathfrak{F}) = \pi(\mathrm{sing}(\tilde{\mathfrak{F}})) \cup \mathrm{sing}(S)$.

We observe that, since S is a solution of $\mathfrak{F}$, then $\mathrm{sing}(S) \subset \mathrm{sing}(\mathfrak{F})$. However, it can happens that $\tilde{\mathfrak{F}}$ has no singularities in some of the divisors. (This happens if the divisor corresponds to a point $p_i \in S$ which is a "dicritical" singularity of $\mathfrak{F}$. See [8] for the definition).

4. π sends leaves of $\tilde{\mathfrak{F}}|(\tilde{M}-(\mathrm{sing}(\tilde{\mathfrak{F}})\cup D))$ onto leaves of $\mathfrak{F}|(M-\mathrm{sing}(\mathfrak{F}))$.

Since $c(\tilde{S},\tilde{\mathfrak{F}})$ does not depend on the foliation $\tilde{\mathfrak{F}}$ (Camacho-Sad Thm.), Theorem C will be proved after the following:

<u>Lemma 3</u>. <u>Let</u> $S \subset M$, $\tilde{S} \subset \tilde{M}$ <u>and</u> $\pi\colon \tilde{M} \to M$ <u>be as above. Then there exists a positive integer</u> k, <u>which depends only on</u> S, M <u>and</u> $\pi\colon \tilde{M} \to M$, <u>such that for any foliation</u> $\mathfrak{F}$, <u>for which</u> S <u>is a solution of</u> $\mathfrak{F}$, <u>we have</u>

$$c(S,\mathfrak{F}) = c(\tilde{S},\tilde{\mathfrak{F}}) + k,$$

<u>where</u> $\tilde{\mathfrak{F}}$ <u>is the foliation satisfying</u> 3 <u>and</u> 4 <u>of</u> (iii).

<u>Proof</u>: Let $p \in S$ be a singularity of $\mathfrak{F}$. If p is not a singularity of S, then S has just one local branch through p, say B. In this case, since $\pi|\tilde{M}-D: \tilde{M}-D \to M\text{-sing}(S)$ is a biholomorphism, $\tilde{S}$ has just one local branch $\tilde{B} = \pi^{-1}(B)$ through $\tilde{p} = \pi^{-1}(p)$. Moreover $i(\tilde{B},\tilde{\mathfrak{F}}) = i(B,\mathfrak{F})$.

Let us suppose $p = p_j \in \text{sing}(S)$. In this case $\pi^{-1}(p_j) = D_j$ is a divisor. Since $\tilde{S}$ has no singularities and intersects D_j, outside the corners, it follows that for each local branch B of S through p_j, there exist a point $q = q(B) \in D_j$ and a local branch $\tilde{B} = \tilde{B}(B)$ through q, such that $B = \pi(\tilde{B})$. This local branch intersects D_j transversely outside its corners. It follows that there exists a coordinate system $\tilde{\varphi} = (u,v): \tilde{U} \to \tilde{\varphi}(\tilde{U}) \subset \mathbb{C}^2$, such that $\tilde{U} \cap D = \{q \in \tilde{U};\ u(q) = 0\}$ and $\tilde{U} \cap \tilde{B} = \{q \in \tilde{U};\ v(q) = 0\}$. If we take a coordinate system $\psi = (x,y): V \to \psi(V) \subset \mathbb{C}^2$ in a neighborhood of p, such that $\psi(p) = (0,0)$, and $\pi(\tilde{U}) \subset V$, then we have a local expression for $\pi|\tilde{U}$ in these coordinate systems of the form $\pi(u,v) = (x(u,v),y(u,v))$, where $x(0,v) \equiv 0$ and $y(0,v) \equiv 0$, because $\pi(\tilde{U}\cap D) = \{p\}$. This implies that $x(u,v) = u^m X(u,v)$ and $y(u,v) = u^n Y(u,v)$, where $m,n \geq 1$. Observe that $u \longmapsto (x(u,0),y(u,0))$ is a Puiseaux's parametrization of the local branch B. Let $f: V \to \mathbb{C}$, $f(0,0) = 0$, be an irreducible equation of B. Let $\tilde{f} = f\circ\pi: U \to \mathbb{C}$. Since $B = \pi(\tilde{B})$, we must have $\tilde{f}(u,0) \equiv 0$, which implies that $\tilde{f}(u,v) = v^r F_1(u,v)$, where $F_1(u,0) \not\equiv 0$ and $r \geq 1$. In fact, since f is irreducible we must have $r = 1$, so that $\tilde{f}(u,v) = vF_1(u,v)$. On the other hand, $\tilde{f}(0,v) \equiv 0$, because $D_j = \pi^{-1}(p_j)$. This implies that $\tilde{f}(u,v) = u^{\ell}.v.F(u,v)$, where $\ell = \ell(B) \geq 1$ and $F(0,0) \neq 0$. If we take a smaller $\tilde{U}$, we can suppose that $F(u,v) \neq 0$ for all $(u,v) \in \tilde{U}$.

<u>Assertion</u>. <u>Let</u> $\mathfrak{F}$ <u>be a foliation on</u> M, <u>for which</u> S <u>is a solution</u>. <u>Let</u> $\tilde{\mathfrak{F}}$, B, $\tilde{B}$, f, $\tilde{f}$ <u>and</u> $\ell(B)$ <u>be as above</u>. <u>Then</u>

$$i(\tilde{B},\tilde{\mathfrak{F}}) = i(B,\mathfrak{F}) - \ell(B).$$

<u>Proof of the assertion</u>: Let $\omega = 0$ be an equation of $\mathfrak{F}|V$. Consider a decomposition $k.\omega = hdf + f\alpha$ as in (5) of §1.3. We can write

$$\pi^*(k.\omega) = (k\circ\pi).\pi^*(\omega) = (h\circ\pi).d(f\circ\pi) + f_o\pi.\pi^*(\alpha)$$
$$= (h\circ\pi).d\tilde{f} + \tilde{f}.\pi^*(\alpha).$$

Now, $\tilde{f}(u,v) = u^{\ell}v\, F(u,v)$, implies that

$$(k\circ\pi).\pi^*(\omega) = ((h\circ\pi).u^{\ell}.F)dv + v(\ell u^{\ell-1}.(h\circ\pi)dF +$$
$$+ u^{\ell}.(h\circ\pi)dF + u^{\ell}.F.\pi^*(\alpha)) =$$
$$= h^*dv + v\alpha^*.$$

On the other hand $\pi^*(\omega) = u^s\omega^*$, where $s \geq 1$, u does not divides both components of ω^* and $\omega^* = 0$ defines $\tilde{\mathfrak{F}}$ in $\tilde{U}$. It follows that $k^*.\omega^* = h^*dv + v\alpha^*$ is a decomposition for $\tilde{\mathfrak{F}}$, relative to $\tilde{B}$, where $k^* = u^s.(k\circ\pi)$. Let $\gamma(\theta) = (re^{i\theta},0)$, $0 \leq \theta \leq 2\pi$, $r > 0$ small. From the definition, we have:

$$i(\tilde{B},\tilde{\mathfrak{F}}) = -\frac{1}{2\pi i}\int_\gamma \frac{\alpha^*}{h^*} = -\frac{1}{2\pi i}\int_\gamma \left(\ell\,\frac{du}{u} + \frac{dF}{F} + \frac{\pi^*(\alpha)}{h\circ\pi}\right) =$$

$$= -\frac{1}{2\pi i}\int_\gamma \pi^*\left(\frac{\alpha}{h}\right) - \frac{1}{2\pi i}\int_\gamma \ell\,\frac{du}{u} - \frac{1}{2\pi i}\int_\gamma \frac{dF}{F} = -\frac{1}{2\pi i}\int_{\pi\circ\gamma} \frac{\alpha}{h} - \ell =$$

$$= i(B,\mathfrak{F}) - \ell .$$

This proves the assertion. When B is a branch of S through a singularity of $\mathfrak{F}$, which is not a singularity of S, we put $\ell(B) = 0$. Let $k = \sum_{B\in\mathcal{B}} \ell(B)$. From the assertion, we must have

$$c(\tilde{S},\tilde{\mathfrak{F}}) = \sum_{B\in\mathcal{B}} i(\tilde{B},\tilde{\mathfrak{F}}) = \sum_{B\in\mathcal{B}} (i(B,\mathfrak{F})-\ell(B)) = c(B,\mathfrak{F}) - k.$$

This proves Lemma 3 and Theorem C. ■

2.2 - Proof of Theorem A

Let $S \subset \mathbb{C}P(2)$ be an algebraic irreducible curve. From Theorem C, $c(S,\mathfrak{F})$ does not depends on the singular foliation $\mathfrak{F}$, for which S is an algebraic solution. Let us choose an affine coordinate system $(x,y)\colon U \to \mathbb{C}^2$ with the property that the line at infinite $L_\infty = \mathbb{C}P(2)-U$ meets S transversely in $dg(S) = k$ points. Let $f\colon \mathbb{C}^2 \to \mathbb{C}$ be an irreducible polynomial of degree k, such that $f = 0$ is the equation of S in this affine coordinate system. Let $\mathfrak{F}$ be the compactification in $\mathbb{C}P(2)$ of the foliation in $\mathbb{C}^2$ whose leaves are the level surfaces of f. An equation for $\mathfrak{F}|\mathbb{C}^2$ is of course $df = 0$. We can divide the singularities of $\mathfrak{F}$ in S in two parts:

1st. Infinite singularities - Corresponding to the intersections of S with L_∞.

2nd. Finite singularities - Corresponding to the singularities of S. These singularities, by the choice of the affine coordinate system, are all in $\mathbb{C}^2 \simeq U$.

Observe that, since the points of $L_\infty \cap S$, are smooth points of S, then to each $p \in L_\infty \cap S$, corresponds an unique local branch. We will denote this branch by $B(p)$. Let us compute $i(B(p),\mathfrak{F})$.

Let us suppose, without lost of generality, that $(1:0:0) \in S \cap L_\infty$, where the affine coordinate system (x,y) is such that $(x,y) \simeq (x:y:1)$. Consider the change of affine coordinate systems $u = y/x$, $v = 1/x$, so that $(1:0:0)$ corresponds to $u = v = 0$ if $(u,v) \simeq (1:u:v)$. Since $f(x,y)$ has degree k, we can write

$$f(1/v,\ u/v) = v^{-k}\,\tilde{f}(u,v)$$

where $\tilde{f}$ is irreducible and $\tilde{f} = 0$ represents S in the affine coordinate system (u,v). Moreover, since $p = (1:0:0) \in S \cap L_\infty$, we must have $\tilde{f}(0,0) = 0$. On the other hand S has just one local branch through p, which is transverse to L_∞ at p. Since $v = 0$ is the equation of L_∞, the local equation of $B(p)$ must be of the form $u = \varphi(v)$, where $\varphi(0) = 0$, so that we can write locally $\tilde{f}(u,v) = (u-\varphi(v)).F(u,v)$, where $F(0,0) \neq 0$. Now, the leaves of the foliation $\mathfrak{F}$ near p, are the level surfaces of the meromorphic function $(u-\varphi(v)).F(u,v)/v^k$, hence $\mathfrak{F}$ can be represented in a neighborhood of p by the differential equation

$$v^{d+1}\, d\left(\frac{\tilde{f}}{v^d}\right) = v d\tilde{f} - k\tilde{f} dv = P dv - Q du = 0,$$

or by the differential equation $\dot{u} = P(u,v)$, $\dot{v} = Q(u,v)$. The Jacobian of (P,Q) at $(0,0)$ is given by the matrix:

$$\begin{pmatrix} \frac{\partial P}{\partial u}(0,0) & \frac{\partial P}{\partial v}(0,0) \\ \frac{\partial Q}{\partial u}(0,0) & \frac{\partial Q}{\partial v}(0,0) \end{pmatrix} = \begin{pmatrix} -kF(0,0) & (k-1)\varphi'(0)F(0,0) \\ 0 & -F(0,0) \end{pmatrix}$$

This matrix has eigenvalues $\lambda_1 = -kF(0,0)$ and $\lambda_2 = -F(0,0)$, where λ_1 corresponds to the direction of L_∞ and λ_2 to the direction of $B(p)$ at $(0,0)$. Therefore, from the particular case 1 of §1.3, we must have $i(B(p),\mathfrak{F}) = k$. This implies that the contribution of each infinite singularity for $c(S,\mathfrak{F})$ is k. Since the number of such singularities is k, the total contribution of the infinite singularities for $c(S,\mathfrak{F})$ is k^2.

Now, let $\operatorname{sing}(S) = \{p_1,\dots,p_m\}$ be the finite singularities of S. Suppose that $f = f_1^i \dots f_{n_i}^i$ is a local decomposition of f in irreducible factors, in a small neighborhood of p_i. Let B_j^i be the local branch of S through p_i, corresponding to f_j^i. Since $\mathfrak{F}$ is represented by $df = 0$ near p_i, we get from the particular case 2 in §1.3, that

$$i(B_j^i,\mathfrak{F}) = -\sum_{\substack{n=1\\ n\neq j}}^{n_i} [f_n^i, f_j^i]_{p_i}, \quad \text{if } n_i>1, \text{ or } i(B_1^i,\mathfrak{F}) = 0, \text{ if } n_i=1.$$

This implies that the contribution of all local branches of S through p_i, for $C(S,\mathfrak{F})$ is $-\sum_{j=1}^{n_i}\sum_{\substack{n=1\\ n\neq j}}^{n_i} [f_n^i, f_j^i]_{p_i} =$

$= -\sum_{n\neq j} [f_n^i, f_j^i]_{p_i}$ (we make the convention that the last sum is zero, if $n_i = 1$). Therefore,

$$(7) \qquad C(S,\mathcal{F}) = k^2 - \sum_{i=1}^{m} \sum_{n \neq j} [f_n^i, f_j^i]_{p_i} .$$

Now we will compute $\chi(S)$. Let $\pi: \tilde{M} \to \mathbb{C}P(2)$ be a desingularization of S, as in the proof of Theorem C. Let $D_i = \pi^{-1}(p_i)$, $i=1,\ldots,m$, and $\tilde{S} \subset \tilde{M}$ be the compact Riemann surface such that $\pi(\tilde{S}) = S$. We set $h = \pi|\tilde{S}: \tilde{S} \to S$. According to our definition $\chi(S) = \chi(\tilde{S})$. The idea is to compute $\chi(\tilde{S})$ by using Poincaré-Hopf's index Theorem. We will construct a real C^∞ vector field Y in $\tilde{S}$ with a finite number of singularities and then we will compute the sum of the Poincaré-Hopf indexes of these singularities.

Let $Z(x,y) = (-\frac{\partial f}{\partial y}(x,y), \frac{\partial f}{\partial x}(x,y))$, be considered as a real vector field on $\mathbb{C}^2$. It is not difficult to see that Z is tangent to $S \cap \mathbb{C}^2$. In order to extend Z to L_∞, we multiply it by a C^∞ real function $\sigma: \mathbb{C}P(2) \to$ $\to [0,+\infty)$, such that $\sigma|\mathbb{C}^2 > 0$, $\sigma|L_\infty \equiv 0$ and σ is flat of infinite order at L_∞. It is not difficult to see that $\sigma.Z$ extends to a C^∞ vector field on $\mathbb{C}P(2)$, such that $\sigma.Z|S$ has a finite number of singularities (since $S \cap L_\infty$ is finite). Let us prove that there exists a C^∞ vector field Y on $\tilde{S}$, such that $(h|(\tilde{S}-D))_*(Y) = \sigma.Z$, where $D = \bigcup_{i=1}^{m} D_i$.

Observe first that, since $h|(\tilde{S}-D): \tilde{S}-D \to S\text{-sing}(S)$ is a diffeomorphism, then we can define $Y|(\tilde{S}-D)$ as $(h^{-1})_*(\sigma.Z|(S\text{-sing}(S))$. Let us prove that Y extends to

$\tilde{S} \cap D$. Recall that $\tilde{S} \cap D_i = \{p_1^i,\ldots,p_{n_i}^i\}$, where for each $j \in \{1,\ldots,n_i\}$, $\tilde{S}$ has an unique smooth branch $\tilde{B}_j^i$ through p_j^i, such that $h(\tilde{B}_j^i) = B_j^i$. Moreover, for each $1 \le j \le n_i$, there is a coordinate system $(u,v)\colon \tilde{U} \to \mathbb{C}^2$, around p_j^i, such that $u(p_j^i) = v(p_j^i) = 0$, $\tilde{B}_j^i \cap \tilde{U} = \tilde{S} \cap \tilde{U} = \{v=0\}$ and $D_i \cap \tilde{U} = D \cap \tilde{U} = \{u=0\}$. We have also seen that $h(u) = \pi(u,0) = (x(u,0),y(u,0))$ is a Puiseaux's parametrization of B_j^i. This implies that $h^*(\sigma.Z)$ is given in a punctured neighborhood of 0 in the u-plane, by

$$u \mapsto -\sigma\circ h(u)\ \frac{\partial f}{\partial y}(h(u))/\frac{\partial x}{\partial u}(u,0) = \sigma\circ h(u)\ \frac{\partial f}{\partial x}(h(u))/\frac{\partial y}{\partial u}(u,0).$$

We have $\frac{\partial f}{\partial x}\frac{\partial x}{\partial u} + \frac{\partial f}{\partial y}\frac{\partial y}{\partial u} = 0$, so that the two expressions above are equal, if $\frac{\partial x}{\partial u} \not\equiv 0$ and $\frac{\partial y}{\partial u} \not\equiv 0$. We will suppose, without lost of generality, that $x(u,0)$ is not constant, which means that $B_j^i \neq \{x=x(0,0)\}$. In §4 of Chapter I of [8], this type of vector field was considered, and it was proved that the function $-\frac{\partial f}{\partial y}(h(u))/\frac{\partial x}{\partial u}(u,0)$ extends to a holomorphic function in $u = 0$. Moreover, it was shown that the Poincaré-Hopf's index of 0 with respect to the vector field $u \mapsto -\frac{\partial f}{\partial y}(h(u))/\frac{\partial x}{\partial u}(u,0)$ is the order at 0 of this function. Since the Poincaré-Hopf index does not change, if we multiply the vector field by a positive function, we get

$$I(Y,p_j^i) = \operatorname{ord}\left(-\frac{\partial f}{\partial y}(h(u))/\frac{\partial x}{\partial u}(u,0),0\right) =$$

$$= \operatorname{ord}\left(\frac{\partial f}{\partial y}(h(u)),0\right) - \operatorname{ord}\left(\frac{\partial x}{\partial u}(u,0),0\right),$$

where $I(Y,p_j^i)$ denotes the Poincaré-Hopf index of Y at p_j^i. On the other hand, we have

$$\operatorname{ord}(\frac{\partial f}{\partial y}(h(u)),0) = [\frac{\partial f}{\partial y},f^i_j]_{p_i}$$

$$\operatorname{ord}(\frac{\partial x}{\partial u}(u,0),0) = \operatorname{ord}(x(u,0)-x(0,0),0)-1 = [x-x(p_i),f^i_j]_{p_i}-1.$$

Since $\frac{\partial f}{\partial y} = \sum_{n=1}^{n_i} \frac{\partial f^i_n}{\partial y} f^i_1 \dots \hat{f}^i_n \dots f^i_{n_i}$, in a neighborhood of p_i, we get from the properties of the intersection number that,

$$[\frac{\partial f}{\partial y},f^i_j]_{p_i} = [\frac{\partial f^i_j}{\partial y} f^i_1 \dots \hat{f}^i_j \dots f^i_{n_i},f^i_j] = [\frac{\partial f^i_j}{\partial y},f^i_j]_{p_i} + \sum_{\substack{n=1\\ n\neq j}}^{n_i} [f^i_n,f^i_j]_{p_i}$$

Therefore,

$$I(Y,p^i_j) = \sum_{\substack{n=1\\ n\neq j}}^{n_i} [f^i_n,f^j_n]_{p_i} + \mu^i_j$$

where $\mu^i_j = [\frac{\partial f^i_j}{\partial y},f^i_j]_{p_i} - [x-x(p_i),f^i_j]_{p_i}+1$. At the end of this section we will prove that $\mu^i_j = \mu(B^i_j)$, the Milnor number of the local branch B^i_j. If we suppose this fact, then the contribution of all singularities of Y in $D \cap \tilde{S}$ for the sum of Poincaré-Hopf's indexes, will be

$$\sum_{i,j} I(Y,p^i_j) = \sum_{i=1}^{m} \sum_{n\neq j} [f^i_n,f^i_j]_{p_i} + \sum_{i,j} \mu(B^i_j).$$

Now let us compute $I(Y,\tilde{p})$, where $\pi(\tilde{p}) \in S \cap L_\infty$. We can suppose without lost of generality, that $\pi(\tilde{p}) = (1:0:0)$, where $(x,y) \simeq (x:y:1)$. Consider the change of coordinates $u = y/x$, $v = 1/x$, so that $(1:0:0)$ corresponds to $u = v = 0$. As before, we have

$$f(1/v,\ u/v) = v^{-k}\,\tilde{f}(u,v),$$

where, locally we can write $\tilde{f}(u,v) = (u-\varphi(v)).F(u,v)$, $F(0,0) \neq 0$ and $u = \varphi(v)$ is the equation of the local branch of S through $(1:0:0)$. In order to simplify we consider the local change of variables $\alpha = u-\varphi(v)$, $\beta = v$, so that in the coordinate system $(\alpha,\beta): U \to \mathbb{C}^2$, we have $L_\infty \cap U = \{\beta=0\}$ and $S \cap U = \{\alpha=0\}$. A straightforward computation shows that $\sigma.z$ can be written it this coordinaty system as, $\sigma.z(\alpha,\beta) = = \sigma(\alpha,\beta)(\alpha\varphi_1(\alpha,\beta),\varphi_2(\alpha,\beta))$, where φ_1 and φ_2 are C^∞ and

$$\varphi_2(\alpha,\beta) = \frac{1}{\beta^{k-3}}[F(\alpha+\varphi(\beta),\beta) + \alpha\frac{\partial F}{\partial u}(\alpha+\varphi(\beta),\beta)].$$

This implies that Y is represented in the coordinate system $\beta \sim (0,\beta)$, of $S \cap U$, by

$$Y(\beta) = \frac{\sigma(0,\beta)}{\beta^{k-3}} \cdot F(\varphi(\beta),\beta) = \frac{\sigma(0,\beta)}{|\beta|^{2(k-3)}} \cdot F(\varphi(\beta),\beta)\cdot(\bar{\beta})^{k-3}.$$

Since $\sigma(0,\beta)/|\beta|^{2(k-3)}$ is positive in a punctured neighborhood of $u=0$, we have

$$I(Y,\tilde{p}) = I(F(\varphi(\beta),\beta)\cdot(\bar{\beta})^{k-3},0).$$

If we set $F(\varphi(\beta),\beta)\cdot(\bar{\beta})^{k-3} = \psi(\beta)$, $\gamma(\theta) = re^{i\theta}$, $r > 0$ small, then $I(\psi(\beta),0)$ can be calculated as

$$I(\psi(\beta),0) = \frac{1}{2\pi i}\int_\gamma \frac{d\psi}{\psi} = \frac{k-3}{2\pi i}\int_\gamma \frac{d\bar{\beta}}{\bar{\beta}} = -(k-3).$$

Since we have k point in $L_\infty \cap S$, we get finally

$$\chi(S) = \chi(\tilde{S}) = \sum_{i=1}^{m} \sum_{n\neq j} [f_n^i,f_j^i]_{p_i} + \sum_{i,j} \mu(B_j^i) - k(k-3).$$

Therefore, from (7), we have

$$\chi(S) + c(S,\mathcal{F}) = \sum_{i,j} \mu(B_j^i) + 3k,$$

which proves Theorem A. It remains to prove that $\mu_j^i = \mu(B_j^i)$.

In order to simplify the notations, let us put $f^i_j = \varphi$, $p=0$. We want to prove that, if the germ of φ is irreducible at 0, $\varphi(0) = 0$ and $\{\varphi=0\} \neq \{x=0\}$, then

$$[\frac{\partial\varphi}{\partial x},\frac{\partial\varphi}{\partial y}]_o = \mu(\varphi,0) = [\frac{\partial\varphi}{\partial y},\varphi]_o - [\varphi,x]_o + 1. \tag{8}$$

Let us suppose first that $\frac{\partial\varphi}{\partial y}(0) \neq 0$. In this case $[\frac{\partial\varphi}{\partial x},\frac{\partial\varphi}{\partial y}]_o = [\frac{\partial\varphi}{\partial y},\varphi]_o = 0$ and $\varphi(x,y) = ax+by+\ldots$, where $b\neq 0$, so that $[\varphi,x]_o = 1$, which proves (8). Suppose now that $\frac{\partial\varphi}{\partial y}(0) = 0$. Let $\frac{\partial\varphi}{\partial y} = \psi_1^{r_1}\ldots\psi_\ell^{r_\ell}$ be a decomposition of $\frac{\partial\varphi}{\partial y}$ in irreducible factors. Let $\alpha_j = (x_j,y_j)\colon D \to \mathbb{C}^2$, be a Puiseaux's parametrization of $\{\psi_j=0\}$, where $D = \{T \in \mathbb{C};\ |T| < 1\}$ and $\alpha_j(0) = 0$. We have $\frac{d}{dT}(\varphi\circ\alpha_j) = \frac{\partial\varphi}{\partial x}\circ\alpha_j\cdot x'_j + \frac{\partial\varphi}{\partial y}\circ\alpha_j\cdot y'_j = \frac{\partial\varphi}{\partial x}\circ\alpha_j\cdot x'_j$, because $\frac{\partial\varphi}{\partial y}\circ\alpha_j \equiv 0$. Since $\varphi\circ\alpha_j(0) = 0$, and $x_j(0) = 0$, we have

$$[\varphi,\psi_j]_o - 1 = \operatorname{ord}(\varphi\circ\alpha_j,0)-1 = \operatorname{ord}(\tfrac{d}{dT}(\varphi\circ\alpha_j),0) = \operatorname{ord}(\tfrac{\partial\varphi}{\partial x}\circ\alpha_j\cdot x'_j,0) =$$
$$= \operatorname{ord}(\tfrac{\partial\varphi}{\partial x}\circ\alpha_j,0) + \operatorname{ord}(x'_j,0) = [\tfrac{\partial\varphi}{\partial x},\psi_j]_o + [\psi_j,x]_o - 1.$$

This implies that,

$$[\frac{\partial\varphi}{\partial x},\frac{\partial\varphi}{\partial y}]_o = [\frac{\partial\varphi}{\partial x},\prod_{j=1}^{\ell}\psi_j^{r_j}]_o = \sum_{j=1}^{\ell} r_j[\frac{\partial\varphi}{\partial x},\psi_j]_o =$$
$$= \sum_{j=1}^{\ell} r_j([\varphi,\psi_j]_o - [\psi_j,x]_o) = [\varphi,\frac{\partial\varphi}{\partial y}]_o - [\frac{\partial\varphi}{\partial y},x]_o.$$

On the other hand,

$$[\frac{\partial\varphi}{\partial y},x]_o = \operatorname{ord}(\frac{\partial\varphi}{\partial y}(0,y),y=0) = \operatorname{ord}(\varphi(0,y),y=0)-1 = [\varphi,x]_o - 1.$$

This proves (8) and finishes the proof of Theorem A. ∎

§3. Proof of Theorem B

3.1 - Examples of singular foliations on $\mathbb{C}P(2)$ without algebraic solutions

Definition. Let $\mathfrak{F}$ be a singular foliation on M and $p \in M$ a singularity of $\mathfrak{F}$. Let us suppose that $\mathfrak{F}$ is represented, in local coordinate system $(x,y): U \to \mathbb{C}^2$ around p, where $x(p) = y(p) = 0$, by a differential equation of the form

$$(9) \qquad \begin{cases} \dot{x} = P(x,y) \\ \dot{y} = Q(x,y) \end{cases}$$

where $P(0,0) = Q(0,0) = 0$. Let λ_1 and λ_2 be the eigenvalues of the Jacobian matrix of (P,Q) at $(0,0)$. We say that p is a nondegenerated singularity of $\mathfrak{F}$, if $\lambda_1 \neq 0 \neq \lambda_2$. If p is nondegenerated, we say that it is a simple singularity, if $\lambda_1/\lambda_2 \notin \mathbb{Q}_+$ (positive rational numbers). We say that p is of Poincaré type, if $\lambda_1/\lambda_2 \notin \mathbb{R}_+$. It can be shown that these conditions are independent of the differential equation (9), which represents $\mathfrak{F}$ in a neighborhood of p.

Let us recall some known facts about nondegenerated singularities. Let λ_1 and λ_2 be as above. We suppose $p = 0 \in \mathbb{C}^2$.

1st. Suppose that $\mathfrak{F}$ has a smooth local separatrix B through 0. Then the tangent space T_pB coincides with one of the eigenspaces of the Jacobian of (P,Q) at 0. If the eigenvalue associated to this eigenspace is $\lambda_1 \neq 0$ then $i(B,\mathfrak{F}) = \lambda_2/\lambda_1$ (see §1.3).

2nd. If $\lambda_1/\lambda_2 \notin \mathbb{N}$ and $\lambda_2/\lambda_1 \notin \mathbb{N}$, then $\mathcal{F}$ has exactly two local smooth separatrices through 0. Moreover, if $\lambda_1/\lambda_2 \notin \mathbb{Q}_+$ (that is 0 is simple), then 0 has no other separatrices through it. These facts follow from Poincaré's linearization theorem, if $\lambda_1/\lambda_2 \notin \mathbb{R}$ or $\lambda_1/\lambda_2 \in \mathbb{R}_+ - \mathbb{Q}$ (cf. [9]). If $\lambda_1/\lambda_2 \in \mathbb{Q}_+$, but $\lambda_1/\lambda_2 \notin \mathbb{N}$, then the proof follows from Poincaré-Dulac's normal form (cf. [9]). If $\lambda_1/\lambda_2 < 0$, then the proof follows from the invariant manifold theorem for real vector fields (cf. [10]).

Definition. We will say that a foliation $\mathcal{F}$ is nondegenerated, simple, or of Poincaré type if all singularities of $\mathcal{F}$ are respectively, nondegenerated, simple, or of Poincaré type. We will use the following notations:

$\mathcal{N}_n$ = set of nondegenerated foliations of degree n in $\mathbb{C}P(2)$.

$\mathcal{S}_n$ = set of simple foliations of degree n in $\mathbb{C}P(2)$.

$\mathcal{P}_n$ = set of Poincaré type foliations of degree n in $\mathbb{C}P(2)$.

Observe that $\mathcal{P}_n \subset \mathcal{S}_n \subset \mathcal{N}_n$. Let $\mathcal{F} \in \mathcal{S}_n$ with N singularities, say $p_1,\ldots,p_N$ (later on we will see that $N = n^2+n+1$). Let B_j^1, B_j^2 be the two local separatrices of $\mathcal{F}$ through p_j, $1 \le j \le N$. Let $\sigma(j,k) = i(B_j^k,\mathcal{F})$, $k=1,2$. If $A \subset \{(j,k);\ 1 \le j \le N,\ k=1,2\}$, then set $\sigma(A,\mathcal{F}) = \sum_{(j,k)\in A} \sigma(j,k)$. We have the following:

Theorem D. Let $n \geq 2$ and $\mathfrak{F} \in \mathcal{S}_n$ be such that for any proper non empty subset A of $\{(j,k);\ 1 \leq j \leq N,\ k=1,2\}$, the number $\sigma(A,\mathfrak{F})$ is not a positive integer. Then $\mathfrak{F}$ has no algebraic solution.

Proof: Suppose by contradiction that $\mathfrak{F}$ has an algebraic solution S. By Remark 3 in §1.2, S contains at least one singularity of $\mathfrak{F}$. Let $p_1,\ldots,p_N$ and B_j^k, $1 \leq j \leq N$, $k = 1,2$, be as before. If $p_j \in S$, then S has at most two local branches through p_j, namely B_j^1 and B_j^2. This follows from the 2nd remark of this section. Let $A =$ $= \{(j,k);\ 1 \leq j \leq N$ and B_j^k is a local branch of S at $p_j\}$. A cannot be a proper subset of $\{(j,k);\ 1 \leq j \leq N,\ k=1,2\}$, because if it was, then we would have $c(S,\mathfrak{F}) = \sigma(A,\mathfrak{F})$, which contradicts Theorem A, since $\sigma(A,\mathfrak{F})$ is not a positive integer. It remains to consider the case
$A = \{(j,k);\ 1 \leq j \leq N,\ k=1,2\}$. We need a lemma:

Lemma 4. Let $\mathfrak{F} \in \mathcal{H}_n$. Then $\mathfrak{F}$ has $N = n^2+n+1$ singularities. Moreover, if $p_1,\ldots,p_N$ are these singularities and $\sigma(j,1),\sigma(j,2) = \sigma(j,1)^{-1}$ are the quotient of the eigenvalues of the Jacobian of some vector field representeing $\mathfrak{F}$ near p_j, $1 \leq j \leq N$, then

$$(10) \qquad \sum_{j=1}^{N} (\sigma(j,1)+\sigma(j,2)) = -n^2 + 2n + 2.$$

Proof: This lemma follows from Baum-Bott's Theorem (cf.[11]) for foliations with nondegenerated singularities. Observe that a foliation $\mathfrak{F}$ of degree n on $\mathbb{C}P(2)$, defines a holomorphic section on the vector bundle $L^{n-1} \otimes T(\mathbb{C}P(2))$, where $T(\mathbb{C}P(2))$ is the tangent bundle and L is the line bundle associated to a linear divisor in $\mathbb{C}P(2)$. This fact can be easily verified by changing affine coordinates in a vector field of the form $X = (P+xg)\partial/\partial x + (Q+yg)\partial/\partial y$, where P, Q and g are as in (a), (b), (c) and (d) of the corollary of Lemma 2 of §1.4. We leave the details for the reader.

Baum-Bott's Theorem asserts that if $\varphi\colon M(2\times 2) \to \mathbb{C}$ is an invariant homogeneous polynomial of degree 2, where $M(2\times 2)$ is the vector space of complex 2×2 matrices (that is $\varphi(P^{-1}AP) = \varphi(A)$), then the sum

$$(11) \qquad \sum_{p\in \operatorname{sing}(\mathcal{F})} \frac{\varphi(J_p)}{\det(J_p)}$$

equals certain Chern class associated to φ and $L^{n-1} \otimes$ $\otimes\, T(\mathbb{C}P(2))$ where J_p is the Jacobian matrix at p of any vector field representing $\mathcal{F}$ near p. In particular the sum in (11) does not depend on the particular nondegenerated foliation of degree n chosen. It follows that we can obtain the sum by calculating it in some particular example. In order to calculate (11) we use

Jouanolou's example (see [4]), which is given in some affine coordinate system by the vector field $(1-xy^n)\partial/\partial x + (x^n-y^{n+1})\partial/\partial y$. According to our definition, the foliation $\mathcal{F}_o$ represented by the differential equation $\dot{x} = 1-xy^n$, $\dot{y} = x^n-y^{n+1}$ has degree n.

Let us consider first $\varphi(A) = \det(A)$. In this case (11) represents the number of singularities of the foliation. If we calculate this number for $\mathcal{F}_o$ above the we get $N = n^2+n+1$ (see the next section).

Now set $\varphi(A) = (\mathrm{tr}(A))^2$. In this case, if p is a nondegenerated singularity of $\mathcal{F}$, J_p is the Jacobian matrix at p of some vector field representeing $\mathcal{F}$ near p, and λ_1, λ_2 are the eigenvalues of V_p, then

$$\frac{\varphi(J_p)}{\det(J_p)} = \frac{(\mathrm{tr}(J_p))^2}{\det(J_p)} = \frac{(\lambda_1+\lambda_2)^2}{\lambda_1\lambda_2} = \frac{\lambda_1}{\lambda_2} + \frac{\lambda_2}{\lambda_1} + 2 = \sigma(p,1)+\sigma(p,2) + 2.$$

Therefore,

$$\sum_{p\in \mathrm{sing}(\mathcal{F})} \frac{\varphi(J_p)}{\det(J_p)} = \sum_{p\in \mathrm{sing}(\mathcal{F})} (\sigma(p,1)+\sigma(p,2)) + 2N.$$

On the other hand, if we calculate (11) in this case for $\mathcal{F}_o$ above, we get

$$\sum_{p\in \mathrm{sing}(\mathcal{F}_o)} \frac{\varphi(J_p)}{\det(J_p)} = (n+2)^2 = n^2+4n+4 \quad \text{(see the next section).}$$

This implies (10), because

$$\sum_{p\in \mathrm{sing}(\mathcal{F})} (\sigma(p,1)+\sigma(p,q)) = (n+2)^2-2N = -n^2+2n+2.$$

This finishes the proof of the lemma. ■

Now let us finish the proof of Theorem D. Let us suppose by contradiction that $\mathcal{F}$ has an algebraic solution S, which contains all local branches B_j^k, $1 \le j \le N$, $k=1,2$. Since B_j^1 and B_j^2 are the only local analytic branches through p_j, $1 \le j \le N$, we get $c(S,\mathcal{F}) = -n^2+2n+2$. If $n \ge 3$ the number $-n^2+2n+2$ is negative and so we must have $n = 2$, $c(S,\mathcal{F}) = 2$. On the other hand by Theorem A, this implies that

$$2 = 3.2 - \chi(S) + \sum_{\substack{1\le j\le N\\ k=1,2}} \mu(B_j^k).$$

Since B_j^k is smooth, we have $\mu(B_j^k) = 0$ for all $1 \le j \le N$, $k=1,2$. This implies that $\chi(S) = 6-2 = 4$, which is impossible, because $\chi(S) \le 2$ for all compact Riemann surfaces. This ends the proof of Theorem D. ■

3.2 - Jouanolou's example

In this section we prove that Jouanolou's example has no algebraic solution on $\mathbb{C}P(2)$. In [4], Jouanolou proves the same fact, but I think that our proof is simpler than his proof.

Let us consider the foliation $\mathcal{F}_0$, defined in some affine coordinate system by the differential equation:

$$\dot{x} = 1-xy^n, \quad \dot{y} = x^n-y^{n+1}.$$

It is not difficult to see that $\mathcal{F}_o$ has no singularities in the line at ∞. The finite singularities of $\mathcal{F}_o$ are the solutions of $xy^n = 1$, $y^{n+1} = x^n$, that is, the points $p_1,\ldots,p_N \in \mathbb{C}^2$, where $N = n^2+n+1$, $p_j = (x_j,y_j)$, $x_j = y_j^{-n}$, $y_j = \exp(2\pi ij/N)$, $1 \le j \le N$.

Now the Jacobian matrix of $(1-xy^n)\partial/\partial x + (x^n-y^{n+1})\partial/\partial y$ at the point p_j is

$$J_j = J_{p_j} = \begin{pmatrix} -y_j^n & -nx_jy_j^{n-1} \\ nx_j^{n-1} & -(n+1)y_j^n \end{pmatrix}.$$

As it can be easily seen, the eigenvalues of J_j are

$$\lambda_j^1 = \frac{-(n+2)+\sqrt{3}\,ni}{2} \cdot y^n, \qquad \lambda_j^2 = \frac{-(n+2)-\sqrt{3}\,ni}{2} \cdot y^n.$$

In particular $\mathcal{F}_o$ is of Poincaré type, because

$$\sigma(j,1) = \frac{\lambda_j^2}{\lambda_j^1} = \frac{-n^2+2n+2}{2N} + \frac{\sqrt{3}\,n(n+2)i}{2N} \notin \mathbb{R}$$

and

$$\sigma(j,2) = \frac{\lambda_j^1}{\lambda_j^2} = \overline{\sigma(j,1)} \notin \mathbb{R}.$$

Let B_j^k be the local separatrix of $\mathcal{F}_o$ tangent at p_j to the eigenspace relative to λ_j^k. We have seen that $i(B_j^k,\mathcal{F}_o) = \sigma(j,k)$. Furthermore, from Theorem D, in order to prove that $\mathcal{F}_o$ has no algebraic solution, it is enough to see that for any proper subset A of $\{(j,k);\ 1 \le j \le N,\ k=1,2\}$ the sum $\sigma(A,\mathcal{F}) = \sum_{(j,k)\in A} \sigma(j,k)$ is not a positive integer.

Let us suppose that $\sigma(A,\mathcal{F})$ is real for some $A \subset \{(j,k);\ 1 \le j \le N,\ k=1,2\}$. This implies that if $(j,1) \in A$ for some $j \in \{1,\ldots,N\}$, then there exists

$\ell \in \{1,\dots,N\}$ such that $(\ell,2) \in A$, because $\sigma(r,1) = \overline{\sigma(s,2)} \notin \mathbb{R}$, for all $r,s \in \{1,\dots,N\}$. This implies that, $m = \#\{(j,1);\ (j,1) \in A\} = \#\{(j,2);\ (j,2) \in A\}$ and moreover that

$$\sigma(A,\mathcal{F}_o) = \frac{m}{N}(-n^2+2n+2).$$

Now, observe that if $n \geq 3$, then $-n^2+n+2 < 0$, which implies that in this case $\sigma(A,\mathcal{F}_o)$ is never a positive integer. On the other hand, if $n = 2$, we have $N = 7$, $-n^2+2n+2 = 2$, so that

$$\sigma(A,\mathcal{F}_o) = \frac{2m}{7}.$$

Since $m = \#\{(j,1);\ (j,1) \in A\} < N = 7$ (because A is a proper subset of $\{(j,k);\ 1 \leq j \leq N,\ k=1,2\}$), we get $0 < \sigma(A,\mathcal{F}_o) < 2$ in this case. Moreover, since m is an integer, we must have $\sigma(A,\mathcal{F}_o) \neq 1$. This finishes the proof.

3.3 - Proof of Theorem B

We prove first the following:

Lemma 5. *$\mathcal{P}_n$ and $\mathcal{h}_n$ are open, dense and connected subsets of $\mathfrak{X}_n$. Moreover, given $\mathcal{F}_o \in \mathcal{h}_n$, with singularities $p_1,\dots,p_N$, then there are neighborhoods U_o of $\mathcal{F}_o$ in $\mathfrak{X}_n$, V_j of p_j in $\mathbb{C}P(2)$, and analytic functions $\varphi_j\colon U_o \to V_j$, $j = 1,\dots,N$, such that $V_i \cap V_j = \emptyset$ if $i \neq j$, and for any $\mathcal{F} \in U_o$, $\varphi_j(\mathcal{F})$ is the unique singularity of $\mathcal{F}$ in V_j (in particular $\varphi_j(\mathcal{F}_o) = p_j$), $1 \leq j \leq N$.*

Proof: Let us prove first the second assertion. Let $\mathcal{F}_o \in \mathcal{h}_n$ and $p_1,\dots,p_N$ be the singularities of $\mathcal{F}_o$, $N = n^2+n+1$.

Consider an affine coordinate system (x,y), such that $p_1,\dots,p_N \notin L_\infty$, the line at infinite. Suppose that $\mathfrak{F}_o$ is represented in this coordinate system by the vector field $(P_o(x,y)+xg_o(x,y))\partial/\partial x + (Q_o(x,y)+yg_o(x,y))\partial/\partial y$, where P_o, Q_o, g_o are polynomials of degree n, g_o homogeneous. We suppose also that some of the coefficients of P_o, Q_o or g_o is 1 (say the coefficient of y^n in g_o). Under this condition, there is a neighborhood $\mathcal{V}$ of $\mathfrak{F}_o$, such that any $\mathfrak{F} \in \mathcal{V}$ has no singularities at L_∞ and has an unique representation in the coordinate system (x,y) of the form $(P(x,y)+xg(x,y))\partial/\partial x + (Q(x,y)+xg(x,y))\partial/\partial y$, where P, Q, g are polynomials of degree n, g homogeneous, and the same coefficient of P, Q or g is 1. Let $F\colon \mathcal{V}\times\mathbb{C}^2 \to \mathbb{C}^2$ and $D\colon \mathcal{V}\times\mathbb{C}^2 \to \mathbb{C}$ be defined by

$$F(\mathfrak{F},x,y) = F(P,Q,g,x,y) =$$

$$(P(x,y)+xg(x,y),Q(x,y)+yg(x,y)) = (F_1(P,Q,g,x,y),F_2(P,Q,g,x,y))$$

$$D(\mathfrak{F},x,y) = D(P,Q,g,x,y) = \left(\frac{\partial F_1}{\partial x}\cdot\frac{\partial F_2}{\partial y} - \frac{\partial F_1}{\partial y}\cdot\frac{\partial F_2}{\partial x}\right)(P,Q,g,x,y).$$

The condition that $p_j = (x_j,y_j)$ is a nondegenerated singularity of $\mathfrak{F}_o$ is equivalent to $F(P_o,Q_o,g_o,x_j,y_j) = 0$ and $D(P_o,Q_o,g_o,x_j,y_j) \neq 0$. This implies that we can apply the implicit function theorem to F at $p_1,\dots,p_N$, to obtain the functions $\varphi_j\colon U_o \to V_j$, $j = 1,\dots,N$. Observe that this implies that $\mathcal{N}_n$ is open. Furthermore, if $\mathcal{V}$ is as above, then $\mathfrak{F} \in \mathcal{V} - \mathcal{N}_n$ if and only if there is $(x,y) \in \mathbb{C}^2$ such that $F(\mathfrak{F},x,y) = 0$ and $D(\mathfrak{F},x,y) = 0$. This implies that $\mathcal{V} - \mathcal{N}_n$ is an analytic subset of codimension ≥ 1 of $\mathcal{V}$, since $\mathcal{H}_n$ is not empty (see Jouanolou's example in §3.2). Therefore $\mathcal{H}_n \cap \mathcal{V}$ is open, dense and connected.

Let us prove that $\mathcal{P}_n$ is open and dense. Let $\mathcal{V}$, F and D be as above. If $\mathfrak{F} \in \mathcal{V}$, set

$$T(\mathfrak{F},x,y) = T(P,Q,g,x,y) = \left(\frac{\partial F_1}{\partial x} + \frac{\partial F_2}{\partial y}\right)(P,Q,g,x,y).$$

Let $\sigma(\mathfrak{F},x_o,y_o)$ and $(\sigma(\mathfrak{F},x_o,y_o))^{-1}$ be quotients of the eigenvalues of the Jacobian matrix, with respect to (x,y), of $F = (F_1,F_2)$ at (x_o,y_o). It is easy to see that σ and σ^{-1} satisfy the equation:

$$(*) \qquad \lambda^2 + (2-T^2/D)\lambda+1 = 0.$$

Observe that the roots of this equation are analytic functions of T^2/D in a neighborhood of T_o^2/D_o, if $T_o^2/D_o \neq 4$. This implies that, if $\mathfrak{F}_o \in \mathcal{P}_n$ and $\varphi_1,\ldots,\varphi_N$ are as before, then $\sigma(\mathfrak{F},\varphi_j(\mathfrak{F})) \notin \mathbb{R}$, $1 \le j \le N$, if $\mathfrak{F}$ is in a neighborhood of $\mathfrak{F}_o$. Therefore $\mathcal{P}_n$ is open.

On the other hand, the roots of (*) are real and non negative if, and only if $T^2/D \ge 4$. This implies that $\mathfrak{F} \in \mathcal{V}-\mathcal{P}_n$ if, and only if, there exists $(x,y) \in \mathbb{C}^2$ such that

$$F(\mathfrak{F},x,y) = 0 \quad \text{and} \quad \frac{T^2(\mathfrak{F},x,y)}{D(\mathfrak{F},x,y)} \ge 4.$$

Therefore $\mathcal{V}-\mathcal{P}_n$ is a semi-analytic subset of $\mathcal{V}$ of real codimension ≥ 1. This implies that $\mathcal{P}_n$ is dense in $\mathfrak{X}_n$. We leave the proof of the connectedness of $\mathcal{P}_n$ for the reader. ∎

The idea of the proof of Theorem B, is to construct an open and dense subset $\mathcal{G}_n \subset \mathcal{P}_n$ such that any $\mathfrak{F} \in \mathcal{G}_n$ has no algebraic solutions.

Fix $\mathfrak{F}_o \in \mathcal{P}_n$, with singularities $p_1,\dots,p_N$. Let $\mathcal{U}_o$ and $\varphi_j: \mathcal{U}_o \to V_j$ be as in Lemma 5, $1 \le j \le N$. We suppose also that $\mathcal{U}_o \subset \mathcal{P}_n$. In order to define $\mathcal{G}_n$ it is sufficient to say what is $\mathcal{G}_n \cap \mathcal{U}_o$. Let $j \in \{1,\dots,N\}$ and consider the equation

$$\lambda^2 + \left(2 - \frac{T^2(\mathfrak{F},\varphi_j(\mathfrak{F}))}{D(\mathfrak{F},\varphi_j(\mathfrak{F}))}\right) \lambda + 1 = 0.$$

Since $T^2(\mathfrak{F},\varphi_j(\mathfrak{F})) \ne 4D(\mathfrak{F},\varphi_j(\mathfrak{F}))$, for any $\mathfrak{F} \in \mathcal{U}_o$, we can write the solutions of this equation as analytic functions $\sigma(j,1): \mathcal{U}_o \to \mathbb{C}$ and $\sigma(j,2) = (\sigma(j,1))^{-1}$ (we suppose, for instance, that $\mathcal{U}_o$ is simply connected). Now, if $A \subset \{(j,k);\ 1 \le j \le N,\ k=1,2\}$, is a proper subset, define $\sigma(A) = \sum_{(j,k)\in A} \sigma(j,k)$. According to Theorem D, if $\mathfrak{F} \in \mathcal{U}_o$ has an algebraic solution, then $\sigma(A)(\mathfrak{F})$ is a positive integer, for some proper non empty subset A of $\{(j,k);\ 1 \le j \le N,\ k=1,2\}$. Let S be the set of all proper non empty subsets of $\{(j,k);\ 1 \le j \le N,\ k=1,2\}$. Then $\# S = 2N-1$. Observe that for any $A \in S$, $\sigma(A)$ can be extended as a multivalued function in $\mathcal{P}_n$. For Jouanolou's example, $\mathfrak{F}_1$, we have seen that $\sigma(A)(\mathfrak{F}_1) \notin \mathbb{N}$ for any $A \in S$. This implies that for all $A \in S$,

$$\{\mathfrak{F} \in U_o;\ \sigma(A)(\mathfrak{F}) \in \mathbb{N}\} = (\sigma(A))^{-1}(\mathbb{N})$$

is closed in U_o and has empty interior (because $\sigma(A)$ is analytic). Therefore the set $\bigcap_{A\in S} (\sigma(A))^{-1}(\mathbb{N}) = B$ is closed and has empty interior. Let $\mathcal{G}_n \cap \mathcal{U}_o = \mathcal{U}_o - B$. It follows that $\mathcal{G}_n \cap \mathcal{U}_o$ is open and dense in $\mathcal{U}_o$. Furthermore, if

$\mathcal{F} \in \mathcal{G}_n \cap U_o$, then $\mathcal{F}$ has no algebraic solutions. This ends the proof of Theorem B. ∎

3.4 - Some comentaries about the real case

If we consider real polynomial differential equations in two variables, instead of complex equations, then there is a problem to prove a result analogous to Theorem B: the set of Poincaré foliations of degree n is not dense in the set of all real foliations of degree n. The reason is that a foliation $\mathcal{F}$ could have a real singularity such that the quotient of its eigenvalues is real positive, different from 1, and this situation is persistent under small perturbations of $\mathcal{F}$. In fact, the problem is when this quotient is a positive rational number p/q, where p, q are relatively primes and $p,q \neq 1$. In this case, from Poincaré-Dulac's Theorem, it is possible to linearize $\mathcal{F}$ near the singularity. Therefore we can suppose that $\mathcal{F}$ is represented near the singularity by one of the differential equations bellow:

$$\begin{cases} \dot{x} = px \\ \dot{y} = qy \end{cases} \qquad \text{or} \qquad pxdy - qydx = 0.$$

Observe that these differential equations have a meromorphic first integral $y^p/x^q = c$. This implies that $\mathcal{F}$ has an infinite number of local analytic separatrices through $(0,0)$. Therefore we cannot use Theorem D in this case. However, with the same argument of Theorem B, we can prove the following:

Theorem B'. Let $\mathcal{S}_n(\mathbb{R}) = \mathcal{S}_n \cap \mathfrak{X}_n(\mathbb{R})$, where $\mathfrak{X}_n(\mathbb{R})$ is the set of all real foliations of degree n in $\mathbb{R}P(2)$. Then $\mathcal{S}_n(\mathbb{R})$ is generic in $\mathfrak{X}_n(\mathbb{R})$. Furthermore, there is a dense subset $\mathcal{G}_n(\mathbb{R}) \subset \mathcal{S}_n(\mathbb{R})$, which is relatively open in $\mathcal{S}_n(\mathbb{R})$, such that any $\mathcal{F} \in \mathcal{G}_n(\mathbb{R})$, has no algebraic solutions.

References

[1] P. Griffiths & J. Harris - Principles of Algebraic Geometry, Wiley-Interscience, New York, 1978.

[2] Peter Orlik - The Multiplicity of a Holomorphic Map at an Isolated Critical Point, Real and Complex Singularities, Proc. ninth Nordic Summer School/NAVF Sympos. Math., Oslo (1976) pp. 405-474.

[3] C. Camacho & P. Sad - Invariant Varieties through Singularities of Holomorphic Vector Fields, Ann. of Math., 115 (1982) pp. 579-595.

[4] J.P. Jouanolou - Equations de Pfaff Algebriques, Lecture Notes in Math. #708, Springer-Verlag.

[5] A. Lins Neto - Construction of Singular Holomorphic Vector Fields and Foliations in Dimension Two, to appear in the Journal of Diff. Geometry.

[6] E. Picard - Traité d'Analyse II, chap. XIII, Gauthier-Villars, Paris, 1893.

[7] H.B. Laufer - Normal Two-Dimensional Singularities, Princeton University Press and University of Tokyo Press, Princeton 1971.

[8] C. Camacho, A. Lins Neto & P. Sad - Topological Invariants and Equidesingularization for Holomorphic Vector Fields, Jr. of Diff. Geometry 20 (1984) pp. 143-174.

[9] V. Arnold - Chapitres Supplémentaires de la Théorie des Équations Differentielles Ordinaires, Edt. MIR, Moscow (1980).

[10] J. Palis & W. de Melo - Geometric Theory of Dynamical Systems, Springer-Verlag (1983).

[11] P.F. Baum & R. Bott - On the Zeroes of Meromorphic Vector Fields, Essays on Topology and Related Topics (Mémoires dediés à Georges de Rham), pp. 29-47, Springer, N.Y. (1970).

Alcides Lins Neto

Instituto de Matemática Pura e Aplicada
Estrada Dona Castorina 110
CEP 22460 - Rio de Janeiro, RJ - Brasil

THE SPACE OF SIEGEL LEAVES OF A HOLOMORPHIC VECTOR FIELD

Santiago López de Medrano
Universidad Nacional de Mexico
D.F. Mexico

We will consider some topological questions arising in the study of complex dynamical systems, namely systems of the form

$$\dot{z} = Az, \quad z \in \mathbb{C}^n \tag{1}$$

where A is a constant $n \times n$ matrix. The results can also be applied to the case of holomorphic dynamical systems that can be locally linearized around an equilibrium point.

Let $\lambda_1, \ldots, \lambda_n$ be the eigenvalues of A. To study the dynamics of this system one usually assumes the following generic *Hyperbolicity Hypothesis*: $\lambda_i \notin \mathbb{R}\lambda_j \quad i \neq j$.

That is, the eigenvalues of A are pairwise independent over the reals.

Under this hypothesis all λ_i are different and non-zero, and A can be assumed to be diagonal. The only equilibrium solution of (1) is then the origin, and all the other solutions are complex 1-dimensional leaves.

One looks then at the configuration of points λ_i in the complex plane, and makes the following distinction: One says that the matrix A (or the system (1)) is in the *Poincaré domain* if the origin is not in the convex hull of the λ_i. Otherwise, one says that A is in the *Siegel domain*. These two types of systems have different dynamical properties, of which we can only mention one: For a system in the Poincaré domain, all leaves get arbitrarily close to the origin. In the Siegel domain, we have some leaves with this property (called *Poincaré leaves*) and also leaves that are bounded away from the origin (called *Siegel leaves*). The analogous situations in the real case, which we can draw, would be a sink (or a source) and a saddle point

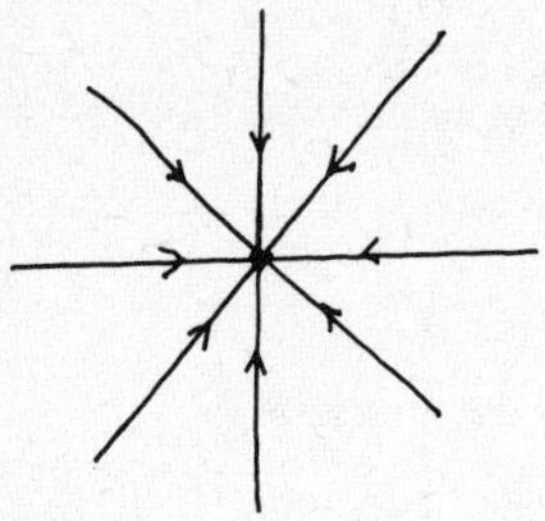

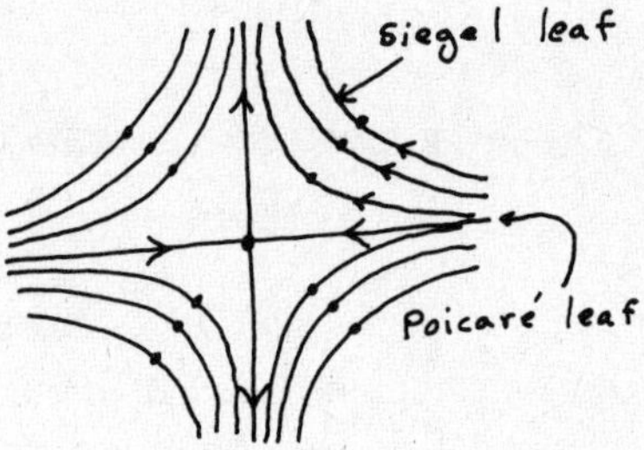

in the Poincaré domain in the Siegel domain

These pictures do not reflect, however, the intricacies of the phase portrait in the complex case. In fact, Camacho, Kuiper and Palis [CKP] have given a *complete topological classification* of these systems, showing that in the Siegel domain there are always *many continuous topological invariants* (or *moduli*).

One problem left open in [CKP] is that of describing the components of the Siegel domain, which is the same as describing the *discrete* topological invariants of these systems. Technically, one considers the open set of unordered configurations $\{\lambda_i\}$ satisfying the hyperbolicity hypothesis, and looks at its components. One of them is the Poincaré domain. (In [CKP], page 9, footnote, the number of components of the Siegel domain is given for $n = 3,4,5$).

PROBLEM 1. DESCRIBE THE COMPONENTS OF THE SIEGEL DOMAIN.

The Siegel leaves by themselves do have a nice disposition. In fact, on every Siegel leaf there is a unique point which is closest to the origin, and the set M of such points is given by the equation

$$\Sigma \lambda_i z_i \bar{z}_i = 0 \tag{2}$$

(without the origin). M can be identified with the quotient space of Siegel leaves (being actually a section) so this space is Hausdorff and in fact a smooth manifold. Let M_1 be the intersection of M with the unit sphere S^{2n-1}. Then M is diffeomorphic with $M_1 \times \mathbb{R}$. Another problem left open in [CKP], and raised by Xavier Gómez Mont and Alberto Verjovsky in various seminars in 1983-84, is the following:

PROBLEM 2. DESCRIBE THE TOPOLOGY OF M_1.

These two problems are closely related, since the topology of M_1 clearly does not change whitin each component of the Siegel domain. The solutions will be given in Theorems 1 and 2 below.

The equation defining M is not holomorphic, so it is better to consider it as two real equations (where $\lambda_i = \alpha_i + i\beta_i$):

$$\left.\begin{aligned} \Sigma\,\alpha_i(x_i^2 + y_i^2) &= 0 \\ \Sigma\,\beta_i(x_i^2 + y_i^2) &= 0 \end{aligned}\right\} \qquad (3)$$

that is, as the intersection of two real quadrics, so now we temporarily change to the field of real numbers. (Nevertheless, X. Gómez Mont has pointed out that M has a natural complex structure).

THE INTERSECTION OF REAL QUADRICS.

We consider now, more generally, the variety M given by equations

$$\left.\begin{aligned} a_1x_1^2 + \cdots + a_nx_n^2 &= 0 \\ b_1x_1^2 + \cdots + b_nx_n^2 &= 0 \end{aligned}\right\} \qquad (4)$$

and M_1 its intersection with the sphere S^{n-1}. One would like to know what are the "non-degenerate" cases, how many such cases there are, what is the topology of M_1 in the non-degenerate cases (to start with), etc.

These look like very classical questions, questions that should have been asked (and answered) 40 or 50 year ago. In fact we consulted lots of literature and experts in the fields of dynamical systems, topology and algebraic geometry, and we could find no reference to these questions. Only after we had obtained the results described in this paper, did we learn from professor C.T.C. Wall that a good number of them had been obtained by him and published in 1980 in [W]. In this paper he arrives at these questions in the study of topological stability of singularities of smooth mappings, and obtains essentialy our theorem 1 and a weaker version of our theorem 2 below. We will follow in the rest of this paper, however, our own point of view and development of the subject.

We first recall the classical results concerning one quadric:

$$\Sigma a_i x_i^2 = 0$$

In this case the non-degenerate situation is when all a_i are non-zero and the different non-degenerate cases are given by the partition of the a_i into positive and negative ones: $n = p + q$, which is known classically as the Sylvester signature or Morse index of the quadratic form.

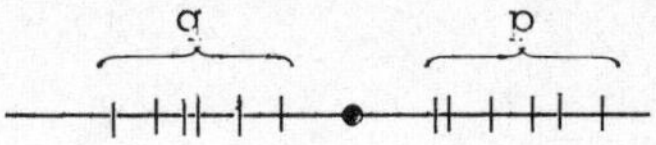

$n = p + q$

The number of cases is then $n + 1$, and each one can be put in normal form by deforming the p points $a_i > 0$ into a multiple point located at $+1$, and the q points $a_i < 0$ similarly into -1. The manifold obtained by intersecting with the unit sphere is then $S^{p-1} \times S^{q-1}$, a fact that follows immediately from the normal form by breaking the equations into two involving disjoint sets of variables (so in fact, for the normal form, the manifold is *equal* to a product of spheres).

For the case of 2 quadrics given by the equations (4) we consider the points $A_i = (a_i, b_i)$ in $\mathbb{R}^2$. The natural non-degeneracy condition to ask is that the two equations (4) be independent at every point of M, and this turns out to be equivalent to the following

WEAK HYPERBOLICITY HYPOTHESIS: $A_i \notin \mathbb{R}^- A_j$

in other words, that the origin is not in the convex hull of any pair A_i, A_j. (The names are borrowed from Chaperon [Ch], where this hypothesis had been previously considered).

So we have a configuration of n points $A_i \in \mathbb{R}^2$ where the origin is not in any of the segments connecting them. For $n = 3$ we have only two possibilities: if the origin is outside the triangle $A_1A_2A_3$ we are in the Poincaré domain, if not we are in the Siegel domain. For $n = 4$ we have again essentially two possibilities, but for $n = 5$ we get 4:

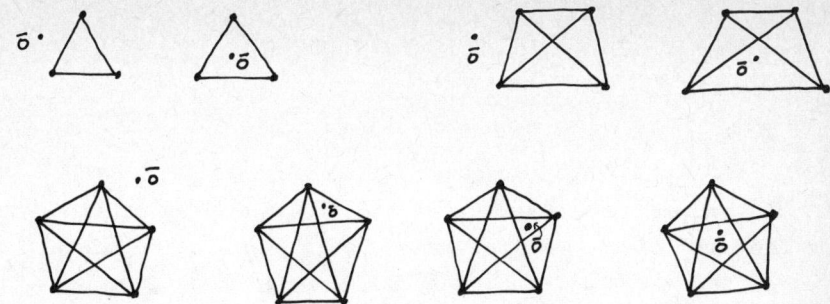

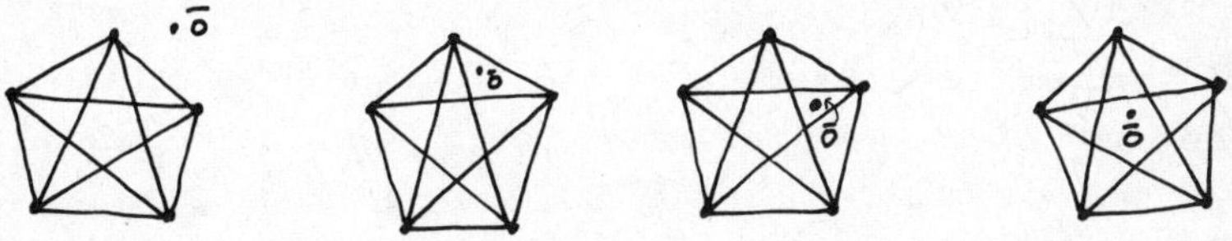

For $n > 5$ the pictures become more complicated, until one finds (after a lot of drawing) that the points A_i actually come in bunches, where each bunch can be concentrated into one multiple point without breaking the weak hyperbolicity condition. For example when $n = 4$ either all points can be concentrated into one, or only one pair can be concentrated into a double point:

In fact we have an equivalence relation, and a standard graph theory argument shows that the number of classes is always odd (as in the case $n = 4$ we just saw). So each class can be deformed into a multiple point, each multiple point can be pushed radially into the unit circle (as pointed out already in [CKP]) and finally one can push them into the vertices of a regular polygon. (For the details of this process see [L]).

We end up with a normal form consisting of the $k = 2\ell + 1$ vertices of a regular k-gon, each having a multiplicity n_i, $i = 1,\dots,k$.

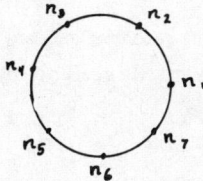

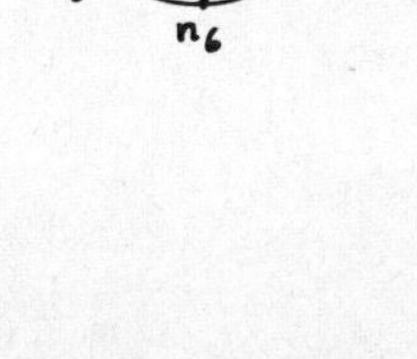

One should consider that two such configurations differing only by a rotation give the same normal form. This normal form is completely defined by the partition

$$n = n_1 + \cdots + n_k$$

into an odd number k of positive integers, where we consider two such as the same if they differ by a cyclic permutation of the n_i. If we call these *odd cyclic partitions*, and think again of the configuration of the A_i as an unordered set of points with multiplicity, we have shown that:

The connected components of configurations of n *points in* $\mathbb{R}^2$ *satisfying the weak hyperbolicity hypothesis are in one-to-one correspondence with the odd cyclic partitions of* n.

(In fact, the proof actually shows that each components is homotopy equivalent to S^1).

So the partition $n = n_1 + \cdots + n_k$ plays the same role as the signature $n = p + q$ in the case of one quadratic form. The number of such partitions can actually be computed: it is the same as the number of ways of dividing a pizza pie with n slices into an odd numbers of connected pieces

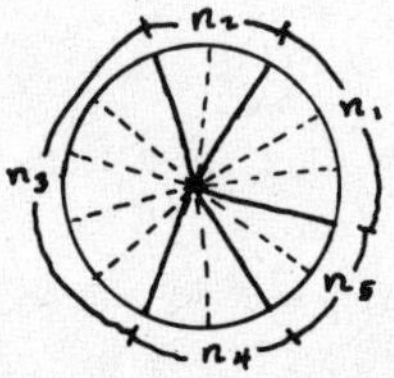

The set of cut points is one of the 2^n subsets of a set of n points, but only half of these are odd. Then one has to consider the n rotations of such a set as giving the same cyclic partition, so we get the number $2^{n-1}/n$. This is actually a *lower bound*, since a given set of cut points could be invariant under some rotations. This shows that the number of components grows exponentially (in contrast with the case of one quadric where the number of cases grows linearly with n).

Now, to describe the topology of M, one should be able to read all the topological invariants of M from the odd cyclic partition $n = n_1 + n_2 + \cdots + n_k$.

The first cases we considered were those coming from the original problem, that is, those having n and all n_i even: $n = 2m$, $n_i = 2m_i$ as in equations (3). In those cases there is a natural action of the torus T^n on M_1, defined by the action of each factor S^1 on a pair (x_i, y_i). The quotient of this action is a spherical convex polytope K lying in M_1 (so it is, in fact, a section of the action). The faces of K correspond with points of M_1 where one or more S^1 factors act trivially, so in principle knowing K one can reconstruct M^1 as a quotient of $T^n \times K$.

Therefore if we describe K, and if we know a similar action of T^n on a manifold M_1^1 whose quotient-section is K, we can conclude that M_1 is homeomorphic to M_1^1. For example, for the case $6 = 2 + 2 + 2$, K is a point, and since $T^3 = S^1 \times S^1 \times S^1$ acts on itself with quotient a single point, we can conclude that M_1 is homeomorphic to $S^1 \times S^1 \times S^1$. (This had been shown previously by X. Gómez Mont using complex variables arguments). More generally, in the case $n = 2p + 2q + 2r$ one gets that K is a product of simplices, at that M_1 is homeomorphic to $S^{2p-1} \times S^{2q-1} \times S^{2r-1}$, since we know the natural action of T^p on S^{2p-1} with quotient a $(p-1)$-simplex, etc.

For the case $10 = 2 + 2 + 2 + 2 + 2$, one gets an action on T^5 on M^7, having as quotient a pentagon. It is easy to show that M_1^7 is simply connected. By looking in the literature one finds that an action like this has been studied, and that D. McGavram [M] has shown that the corresponding manifold is homeomorphic to the connected sum of 5 copies of $S^3 \times S^4$. Therefore, in this case

$$M_1 = \underset{5}{\#}\, S^3 \times S^4$$

which is *not* a product of spheres, a very surprising result! (The story of these actions is interesting: first the experts on torus actions thought that T^n could not act on a simply connected manifold M^{n+2}, then they found some such actions, then they gave a complete list of them, andfinally McGavran showed that the corresponding manifolds were complicated connected sums of products of spheres. See [Mc] and its references).

Unfortunately, there are not many such good catalogs of torus actions, and this method does not give more results.

For the general case (4), we only have a $\mathbb{Z}_2^n$-action, where the i factor acts by sending x_i to $-x_i$. The quotient in this case is the intersection K of M_1 with the first orthant $(\mathbb{R}^+)^n$ which is again a spherical convex polyhedron which can be flattened to the linear convex polyhedron K^1 given by

$$a_1x_1 + \cdots + a_nx_n = 0$$

$$b_1x_1 + \cdots + b_nx_n = 0 \qquad x_i \geqslant 0$$

$$x_1 + \cdots \quad + x_n = 1$$

(which is the same as the set of convex combinations of the A_i that are equal to $\bar{0}$). M_1 can be reconstructed from K by reflecting on all hyperplanes $x_i = 0$. Again, from this construction one can tell in some cases the topological type of M_1: In the case $n = p + q + r$, K is a product of simplices and M_1 is $S^{p-1} \times S^{q-1} \times S^{r-1}$. In the case $5 = 1 + 1 + 1 + 1 + 1$, K is a pentagon, M_1 is connected and by computing the Euler characteristic one decides that M_1 is the surface of genus 5:

$$M_1 = \underset{5}{\#}\, S^1 \times S^1$$

So we have an action of $\mathbb{Z}_2^5$ (*not* orientation preserving) on the surface of genus 5 with quotient a pentagon which should have an interesting geometry behind it. This is the analog of McGavran's example, and belongs to a family of examples considered by Hirzebruch [H].

But again we don't know enough explicit examples of $\mathbb{Z}_2^n$ actions on manifolds to continue along these lines.

We have, however, a cell decomposition of M_1 given by K and its $2^n - 1$ reflections, and from this all the homology groups of M_1 can be computed in terms of the partition $n = n_1 + \cdots + n_k$. For example. assume we have a *Siegel* configuration (i.e., $\bar{0}$ is in the convex hull of the A_i, which means $k > 1$) and a point A_r such that when we remove A_r we get a *Poincaré* configuration (i.e. $\bar{0}$ is not on the convex hull of the rest of A_i, which only happens if $k = 3$ and one $n_i = 1$). We say that A_r is *indispensable*. This implies that $M_1 \neq \phi$ but $M_1 \cap \{x_r = 0\} = \phi$. By considering the reflection on $\{x_r = 0\}$ we see that there are parts of M_1 on both sides of this hyperplane:

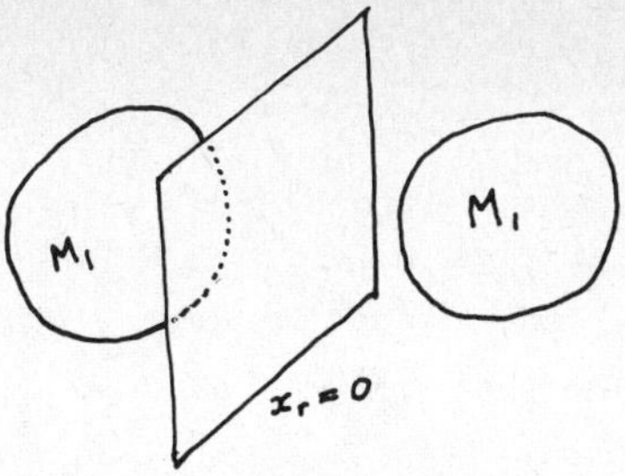

This means that M_1 is disconnected, or that we have detected a non-trivial 0-cycle.

If A_r, A_s is now a pair of points which is indispensable, but neither of them by itself is indispensable, we have a codimension 2 subspace $\{x_r = x_s = 0\}$ that doesn't meet M_1, but such that there are parts of M_1 all around it, and this detects a non-trivial 1-cycle in M_1.

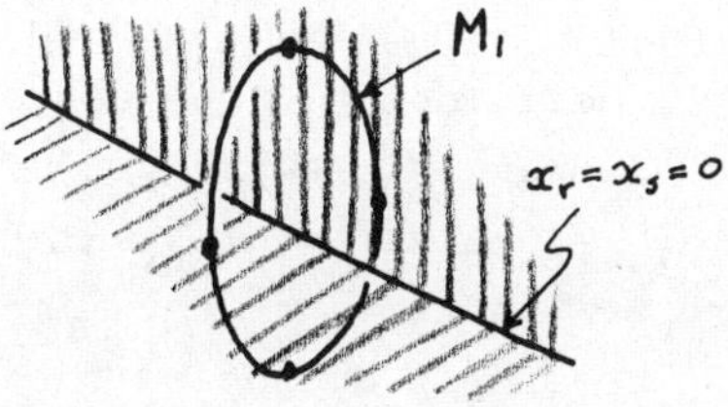

Notice, however that this can only happen if $k = 3$ and some $n_i = 2$, or if $k = 5$ and two consecutive n_i are 1. In fact, the dual cell complex can be identified with a subcomplex of the n-cube, and this shows that only on these cases can M_1 be non-simply connected.

This computation can be carried out to the end to show that the homology of M_1 is always free with one generator for each minimal indispensable subset of A_i's and a dual generator for each complementary subset of A_i's (these are the maximal Poincaré subsets). The minimal indispensable subsets can be easily described: if $k = 2\ell + 1$, then they are the unions of ℓ consecutive classes. Define then the numbers d_i, $i = 1,\dots,k$

$$d_1 = n_1 + n_2 + \cdots + n_\ell$$

$$d_2 = n_2 + n_3 + \cdots + n_{\ell+1}, \text{ etc.}$$

(following the cyclic order), then we have shown that the homology of M_1 is the same as that of the connected sum

$$M_1^1 = \mathop{\#}_{i=1}^{k} S^{d_i - 1} \times S^{n-d_i-2}$$

The proof of these facts is long and can be found in [L]. A different proof had been carried out previously in [W]. Our explicit description of the generators lets us go a couple of steps further to show that *if* $k > 3$ then M_1 looks very much like the connected sum M_1^1: all homology classes of M_1 can be represented by embedded spheres with trivial normal bundle, and further M_1 and M_1^1 are H-cobordant (i.e. there is a cobordism between M_1 and M_1^1 whose homology is that of $M_1 \times I$). See [L].

These facts are definitely not true for $k = 3$, but in this case we know M_1 is a product of spheres. (In fact by playing with the equations in normal form defining M_1 one can show in a completely elementary way that M_1 is *equal* to a product of 3 spheres).

To conclude that M_1 *is actually diffeomorphic to* M_1^1 one can now apply the h-cobordism theorem, but this implies some technical difficulties: M_1 must be simply connected and of dimension greater than 4, which happens in *most* of the cases. (The only cases excluded are $k = 5$ with some $d_i = 2$, and the case $n = k = 7$). Other cases can be covered by different arguments and the proof could be pushed to cover some other ones, so we can safely conjecture that M_1 and M_1^1 are always diffeomorphic. But it seems clear that a different, more elementary, proof is needed. All of the cases (but one) considered in the applications are covered by our proof; in particular all those cases coming from holomorphic vector fields.

(One conjecture we made in the early stages of this work was that the topology of M_1 determined the partition of n up to a rotation and a reflection in $\mathbb{R}^2$, or, in view of the results of [CKP], that dynamical systems with homeomorphic M_1 could be deformed into topologically equivalent ones without breaking the hyperbolicity condition. The examples $8 = 3 + 1 + 2 + 1 + 1$ and $8 = 2 + 2 + 2 + 1 + 1$ and their doubles show that these conjectures were wrong).

Here the main open question would be to extend these result to the case of the intersection of more than 2 quadrics. The main problem seems to be the description of the simple configurations (i.e. the analogs of the regular odd polygons). For the intersection of 3 quadrics, the corresponding manifold can be a product of 4 spheres, a product of a sphere with a connected sum of the type M_1^1 above, and also a connected sum that is not a product. This questions can be applied to the study of linear $\mathbb{R}^k$-actions, as was pointed to us by Chaperon (See [Ch]).

Other important questions would be to extend these results to the intersections of quadrics whose equations are not simultaneously diagonalizable and to the simplest degenerate cases.

SOME APPLICATIONS.

We mention briefly some applications of the previous results and constructions. First we have the manifolds M_1. Unlike the Brieskorn varieties there is nothing exotic or new about them except for the fact that we have connected sums given by very simple equations.

Then we have a lot of group actions. For the case $n = n_1 + \cdots + n_k$ we have a natural action of $O(n_1) \times \cdots \times O(n_k)$ on M_1. In particular we have an interesting family of actions of $\mathbb{Z}_2^n$ on highly connected manifolds. The first example is the action of $\mathbb{Z}_2^5$ on the surface of genus 5, which also belongs to the family studied by Hirzebruch ([H]) which includes actions of $\mathbb{Z}_2^n$ on the surface of genus $2^{n-3}(n - 4) + 1$. In fact Hirzebruch represent these actions as an intersection of $n - 2$ quadrics in $\mathbb{R}^n$. His construction generalizes as follows: Let K be a compact convex polytope of dimension k, which we can assume embedded in $\mathbb{R}^k$, and given by n linear inequalities $L_i \geqslant 0$. The mapping $p \to (L_1(p),\dots,L_n(p))$ defines an embedding of K into $\mathbb{R}^n$ whose image is the intersection of an affine k-plane with the first orthant. If $\Sigma a_{ij}x_j = b_i$ are the equations defining L, $\Sigma a_{ij}x_j^2 = b_j$ defines an intersection of quadrics, with an action of $\mathbb{Z}_2^n$ whose quotient is K. If K is such that its faces are in general position then the intersection of quadrics is non-degenerate and therefore a manifold. This gives a complete correspondence between intersections of quadrics, convex polytopes and certain $\mathbb{Z}_2^n$ actions.

(This in turn gave us the idea to apply our previous results to the classification of generic k-polytopes with $k + 3$ faces, but then we learned from Wall [W] that all this could be found in the literature on convex polytopes, where the above construction is known as the Gale diagram of the polytope. See [G].

We also have the torus actions, with a nice family of T^n actions on highly connected manifolds of dimension $2n-3$, the first one being the example of McGavran [M], so we can represent this action by equations. In fact, from the above constructions we can also get T^n actions with a given polytope as quotient, so we can give all of McGavran's examples (i.e., all T^n actions on simply connected M^{n+2}), in principle, by simple quadratic equations.

When all the n_i are equal we have an additional action of $\mathbb{Z}_k$ by cyclic permutation of coordinates, which restricts to an action on K. If further all the $n_i = 1$ then this action is fixed point free, giving some interesting cell decompositions of some lens spaces.

Another application on which we worked very hard was that of finding new minimal submanifolds of S^n (suggested by A. Verjovsky). A long computation, which really should be checked over, seems to show that the only manifolds, M_1 which can be minimal are certain products of round spheres, and these are in fact well known. (It should be acknowledged here, however, that all the other suggestions of Verjovsky regarding this subject, and there were many, were very fruitful).

BACK TO DYNAMICAL SYSTEMS.

If we observe that the connected componets of configurations satisfying weak hyperbolicity do not change when we restric to hyperbolic ones (but "normal forms" should now be clouds of points in the neighborhood of the vertices of the regular odd polygons) we have

THEOREM 1. *The components of the Siegel domain are in one-to-one correspondence with the odd cyclic partitions* $n = n_1 + \cdots + n_k$ *of* n *into* $k > 1$ *positive integers.*

Now one can check that all the cases arising in this situation satisfy the conditions needed to apply the h-cobordism theorem to the manifolds M_1 and M_1^1 above so we have:

THEOREM 2. *If* M *is the manifold of Siegel leaves of the system* (1) *corresponding with the partition* $n = n_1 + \cdots + n_k$,

(i) *If* $k = 1$, *then* $M = \phi$ *(Poincaré domain)*

(ii) *If* $k = 3$, *then* M *is diffeormorphic to*

$$S^{2n_1-1} \times S^{2n_2-1} \times S^{2n_3-1} \times \mathbb{R}$$

(iii) *If* $k = 2\ell + 1 > 3$ *then* M *is diffeomorphic to*

$$\left(\mathop{\#}_{i=1}^{k} S^{2d_i - 1} \times S^{2n-2d_i-2} \right) \times \mathbb{R}$$

where $d_i = n_i + n_{i+1} + \cdots + n_{i+\ell-1}$.

These theorems give the solution to problems 1 and 2 above.

The main open question here would be to find the implications of these results in the dynamics of the system (1).

REFERENCES

[CKP] C. Camacho, N. Kuiper and J. Palis. The topology of holomorphic flows with singularities. Publications Mathematiques I.H.E.S., 48 (1978), 5-38.

[Ch] M. Chaperon. Géometrie différentielle et singularités de systèmes dynamiques. Asterisque (1986), 138-139.

[G] B. Grünbaum. Convex Polytopes, Wiley, 1967.

[H] F. Hirzebruch. Arragnements of lines and algebraic surfaces, in M. Artin, J. Tate, eds., Arithmetic and Geometry, Papers dedicated to I.R. Shafarevich, Vol. II, Birkhäuser, 1983, 114-140.

[L] S. López de Medrano. Topology of the Intersection of quadrics in $\mathbb{R}^n$. To appear in Proceedings of the Algebraic Topology Conference in honor of E.H. Brown, Arcata 1986. Springer-Verlag Lecture Notes.

[Mc] D. McGavran. Adjacent connected sums and torus actions, Trans. Amer. Math. Soc. 251 (1979), 235-254.

[W] C.T.C. Wall. Stability, pencils and polytopes, Bull. London Math. Soc., 12 (1980), 401-421.

DEFORMATIONS OF SINGULAR HOLOMORPHIC FOLIATIONS ON REDUCED COMPACT $\mathbb{C}$-ANALYTIC SPACES.

by

Geneviève POURCIN

Summary

One can define holomorphic foliations with singularities on a reduced complex space X as a coherent subsheaf T of the tangent sheaf Θ_X stable by the bracket of derivations ([B],[G-M],[P2],[S]) or as a coherent subsheaf Ω of the sheaf of holomorphic 1-forms satisfying an integrability condition ([R],[Su]).

If X is compact the set of all the (singular) foliations on X has an universal analytic structure associated to each definition (vector fields or differential forms) ;these analytic structures are different but coincide on the open subset of regular foliations ,

Moreover one obtains a semi-universal simultaneous deformation of a compact manifold and its foliations .

I thank H.J.REIFFEN , X.GOMEZ-MONT and H.FLENNER for usefull discussions.

I - SINGULAR FOLIATIONS ON A REDUCED COMPLEX SPACE

Let X be a reduced $\mathbb{C}$-analytic space , O_X its structural sheaf , Ω_X the coherent sheaf of holomorphic 1-forms ; the tangent sheaf $\Theta_X = \underline{Hom}_{O_X}(\Omega_X, O_X)$ is the sheaf of the derivations of O_X with values in O_X ([G-R] III §4) .

- If V is a Stein open set of $\mathbb{C}^n$ and X the analytic subspace of V defined by a coherent ideals sheaf I , then Θ_X is the kernel of the map

$$\varphi : O_X^n \longrightarrow \underline{Hom}_{O_X}(I, O_X)$$

defined by

$$\varphi(a_1,\dots,a_n)(f) = \sum_{i=1}^{n} a_i \frac{\partial f}{\partial z_i}\Big|_X$$

(where $(z_1,\dots,z_n)$ are the coordinates of $\mathbb{C}^n$.)

Take $\tau = (a_1,\dots,a_n) \in \ker \varphi(V)$ and $f \in O_X(V)$ one has by definition

(1) $$\tau(f) = \sum_{i=1}^{n} a_i \frac{\partial \tilde{f}}{\partial z_i}\Big|_X$$

for any holomorphic extension $\tilde{f}$ of f to V .

- For any open subset U of X let $m_U : \Theta_X(U) \times \Theta_X(U) \to \Theta_X(U)$ the bracket of derivations :

(2) $$m_U(\tau_1, \tau_2) = \tau_1 \tau_2 - \tau_2 \tau_1$$

and $m : \Theta_X \times \Theta_X \to \Theta_X$ the $\mathbb{C}$-bilinear morphism of sheaves defined by the m_U's. As for any coherent sheaf , for any open set U , $\Theta_X(U)$ has a natural Frechet topology ([C]).

<u>Proposition 1</u>

<u>For any open set</u> U <u>the</u> $\mathbb{C}$<u>-bilinear map</u> m_U <u>is continuous for the natural Frechet topology on</u> $\Theta_X(U)$.

Proposition 1 follows from formulas (1) and (2) and from the Cauchy majorations

- For any coherent O_X-submodule T of Θ_X the restriction of m to TxT induces an O_X-linear map

$$m_T : \Lambda^2 T \to \Theta_X / T$$

(this remark is already done in [G-M] when X is a manifold .)

- Let Ω be a coherent O_X-submodule of Ω_X of generic rank r ; in [R] H.J.REIFFEN considers the $\mathbb{C}$-multilinear map

$$\delta : \Omega_X^{r+1} \to \Lambda^{r+2}\Omega_X / \tau$$

defined by

$$\delta(\omega_o, \omega_1, \dots, \omega_r) = \text{class of } d\omega_o \wedge \omega_1 \wedge \dots \wedge \omega_r$$

(τ being the torsion submodule of $\Lambda^{r+2}\Omega_X$) and the restriction δ_Ω of δ to Ω^{r+1} .

δ_Ω is an O_X-multilinear map .

<u>Definition 1</u>

(1) <u>An holomorphic</u> V-<u>foliation of dimension</u> r <u>on</u> X <u>is a coherent</u> O_X<u>-submodule</u> T <u>of</u> Θ_X <u>of generic rank</u> r , <u>stable by the bracket of derivations (or equivalently such that</u> $m_T = 0$.)

(2) <u>An holomorphic</u> D-<u>foliation of codimension</u> r <u>on</u> X <u>is a coherent</u> O_X<u>-submodule</u> Ω <u>of</u> Ω_X <u>of generic rank</u> r <u>and such that</u> $\delta_\Omega = 0$.

Remark

A V-foliation (resp. D-foliation) induces a regular foliation outside an analytic set .

For any D-foliation Ω , $T = \underline{\mathrm{Hom}}_{O_X}(\Omega_X/\Omega\ ,\ O_X)$ is a V-foliation .For any V-foliation T on a manifold X , $\Omega = \underline{\mathrm{Hom}}_{O_X}(\Theta_X/T\ ,\ O_X)$ is a D-foliation; but the correspondence between D-foliations and V-foliations is not a bijection (see for instance [Su] .)

II - DEFORMATIONS OF FOLIATIONS ON X

For any analytic space S we denote by $p_S : S\times X \rightarrow X$ the projection , $m_S : p_S^*\Theta_X \times p_S^*\Theta_X \rightarrow p_S^*\Theta_X$ and $\delta_S : p_S^*\Omega_X^{r+1} \rightarrow p_S^*(\wedge^{r+2}\Omega_X/\mathcal{C})$ the pull back of m and δ by p_S .

Definition 2 : Flat family of V-foliations on X

Let S be an analytic space , a flat family of V-foliations parametrized by S is a S-flat quotient of $p_S^*\Theta_X$ by a $O_{S\times X}$-coherent m_S-stable submodule T (or equivalently such that the $O_{S\times X}$-linear map $\wedge^2 T \rightarrow p_S^*\Theta_X/T$ induced by m_S is identically zero).

Definition 3 : Flat family of D-foliations on X

Let S be an analytic space , a flat family of D-foliations on X parametrized by S is a S-flat quotient of $p_S^*\Omega_X$ by a $O_{S\times X}$-coherent submodule Ω such that the restriction of δ_S to Ω^{r+1} is identically zero .)

- Let us recall the following notations and results of [P1] : let $Y \rightarrow S$ be a morphism and $\underline{E}$ a coherent sheaf on Y ; for any morphism $Z \rightarrow S$ the pull back of $\underline{E}$ by the projection $Y\times_S Z \rightarrow Y$ is denoted by $\underline{E}_Z$ .Then we have

Proposition 2 ([P1] §2 prop.1)

Let $Y \rightarrow S$ be a morphism of analytic spaces and $\underline{E}$ a coherent sheaf on Y S-proper and S-flat . Let $\underline{F}$ be a coherent quotient of $\underline{E}$ .Then it exists an anlytic subspace T of S such that

(i) $\underline{E}_T = \underline{F}_T$

(ii) any morphism $Z \rightarrow S$ such that $\underline{E}_Z = \underline{F}_Z$ admits an unique factorization through T .

as an easy corollary we get

Proposition 3

Let $Y \to S$ be a morphism of analytic spaces , $\underline{E}$ and $\underline{G}$ two coherent sheaves on X , $\underline{E}$ being S-proper and S-flat . Let $h : \underline{G} \to \underline{E}$ be an O_X-morphism . Then it exists an analytic subspace T of S such that

(i) $h_T : \underline{G}_T \to \underline{E}_T$ is identically zero

(ii) any morphism $Z \to S$ such that the pull back $h_Z : \underline{G}_Z \to \underline{E}_Z$ is identically zero factorizes through T .

proof : apply proposition 2 to $\underline{E}$ and coker h .

- Now we suppose that X is a compact reduced analytic space . Let H_1 be the DOUADY space ([D1]) of the coherent quotients of Θ_X , $p_1 : H_1 \times X \to X$ the projection , R_1 the universal quotient .We can write $R_1 = p_1^* \Theta_X / \underline{T}$ where $\underline{T}$ is a coherent $O_{H_1 \times X}$- submodule of $p_1^* \Theta_X$; R_1 is H_1-proper and H_1-flat and for any $h \in H_1$ the fiber $R_1(h) = \Theta_X / T(h)$ is the quotient of Θ_X represented by h .

The pull back m_{H_1} of the bracket map induces an $O_{H_1 \times X}$-linear map

$$m_T : \Lambda^2 T \to R_1$$

Then we apply proposition 3 to $S = H_1$, $Y = H_1 \times X$, $\underline{E} = R_1$, $\underline{G} = \Lambda^2 T$ and $h = m_T$ and we get

THEOREM 1

Let X be a compact reduced space ; there exist an analytic space H and a flat family $\mathcal{R}$ of V-foliations on X parametrized by H with the following universal property :

for any analytic space S and any flat family R of V-foliations on X parametrized by S it exists an unique morphism $\varphi : S \to H$ satisfying $(\varphi \times I_X)^* \mathcal{R} = R$

H is the set of all the V-foliations on X and it is an analytic subspace of the DOUADY space H_1 .This result has also been obtained by X.GOMEZ-MONT ([G-M] if X is a compact manifold ;see also [P2]) .

Let H_2 be the DOUADY space of the coherent quotients of Ω_X, $p_2 : H_2 \times X \to X$ the projection , $R_2 = p_2^* \Omega_X / \Omega$ the universal quotient and

$$\delta_\Omega : \Omega^{\otimes r+1} \longrightarrow p_2 (\Lambda^{r+2} \Omega_X / \mathcal{I})$$

the $O_{H_2 \times X}$-morphism induced by the pull back of δ. It follows from proposition 3 applied to $S = H_2$, $Y = H_2 \times X$, $\underline{E} = p_2^*(\Lambda^{r+2}\Omega_X/\mathcal{Z})$, $\underline{G} = \Omega^{\otimes r+1}$ and $h = \delta_\Omega$ the following theorem

THEOREM 2

Let X be a compact reduced space and $r \in N$; there exist an analytic space K and a flat family $\mathcal{D}$ of D-foliations of codimension r on X parametrized by K with the following universal property :

for any analytic space S and any flat family $\underline{D}$ of D-foliations of codimension r on X parametrized by S it exists an unique morphism $\Psi : S \to K$ satisfying $(\varphi \times I_X)^* \mathcal{D} = \underline{D}$.

K is the set of all the D-foliations of codimension r on X .
H.J.REIFFEN ([R]) has obtained the reduced structure on K .

III - THE UNIVERSAL SPACE OF REGULAR FOLIATIONS ON A COMPACT MANIFOLD

Let X be a connected compact $\mathbb{C}$-analytic manifold . In [G-H-S] a germ of versal deformation of a given regular V-foliation on X is obtained .
A V-foliation $\underline{R} = \Theta_X/T$ is regular if and only if $\underline{R}$ is locally free ; a D-foliation $\underline{D} = \Omega_X/\Omega$ is regular if and only if $\underline{D}$ is locally free .
Let $n = \dim_{\mathbb{C}} X$.
The correspondence

$$\underline{R} \mapsto \Omega_X / \underline{Hom}_{O_X}(\underline{R}, O_X)$$

$$\underline{D} \mapsto \Theta_X / \underline{Hom}_{O_X}(\underline{D}, O_X)$$

is a bijection between the set V_r of regular V-foliations of dimension r and the set D_r of the regular D-foliations of codimension (n-r). We shall prove that V_r is an open subset of H , D_r an open subset of K ,and that V_r and D_r are canonically isomorphic .
Notations are those of theorems 1 and 2 .

Lemma 1

(i) V_r is an open subset of H

(ii) D_r is an open subset of K

proof : let $h_o \in V_r$ be a regular V-foliation ; then $\mathcal{R}(h_o)$ is locally free of rank r at any point of X . For any $x \in X$ the H-flatness of $\mathcal{R}$ implies the existence of neighborhoods U_x of h_o and V_x of x such that the restriction of $\mathcal{R}$ to $U_x \times V_x$ is free of rank r ; the compacity of X implies that $\mathcal{R}(h)$ is locally free of rank r if h belongs to a suitable neighborhood of h_o ; lemma 1(i) follows ; the proof of (ii) is similar .

Let us denote by $p : V_r \times X \to X$ the projection , again by $\mathcal{R}$ the restriction of $\mathcal{R}$ to V_r and $\mathcal{R} = p^*\Theta_X / T$; then we have a splitted exact sequence of $O_{V_r \times X}$-locally free modules

$$0 \to T \to p^*\Theta_X \to \mathcal{R} \to 0$$

and by duality a splitted exact sequence on $V_r \times X$

$$0 \to \underline{\mathrm{Hom}}\,(\mathcal{R}, O_{V_r \times X}) \to p^*\Omega_X \to \underline{\mathrm{Hom}}\,(T, O_{V_r \times X}) \to 0$$

Then $\Omega_1 = \underline{\mathrm{Hom}}\,(\mathcal{R}, O_{V_r \times X})$ is free of rank (n-r) ; let us verify that

the free module $\underline{D} = \underline{\mathrm{Hom}}\,(T, O_{V_r \times X})$ is a flat family of regular D-foliations ; we have to prove that

$$\delta_{\Omega_1} : \Omega_1^{n-r+1} \longrightarrow p^*(\Lambda^{n-r+2}\Omega_X)$$

is identically zero ; it is a relative version of Frobenius's theorem: If $d_X : p^*\Omega_X \to p^*\Lambda^2\Omega_X$ is the pull back of the differential $\Omega_X \to \Lambda^2\Omega_X$, we have for any sections t_1 and t_2 of $p_1^*\Theta_X$ and any section ω of $p^*\Omega_X$

$$(*) \qquad d_X\omega(t_1,t_2) = t_1(\omega(t_2)) - t_2(\omega(t_1)) - \omega(m(t_1,t_2))$$

and if t_1 and t_2 are sections of T and if ω is a section of Ω_1

$$d_X\omega(t_1,t_2) = 0$$

Now take locally a basis $\{t_1,\dots,t_r,\dots,t_n\}$ of $p^*\Theta_X$ such that $\{t_1,\dots,t_r\}$ is a basis of T ; any (n-r+2)-uple of the t_i's contains at least two sections of T ;then $\delta_{\Omega_1} = 0$ follows .

Let us denote again by $\mathcal{D}$ the restriction of $\mathcal{D}$ to D_r and put $\mathcal{D} = p^*\Omega_X/\Omega$ .We obtain

<u>Proposition 4</u>

<u>With the previous notations it exists an unique morphism $\varphi : V_r \to D_r$ such that $(\varphi \times I_X)^*\mathcal{D} = \underline{\mathrm{Hom}}\,(T, O_{V_r \times X})$</u>
<u>Moreover one has $(\varphi \times I_X)^*\Omega = \underline{\mathrm{Hom}}(\mathcal{R}, O_{V_r \times X})$</u> .

Now ,by duality , the splitted exact sequence of $O_{D_r \times X}$-modules

$$0 \to \Omega \to p^*\Omega_X \to \mathcal{D} \to 0$$

gives a splitted exact sequence

$$0 \to \underline{\mathrm{Hom}}\,(\mathcal{D}, O_{D_r \times X}) \to p^*\Theta_X \to \underline{\mathrm{Hom}}\,(\Omega, O_{D_r \times X}) \to 0$$

If $\{\omega_1, \ldots, \omega_{n-r}, \ldots, \omega_n\}$ is locally a basis of $p^*\Omega_X$ such that $\{\omega_1, \ldots, \omega_{n-r}\}$ is a basis of Ω ; then it results from [Sa] that for any $i \in \{1, \ldots, n-r\}$ the condition $d_X\, \omega_i \wedge \omega_1 \wedge \cdots \wedge \omega_{n-r} = 0$ implies the existence of sections s_{ij} of $p^*\Omega_X$ such that

$$d_X\, \omega_i = \sum_{j=1}^{r} s_{ij} \wedge \omega_j$$

Then it follows easily from ($*$) that $T = \underline{\mathrm{Hom}}\,(\mathcal{D}, O_{D_r \times X})$ satisfies $m_T = 0$. We obtain

Proposition 5

With the previous notations it exists an unique morphism $\psi : D_r \to V_r$ such that $(\psi \times I_X)^* \mathcal{R} = \underline{\mathrm{Hom}}(\Omega, O_{D_r \times X})$

Moreover one has $(\psi \times I_X)^* T = \underline{\mathrm{Hom}}\,(\mathcal{D}, O_{D_r \times X})$.

Therefore ,for free modules of finite type , duality and pull back commute ;then $((\varphi \circ \psi) \times I_X)\, \mathcal{D} = \mathcal{D}$ and $((\psi \circ \varphi) \times I_X)\mathcal{R} = \mathcal{R}$ and the universal property of H and K implies

$$\varphi \circ \psi = \text{identity of } D_r$$

$$\psi \circ \varphi = \text{identity of } V_r$$

Proposition 6

Morphisms φ and ψ of propositions 4 and 5 are reciprocal isomorphisms .

IV - SIMULTANEOUS DEFORMATIONS OF A COMPACT MANIFOLD AND ITS FOLIATIONS

Let X_o be a compact connected manifold and $\pi : X \to S$ its semi-universal deformation ([K] ,[D2]) ; S is a germ of analytic space and π is proper and smooth .One can define the relative tangent sheaf $\Theta_{X/S}$ (it satisfies $\Theta_{X/S}(s) = \Theta_{X(s)}$) and a Lie bracket in the direction of the fibers of π

$$m_S : \Theta_{X/S} \times \Theta_{X/S} \longrightarrow \Theta_{X/S}$$

m_S is an O_S-bilinear map, the restriction to the fibers of which is the classical Lie bracket.

Let Σ be the set of the coherent quotients $\Theta_{X/S}/T$ with compact support such that T is m_S-stable. It follows from [P2] (III-3 Remark) that Σ has an universal analytic structure and it exists on $\Sigma \times X$ an universal quotient $Q = p_\Sigma^* \Theta_{X/S}/T$, p_Σ being the projection of $\Sigma \times X$ on X. Q is Σ-proper and Σ-flat.

Now we follow the argument of [S-T] (§1-5) : Q is an O_S-module via $\pi \circ p_\Sigma$ but Q is not in general $S \times \Sigma$-flat ; the flattening theorem of J. FRISCH [F] implies the existence of a bijective immersion

$$\alpha : \mathfrak{S} \longrightarrow S \times \Sigma$$

such that $F := \alpha^* Q$ is $\mathfrak{S}$-flat and satisfies :

for any morphism $\beta : Z \longrightarrow S \times \Sigma$ such that $\beta^* Q$ is Z-flat, β has an unique factorization through $\mathfrak{S}$.

Let $\underline{X} = X \times_S \mathfrak{S}$, let $\underline{\pi} : \underline{X} \longrightarrow \mathfrak{S}$ and $p : \underline{X} \longrightarrow X$ the projections ; $\underline{\pi}$ is smooth and one has

$$p^* \Theta_{X/S} = \Theta_{\underline{X}/\mathfrak{S}}$$

and the following theorem :

THEOREM 3

($\underline{X}$, $\mathfrak{S}$, $\underline{F}$) is a simultaneous semi-universal deformation of X_0 and its V-foliations, i.e.

(i) $\underline{X} = X \times_S \mathfrak{S} \longrightarrow \mathfrak{S}$ is a deformation of X_0

(ii) $\underline{F} = \Theta_{\underline{X}/\mathfrak{S}}/\mathcal{E}$, $\mathcal{E}$ being a $m_\mathfrak{S}$-stable coherent submodule of $\Theta_{\underline{X}/\mathfrak{S}}$

(iii) $\underline{F}$ is $\mathfrak{S}$-proper and $\mathfrak{S}$-flat

(iv) ($\underline{X}, \mathfrak{S}, \underline{F}$) is semi-universal for these properties (i),(ii),(iii).

(iv) means that for any other $(X', \mathfrak{S}', F')$ satisfying (i),(ii),(iii) it exists a morphism $\gamma : \mathfrak{S}' \longrightarrow \mathfrak{S}$ unique up to the first order such that $X' = \underline{X} \times_\mathfrak{S} \mathfrak{S}'$ and $\gamma^* \underline{F} = F'$.

The semi-universality of ($\underline{X}$, $\mathfrak{S}$, $\underline{F}$) results from the semi-universality of S, the universality of Σ and from the fact that α is an immersion.

Remark

The subset of $\mathfrak{G}$ corresponding to regular V-foliations of a fixed dimension r is an open subset $\mathfrak{G}_r$ of $\mathfrak{G}$.

- For D-foliations we have a similar result .Let $\Omega_{X/S}$ be the sheaf of relative differential forms and

$$d_{X/S} : \Omega_{X/S} \to \Lambda^2 \Omega_{X/S}$$

tne relative differential map ; let us consider an $r \in N$, $n = \dim_{\mathbb{C}} X$ and

$$\delta : (\Omega_{X/S})^{n-r+1} \to \Lambda^{n-r+2} \Omega_{X/S}$$

$$\delta (\omega_o, \omega_1, \ldots, \omega_{n-r}) = d_{X/S}\omega_o \wedge \omega_1 \wedge \cdots \wedge \omega_{n-r}$$

Following [P2] we obtain an universal structure on the set $\tilde{\Sigma}$ of the coherent quotients $\Omega_{X/S}/G$ with compact support , G being a submodule of generic rank (n-r) such that the restriction δ_G of δ to G is identically zero ; moreover it exists on $\tilde{\Sigma} \times X$ an universal quotient $\tilde{Q} = p^*_{\tilde{\Sigma}} \Omega_{X/S} / \Omega$; $\tilde{Q}$ is $\tilde{\Sigma}$ -proper and $\tilde{\Sigma}$ - flat .
Following the previous construction([S-T]) we get a semi-universal simultaneous deformation of X_o and its D-foliations of codimension (n-r), $(\tilde{X}, \tilde{\mathfrak{G}}, \underline{\tilde{F}})$. The subset of $\tilde{\mathfrak{G}}$ corresponding to regular D-foliations is an open subset $\tilde{\mathfrak{G}}_r$ of $\tilde{\mathfrak{G}}$.

From the argument of A.DOUADY given in [G-H-S](§1-6) it follows that the analytic spaces $\mathfrak{G}_r$ and $\tilde{\mathfrak{G}}_r$ are (not canonically) isomorphic.

BIBLIOGRAPHY

[B] - P.BAUM "Structure of foliations singularities" Adv.in Math.15(1975)

[C] - H.CARTAN "Faisceaux analytiques coherents" C.I.M.E.(1963)Ed.Cremonese

[D1]-A.DOUADY "Le problème des modules pour les sous-espaces ..." Ann.Inst. Fourier , 16, (1966) , 1-95

[D2]-A.DOUADY "Le problème des modules locaux .." Ann.Sc. ENS ,7,(1974) 569-602

[F] -J.FRISCH "Aplatissement en géométrie analytique" Ann.SC.Norm.Sup. PISA ,1,(1968),305-312

[G-H-S] - J.GIRBAU -A.HAEFLIGER-D.SUNDARARAMAN "On deformations of transversally holomorphic foliations" J.Reine u.Ang.Math.,345, (1983),122-147

[G-M] - X.GOMEZ-MONT "The transverse dynamics of a holomorphic flow" Pré-Pub.UNAM MEXICO,(1986)

[G-R] -H.GRAUERT-R.REMMERT "Analytische Stellenalgebren"(Springer 1971)

[K] - M.KURANISHI "On the locally complete families of complex analytic structures"Ann.of Math.,75,(1962),536-577

[P1] - G.POURCIN "Théorème de Douady au-dessus de S"Ann.Sc.Norm.Sup.PISA XXIII,(1969),451-459

[P2] - G.POURCIN "Deformations of coherent foliations on a compact normal space" Ann.Inst.Fourier ,2,t.37,(1987) (à paraître)

[R] - H.J.REIFFEN "The variety of moduli of foliations on a compactcomplex space" Pub.Universität OSNABRUCK ,(1986)

[Sa]- K.SAITO "On a generalization of de Rham lemma " Ann.Inst.Fourier XXVI,(1976),2,165-170

[Su]-T.SUWA "Unfoldings of complex analytic foliations with singularities" Japan.J. Math. ,9,1,(1983),181-206 .

Geneviève POURCIN
Département de Mathématiques
Faculté des Sciences
2 Boulevard Lavoisier
49045 ANGERS CEDEX -FRANCE

Product Singularities and quotients

K. Reichard, K. Spallek
Bochum

Summary: In [14] a unique-product-factorisation-theorem is proved for quite arbitrary space germs (as for Whitney-stratified germs). This we are going to apply here for factorisations of quotientsingularities according to the announcement in [14] .

Introduction, statement of results: With [14] we first have:

1) To "any"$^{*)}$ germ ^NA of a reduced N-differentiable space with $N \in \{\infty,$ ω(realanalytic), ω^*(complexanalytic)$\}$ there exists -up to numbering and C^N-diffeomorphisms of the factors- a unique p-irreducible factorisation into germs ^NA_j:

$$^NA \simeq {}^NA_1 \times \cdots \times {}^NA_r,$$

where: $^NA_1 \simeq {}^N(K^\ell)$ with $\ell \geq o$; $K = \mathbb{R}_o$, $N \in \{\infty,\omega\}$ or $K = \mathbb{C}_o$, $N = \omega^*$; and the other germs ^NA_j are singular and p-irreducible (: admit no further product-factorisation).

1') Any other factorisation $^NA \simeq {}^NB_1 \times \cdots \times {}^NB_s$, where $^NB_1 \simeq {}^N(K^m)$ and no other ^NB_j splits up some ^NK, gives: $\ell = m$ and each ^NB_i, $i \neq 1$, is a product of some ^NA_j 's with $j \neq 1$.

In general "^NA p-irreducible" does not imply "$^{N'}A$ p-irreducible" for $N' < N$. However this holds in the following cases.

2) A is semianalytic, $N = \omega$, $N' = \infty$. Moreover then: Given p-irreducible factorisations $^\infty A \simeq {}^\infty A_1 \times \cdots \times {}^\infty A_r \simeq {}^\infty B_1 \times \cdots \times {}^\infty B_r \curvearrowright A_i, B_i$ are up to C^∞-diffeomorphisms semianalytic (: [14] for analytic cases) and we even have modulo numbering $^\omega A_i \simeq {}^\omega B_i$ (: 1) and [8]).

*) "any": locally compact, kurvenreich ([14]), for ex. semianalytic

3) A is complexanalytic and for example (algebraically) irreducible, $N=\omega^*$, $N'\in\{\infty,\omega\}$. Moreover then: Given p-irreducible factorisations $^{\infty}A \simeq {}^{\infty}A_1 \times\cdots\times {}^{\infty}A_r \simeq {}^{\infty}B_1\times\cdots\times {}^{\infty}B_r \curvearrowright$ A_i, B_i are up to C^∞-diffeo's complexanalytic and modulo numbering even pairwise holomorphic or antiholomorphic equivalent (:1), [2] or more generally [14]).

As here, properties of factors A_i in general carry over to the product $\times A_i$ and vers versa. In the following we study such phenomena for quotientsingularities:

Any Liegroup G operating properly on an N-differentiable space X leads to a quotient-N-differentiable space X/G for $N\in\{\infty,\omega,\omega^*\}$ ([7]). Especially if $G\subset GL(n,K)$ and G is a compact (Lie-) group, the quotient space $^N(K^n)/G$ exists ([1],[5],[10]). Note that we consider here K as germ $\mathbb{R}_o$ or $\mathbb{C}_o$ at zero; then also $^N(K^n)/G$ is a germ of a space.
In general, the ω-differentiable space induced by $^{\omega^*}(\mathbb{C}^n_o)/G$ is different from $^{\omega}(\mathbb{C}^n_o)/G$. One has only a map $^{\omega}(\mathbb{C}^n_o)/G \to {}^{\omega^*}(\mathbb{C}^n_o)/G$.

To any $G\subset GL(m,K)$ there is associated the largest group $\max^N G$, that leaves invariant the same C^N-functions as G. We have, respectively we obtain:

4) Any compact group $G\subset GL(n,\mathbb{C})$ is finite (:classical).

5) If $G\subset GL(n,K)$ is finite, then $G=\max{}^{\infty}G=\max{}^{\omega}G$, and if $K=\mathbb{C}_o$ also $G=\max{}^{\omega^*}G$.

6) If $G\subset GL(n,K)$ is compact, so is $\max{}^N G$, $N\in\{\infty,\omega,\omega^*\}$.

From now on let $G\subset GL(n,K)$ be a compact (Lie-) group, $^NA={}^N(K^n)/G$ the quotientgerm with $q: K^n\to {}^NA$ as some quotientmap; and assume always:

if $K=\mathbb{R}$, then $N\in\{\infty,\omega\}$ and G is maximal, i.e.: $G=\max{}^N G$

if $K=\mathbb{C}$, then $N=\omega^*$ and G has no reflections

In case $K=\mathbb{C}$ define $^{\infty}A$ to be the reduced C^∞-space associated to $^{\omega^*}A$. Note, that in case $K=\mathbb{R}$ the germ $^{\infty}(\mathbb{R}^n)/G$ is the reduced C^∞-space associated to $^{\omega}(\mathbb{R}^n)/G$ (see note, p. 5).

The assumptions above on G are necessary to obtain:

7) *Theorem* The following are equivalent

α) ${}^{\infty}A \simeq {}^{\infty}A_1 \times \cdots \times {}^{\infty}A_r$ (not necessarily p-irreducible)

β) $\exists\, n_i \in \mathbb{N}$, $G_i \subset GL(n_i,K)$ compact, maximal if $K = \mathbb{R}$, without reflections if $K = \mathbb{C}$, such that: $n = \sum^r n_i$

$G \simeq G_1 \times \cdots \times G_r$ (up to conjugation in GL(n,K))

${}^{\infty}A_i \simeq {}^{\infty}(K^{n_i})/G_i$ for each i.

In the equivalent situations of 7) we have moreover:

8) *Theorem* α) ${}^{\infty}A_i$ is singular iff $G_i \neq id$

β) ${}^{\infty}A \simeq {}^{\infty}A_1 \times \cdots \times {}^{\infty}A_r$ is p-irreducible iff

$G \simeq G_1 \times \cdots \times G_r$ is irreducible (: has no refined factorisation).

For the next assume that $G \subset GL(n,K)$, $\tilde{G} \subset GL(m,K)$ are finite groups with irreducible factorisations

$$G = G_1 \times \cdots \times G_r, \quad \tilde{G} = \tilde{G}_1 \times \cdots \times \tilde{G}_s ,$$

quotients ${}^{N}A$ for G as above, similarly ${}^{N}\tilde{A}$ for $\tilde{G}$.

9) *Theorem* ${}^{\infty}A \simeq {}^{\infty}\tilde{A}$ iff $n = m$, $r = s$ and

α) $\forall\, i : G_i = \tilde{G}_i$ if $K = \mathbb{R}$ (up to numbering and conjugation)

β) $\forall\, i : G_i = \tilde{G}_i$ or $\bar{G}_i = \tilde{G}_i$ if $K = \mathbb{C}$ (")

Moreover: ${}^{\omega^*}A \simeq {}^{\omega^*}\tilde{A}$ in case $K = \mathbb{C}$ iff $n = m$, $r = s$ and

γ) $\forall\, i : G_i = \tilde{G}_i$ (up to numbering and conjugation)

Here $\bar{G}_i := \{\bar{g} := \bar{a}\, z \mid g = a z \in G_i\}$.

Especially we have now: The C^{∞}- and C^{ω}-p-irreducible (in case $K = \mathbb{C}$ also the C^{ω^*}-p-irreducible) factorisations of K^n/G are all the "same" and correspond exactly to the irreducible factorisations $G \simeq G_1 \times \cdots \times G_s$. Here the G_i' s are

i) up to conjugation if $K = \mathbb{R}$

ii) up to holomorphic or antiholomorphic conj. if $K = \mathbb{C}$

uniquely determined by the C^∞-p-factors of K^n/G.

The proofs require different methods for the case $K = \mathbb{R}$ and for the case $K = \mathbb{C}$. In the more involved case $K = \mathbb{R}$ essentially they are applications of results (for ex. factorisation-theorems) and methods (locally integrable vectorfields) of [13],[14] and of [8],[9], described first by the first author. The case $K = \mathbb{C}$ uses the "Riemannscher Hebbarkeitssatz".

§ 1 Locally integrable vectorfields and trivial factors

From a quotient-singularity we first factor aut trivial factors $^N(K^n)$. For this and further use we need some results from [13],[14].

A (tangent) vectorfield V on an N-differentiable space X is called locally integrable, if through each point $p \in X$ passes an integral curve of V on X. Let $T^i_p X$ denote the set of those tangentvectors $v \in T_p X$ of X at p, for which there exists a locally integrable field V in a neighbourhood of p on X with $V(p) = v$. We have ([13],[14]):

10) α) $T^i_p X$ is a vectorspace. β) A tangentvectorfield V on X is locally integrable iff $V(p) \in T^i_p X \quad \forall\, p \in X$.
γ) $T^i_p X = T^i_{p_1} X_1 \times T^i_{p_2} X_2$ for $p = (p_1, p_2) \in X := X_1 \times X_2$, if the set of manifold-points of X is dense in X. In this situation we also have: δ) A field $V = (V_1, V_2)$ on $X = X_1 \times X_2$ is locally integrable iff $V_1 | X_1 \times \{q_2\}$, $V_2 | \{q_1\} \times X_2$ are locally integrable on X_1 resp. X_2 for each $(q_1, q_2) \in X_1 \times X_2$.
ε) $r := \dim T^i_p X$ is the largest number s such that $X \simeq Y \times K^s$ (locally near p).

Note, that on real or complex analytic sets each differentiable or analytic vectorfield is locally integrable ([13]). However this is not true in general on more general spaces, especially not in general on semianalytic sets or even on quotientsingularities.

For any germ $A \subset K^n$ let ${}^N A$ denote the associated reduced C^N-space(-germ). For a compact Liegroup $G \subset GL(n,K)$ let $q : {}^N(K^n) \to {}^N(K^n)/G$ be a quotient-map (which therefore factors any G-invariant mapping ${}^N(K^n) \to X$). The following known fact gives "$\beta) \curvearrowright \alpha)$" of 7).

11) If $G = G_1 \times G_2$, $G_i \subset GL(n_i,K)$, $n_1 + n_2 = n$, then

$$ {}^N(K^n)/G \simeq {}^N(K^{n_1})/G_1 \times {}^N(K^{n_2})/G_2 $$

$(N \in \{\infty,\omega\}$ if $K = \mathbb{R}$; $N = \omega^*$ if $K = \mathbb{C})$

<u>*Because:*</u> As quotient map $q = (q_1,..,q_\ell) : K^n \to K^\ell$ any finite sequence of G-invariant polynomials q_i can be taken, which generates (as algebra) the set of all G-invariant polynomials on K^n ([1],[5],[10]). Moreover, by taking averages (summing over G) any polynomial q can be turned into a G-invariant polynomial q^*. We have $q = q^*$ iff q is G-invariant. This gives for $q_i = \sum a_{i\sigma\tau} x^\sigma \cdot y^\tau$, x on K^{n_1}, y on K^{n_2}, $G = G_1 \times G_2$:

$$ q_i^* = \sum a_{i\sigma\tau} x^{\sigma *} \cdot y^{\tau *} $$

Here $x^{\sigma *}$ (resp. $y^{\tau *}$) is G_1- (resp. G_2-) invariant. As generating sequence above we therefore may assume $q = (q_1,..,q_\ell)$ to be of the following type: $q^1 := (q_1,..,q_r)$ (resp. $q^2 := (q_{r+1},..,q_\ell)$) are G_1- (resp. G_2-) invariant polynomials on K^{n_1} (resp. K^{n_2}) generating the quotientmappings

$$ q^i : {}^N(K^{n_i}) \to {}^N(K^{n_i})/G_i, \quad q : {}^N(K^n) \to {}^N(K^n)/G . $$

This leads to the required result. ✓

<u>*Note:*</u> The proof shows: If $K = \mathbb{R}$, then ${}^\infty(\mathbb{R}^n)/G$ <u>is</u> the reduced C^∞-differentiable space associated to the reduced C^ω-differentiable space ${}^\omega(\mathbb{R}^n)/G$.

12) <u>*Remark*</u> If G operates properly on a space X, then any G-equivariant vectorfield $V : X \to TX$ pushes down to a vectorfield $V^* : X/G \to T(X/G)$. If V is locally integrable, so is V^*.

<u>*Because:*</u> We have the following commutative diagrams for any $g \in G$:

$$\begin{array}{ccc} X & \xrightarrow{g} & X \\ \downarrow q & & \downarrow q \\ X/G & \xrightarrow{id} & X/G \end{array} \qquad \begin{array}{ccc} TX & \xrightarrow{Tg} & TX \\ \downarrow Tq & & \downarrow Tq \\ T(X/G) & \xrightarrow{id} & T(X/G) \end{array} \curvearrowright \begin{array}{ccc} X & \xrightarrow{V} & TX \\ \downarrow q & & \downarrow Tq \\ X/G & \overset{V^*}{\dashrightarrow} & T(X/G) \end{array}$$

Because V is equivariant ($V \circ g = Tg \circ V$), $Tq \circ V$ is G-invariant and factors therefore over V. The rest is obvious. ✓

Let $G \subset GL(n,K)$ be compact with quotientmap $q : {}^N(K^n) \to {}^N(K^n)/G$ and G without reflections in case $K = \mathbb{C}_o$

13) *Theorem* The following are equivalent for $K = \mathbb{R}_o$:

α) $Tq(o) \neq o$. β) There exists a constant G-equivariant vectorfield $V \neq o$ on ${}^N(K^n)$. γ) $T^i_o({}^N(K^n)/G) \neq o$. δ) ${}^N(K^n)/G \simeq {}^NA \times {}^NK$. ε) There exists a regular curve k on ${}^N(K^n)/G$ through the origin. ζ) $G \subset GL(n-1,K)$ (up to conjugation in $GL(n,K)$).

Proof Without restriction: $q(o) = o$. By assumption: G is compact; any metric on $\mathbb{R}^n$ can be changed into a G-invariant metric (taking average over G). So without restriction: G is an orthogonal group.
"α) $\curvearrowright$ β)": $q = (q_1, .., q_\ell)$, and without restriction: $Tq_1(o) \neq o$. q is G-invariant, so q_1, so its linear part $Tq_1(o)$; i.e. $Tq_1(o)$ considered as a row-vector, $g \in G$ as a matrix gives: $Tq_1(o)^t = (Tq_1(o) \cdot g)^t = g^t \cdot Tq_1(o)^t = g^{-1} \cdot Tq_1(o)^t$ for the transposed „ t ". "β) $\curvearrowright$ γ)": The field V under β) pushes down under Tq to a locally integrable field V^* on the quotient with $V^*(o) \neq o$. "γ) $\leftrightarrow$ δ)": [14]. "δ) $\curvearrowright$ ε)": obvious. "ε) $\curvearrowright$ α)": x^2 on $\mathbb{R}^n$ is G-invariant and factors therefore: $x^2 = H \circ q$. Assume: $Tq(o) = o$, then $|q(x)| \leq c\, x^2$ for some $c > o$, all x near o. Therefore: $|q(x)| \leq c \cdot x^2 = c \cdot H(q(x))$, $|y| \leq c \cdot H(y)$ for all y on the quotient close to o. For any differentiable curve k on the quotient with $k(o) = o$ we obtain $H \circ k(t) \geq o$, therefore $(H \circ k)'(o) = o$, $H \circ k(t) \leq d \cdot t^2$, $|k(t)| \leq c \cdot H \circ k(t) \leq c \cdot dt^2$, therefore $k'(o) = o$ ✓ Finally, the equivalence "β) $\leftrightarrow$ ζ)" holds, because G is assumed to be orthogonal. With this 13) is proved.

13') *Theorem* The following are equivalent for $K = \mathbb{C}_o$:

$\alpha + \beta$) There exist: $L : K^n \to K^\ell$ linear, G-invariant and $V : K^n \to K^n$ constant, G-equivariant vectorfield, with $L(V) \neq o$. γ) $T^i_o({}^N(K^n)/G) \neq o$.

δ) ${}^N(K^n)/G \simeq {}^NA \times {}^NK$. ζ) $G \subset GL(n-1,K)$ (up to conjugation).

Proof "$\gamma) \leftrightarrow \delta)$" holds as above. "$\zeta) \rightarrow \delta)$" follows from 11).
"$\gamma) \rightarrow \alpha + \beta)$": G has no reflections; the set S of those points in K^n, where q has rank $<n$, therefore is of codimension ≥ 2. Any vectorfield V^* on the quotient, $V^*(o) \neq o$, can locally be lifted to a field on K^n outside of S. Because of codim $S \geq 2$ this lifting can uniquely be extended over $K^n \setminus S$ and then extended to a field $\tilde{V}$ on K^n (Hebbarkeitssatz of Riemann). $\tilde{V}$ is G-equivariant with V^* as push-down. Therefore $Tq(o)(\tilde{V}(o)) = V^*(o) \neq o$. Now $\tilde{V}(o)$ can be considered as a constant G-equivariant vectorfield on K^n. With $L := Tq(o)$ statement $\alpha + \beta$) follows. Finally "$\alpha + \beta) \rightarrow \zeta)$" can be seen directly and is left to the reader.

§ 2 Factoring in the real case $K = \mathbb{R}_o$

Due to the general factorisation theorems of our introduction as well as to 10) and 13) we may first factor out trivial factors ${}^\infty(K^n)$ and restrict then to the following situation ($G \subset GL(n,\mathbb{R})$ compact, $G = \max{}^\infty G$, without restriction: G orthogonal):

14) Let ${}^\infty(\mathbb{R}^n_o)/G \simeq {}^\infty A_1 \times {}^\infty A_2$, $Tq(o) = o$ for the associated quotientmap q. Then there exist $G_i \subset GL(n_i, \mathbb{R})$ compact maximal Liegroups with $n = n_1 + n_2$, $G = G_1 \times G_2$ up to conjugation and ${}^\infty A_i \simeq {}^\infty(\mathbb{R}^n_o)/G_i$ for $i = 1,2$.

Proof For $q = (q_1, q_2)$, $q_i : \mathbb{R}^n_o \to A_i$, $A^*_1 := q_2^{-1}(o)$, $A^*_2 := q_1^{-1}(o)$ we obtain step by step a) - e):

a) Each A_i^* is G-invariant; the tangentspaces $T_o A_i^* \subset \mathbb{R}^n$ are G-invariant vectorspaces; ${}^{\infty}A_i^*/G \simeq {}^{\infty}A_i$.

<u>*Because:*</u> For example any G-invariant function on A_i^* can be extended over K^n, then can be made G-invariant over K^n and finally can be pushed down to $A_1 \times A_2$ and restricted onto $A_i \times \{o\}$. This gives the last statement.

Consider now q as a map $\mathbb{R}^n \to \mathbb{R}^s \supset ({}^{\infty}\mathbb{R}^n)/G$ and its differential as a map $Tq : \mathbb{R}^n \ni x \to \mathrm{Hom}(\mathbb{R}^n, \mathbb{R}^s) \simeq \mathbb{R}^{n\cdot s}$. We obtain next:

b) $Tq_1 \mid A_2^* = o, \quad Tq_2 \mid A_1^* = o,$

$$T(Tq_1)(o) \mid T_o A_2^* = o, \quad T(Tq_2)(o) \mid T_o A_1^* = o \quad \text{with } A_i^* := {}^{\infty}A_i^*$$

<u>*Because:*</u> If for example $x \in A_2^*$ gives $Tq_1(x) \neq o$, take a curve h in $\mathbb{R}^n$ with $h(o) = x$ and $q_1 \circ h$ regular. But $q_1 \circ h(o) = o$, so $k := (q_1 \circ h, o)$ would be a regular curve on ${}^{\infty}A_1 \times {}^{\infty}A_2$ through the origin, contradicting $Tq(o) = o$ and 13). So $Tq_1 \mid A_2^* = o$, similarly $Tq_2 \mid A_1^* = o$. The rest then follows.

c) $A_i^* \subset T_o A_i^*$ for $i = 1,2$.

<u>*Because:*</u> The identical vectorfield $V^* : x \to x$ in $\mathbb{R}^n$ is G-equivariant (G is assumed to be linear) and therefore pushes down to a locally integrable vectorfield $V = (V_1, V_2)$ on ${}^{\infty}A_1 \times {}^{\infty}A_2$. Then each $V_i \mid A_i$ x const. is locally integrable on A_i (10), δ) and 12)). Connect a given $x \in A_1^*$ to o by the integralcurve $h(t) = e^{t-1}x$ of V^*. Then $q \circ h =: (k_1, k_2)$ is an integralcurve of V. We have: $k_2(1) = o$; $V_2 \mid A_1 \times o = o$ (by b)); $V_1 \mid A_1 \times o$ locally integrable (10, δ)). Then also (k_1, o) is an integralcurve of V. By uniqueness of integralcurves we now obtain $k_2 = o$; hence $k = (k_1, o)$, and k lies on $A_1 \times o$, h on A_1^*, i.e. $s \cdot x \in A_1^*$ $\forall\ o < s \leq 1$. This gives c) for i=1 and similarly for $i = 2$.

d) $T_o A_1^* \cap T_o A_2^* = o$

<u>*Because:*</u> x^2 is G-invariant and factors therefore by $x^2 = H \circ q$. Considering differential-maps as above and using $q(o) = o$, $Tq(o) = o$ we obtain by Taylor:

$$2x = T(H \circ q)(x) = TH(o)(T(Tq)(o)(x))$$

Therefore: $o = T(Tq)(o)(x) = (T(Tq_1)(o)(x),\ T(Tq_2)(o)(x))$ iff $x = o$. This gives d).

e) $T_o A_1^* + T_o A_2^* = \mathbb{R}^n$

Because: G is orthogonal, there exists therefore a G-invariant vector-space $V_3 \subset \mathbb{R}^n$ with $\mathbb{R}^n = T_o A_1^* \oplus T_o A_2^* \oplus V_3$. Write $V_i := T_o A_i^*$, $\bar{q}_i := q_i \mid V_i$ for $i = 1,2$ and look at these as germs at o. Then $\bar{q}_i : V_i \to A_i$ is surjectiv ($A_i^* \subset V_i$!) and G-invariant. Especially $\bar{q} := (\bar{q}_1, \bar{q}_2)$ factors like follows

$$\begin{array}{ccc} V_1 \times V_2 \times V_3 & \xrightarrow{\text{projection}} & V_1 \times V_2 \\ q \downarrow & & \downarrow \bar{q} \\ {}^{\infty}A_1 \times {}^{\infty}A_2 & \overset{f}{\dashrightarrow} & {}^{\infty}A_1 \times {}^{\infty}A_2 \end{array}$$

$\bar{q}$ is onto, hence also f. But $f \mid A_1^{\infty} \times \{o\} = \mathrm{id}$, $f \mid \{o\} \times A_2^{\infty} = \mathrm{id}$, hence $Tf(o) = \mathrm{id}$ and f is a diffeo. This means, that on $V_1 \times V_2$ and $V_1 \times V_2 \times V_3$ we have the same G-invariant functions. If z are coordinates on V_3, z^2 is an G-invariant function on V_3, so on $V_1 \times V_2 \times V_3$; but z^2 is zero on $V_1 \times V_2$. So $z = o$, so $\dim V_3 = o$, and e) is proved. ✓

Changing linearly coordinates if necessary, we may assume now $\mathbb{R}^n = \mathbb{R}^{n_1} \times \mathbb{R}^{n_2}$, $V_i = \mathbb{R}^{n_i}$. Let $G_i := G \mid \mathbb{R}^{n_i}$, then $G \subset G_1 \times G_2$. Any G_i-invariant function g on V_i can be considered as G-invariant function on $V_1 \times V_2 = \mathbb{R}^n$, therefore can be pushed down by q to ${}^{\infty}A_1 \times {}^{\infty}A_2$. So g can be pushed down by $\bar{q}_i$ on ${}^{\infty}A_i$, therefore $\bar{q}_i : V_i \to {}^{\infty}A_i$ is a quotientmap to G_i, and by 11 (and its proof) also $\bar{q} = (\bar{q}_1, \bar{q}_2)$ is a quotientmap to $G_1 \times G_2$. So by the above diagramm (with $V_3 = o$) G and $G_1 \times G_2$ have the same invariant functions. By $G \subset G_1 \times G_2$ and the maximality of G we obtain: $G = G_1 \times G_2$, and moreover the G_i are maximal themselves. This proves 14). ✓

The theorems 7), 8) now follow for $K = \mathbb{R}_o$:

For "7), $\beta) \rightsquigarrow \alpha)$" see 11). For "7), $\alpha) \rightsquigarrow \beta)$" start with a p-irreducible factorisation ${}^{\infty}A = {}^{\infty}B_1 \times \ldots \times {}^{\infty}B_s$ according to 1), especially ${}^{\infty}B_1 \simeq {}^{\infty}(\mathbb{R}^{\ell})_o$. Using 10), we may apply 13) inductively to obtain $G \subset GL(n-\ell, \mathbb{R})$; i.e.: $G = \mathrm{id} \times \tilde{G}$ with $\tilde{G} \subset GL(n-\ell, \mathbb{R})$, $q = (q^1, q^2)$,

$q^1 : \mathbb{R}^n \to \mathbb{R}^\ell$ projection, $q^2 : \mathbb{R}^{n-\ell} \to \mathbb{R}^s$ quotientmap for $\tilde{G}$ with $Tq^2(o) = o$. Then apply 14) inductively to obtain 7), β) for a p-irreducible factorisation. Now factoring out ${}^\infty A_j = {}^\infty(\mathbb{R}_o^{\ell_j}) \times {}^\infty A_j^*$ with largest possible ℓ_j, 10) gives $\sum \ell_j = \ell$. By 1') each ${}^\infty A_j$ is a product of ${}^\infty(\mathbb{R}_o^{\ell_j})$ with some of the ${}^\infty B_i$'s, $i > o$, and $\underset{j}{\times}{}^\infty(\mathbb{R}_o^{\ell_j}) = {}^\infty B_1$. Gathering similarly on the group-level and using 11) we obtain the required G_i's. To prove 8), α), split up ${}^\infty A_j = {}^\infty(\mathbb{R}_o^{\ell_j}) \times {}^\infty A_j^*$ with largest possible ℓ_j. Then ${}^\infty A_j$ is singular iff ${}^\infty A_j^* \neq \{o\}$ and $T_o^i A_j^* = o$ (use 10)), iff $G_j \neq \mathrm{id}$ (use 13)). 8), β) follows similarly with 1), 11) and 14). ✓

Also theorem 9) follows for $K = \mathbb{R}$:

Let ${}^\infty A \simeq {}^\infty\tilde{A}$, especially $\dim A = \dim \tilde{A}$, therefore $n = m$ (the groups are finite). The irreducible factorisations of G resp. $\tilde{G}$ induce p-irreducible factorisations of ${}^\infty A, {}^\infty\tilde{A}$, which are the "same" (1)). By [9] the corresponding groups are conjugate. This proves one direction, the other direction is obvious. ✓

Finally we prove 5) and 6) for $K = \mathbb{R}$:

We have $G \subset \max^\infty G \subset \max^\omega G$ and ${}^\omega(\mathbb{R}^n)/G \simeq {}^\omega(\mathbb{R}^n)/\max^\omega G$, because G and $\max^\omega G$ have the same invariant analytic functions. By [9]: G and $\max^\omega G$ are conjugate; due to $G \subset \max^\omega G$ and G finite, they are equal then. This proves 5). If G is compact, we may suppose G is orthogonal, then G, hence also $\max^\omega G$ keeps invariant the function x^2 on $\mathbb{R}^n$. Then also $\max^\omega G$ is orthogonal, and obviously also closed, hence compact. The same holds for $\max^\infty G$; but due to the proof of 11) we even have $\max^\infty G = \max^\omega G$. ✓

§ 3 Factoring in the complex case $K = \mathbb{C}_o$

Let $G \subset GL(n,\mathbb{C})$ be a finite group without reflections. Again due to the unique decomposition theorems we may restrict at first to $N = \omega^*$ and the following situation:

15) Let ${}^{\omega^*}(\mathbb{C}^n)/G \simeq {}^{\omega^*}A_1 \times {}^{\omega^*}A_2$. Then there exist $G_i \subset GL(n_i,\mathbb{C})$, without reflections, with $n = n_1 + n_2$, $G = G_1 \times G_2$ up to conjugation and ${}^{\omega^*}A_i \simeq {}^{\omega^*}(\mathbb{C}^n)/G_i$ for $i = 1,2$.

Proof In the following we omit the index ω^*; for ex. A_i means ${}^{\omega^*}A_i$. As in § 2 let $q=(q_1,q_2):\mathbb{C}_o^n \to A_1 \times A_2$ be a (holomorphic) quotientmap, $q(o)=o$. $A_1^* := q_2^{-1}(o)$, $A_2^* := q_1^{-1}(o)$ and their tangentvectorspaces $T_o A_1^*$, $T_o A_2^*$ are G-invariant. q is a finitely branched covering map; the branching germ $S \subset \mathbb{C}_o^n$ is of codimension ≥ 2 (: G has no reflections). Also $\bar{q}_i := q_i \mid A_i^* \to A_i \times \{o\}$ is a branched covering and $A_i^*/G \simeq A_i$. Especially $\dim A_i^* = \dim A_i$ everywhere.

a) If $x \in A_1$ is nonsingular, then each $y \in q_1^{-1}(x) \subset A_1^*$ is nonsingular for A_1^*, and $\bar{q}_1$ at these points is a local bimorphism. A similar result holds for A_2 and $A_1 \times A_2$.

Because: Through any $v \in T_x A_1 \setminus \{o\}$ passes a field V on the germ A_{1x}. V can be considered as a field on $(A_1 \times A_2)_{(x,o)}$, which is constant in the A_2-variable and zero in the A_2-component. V can be lifted therefore to a field V^* on $\mathbb{C}_y^n$ for $q(y)=(x,o)$ (see the proof of 13')). V^* is everywhere different from zero, its integral curves push down to integral curves of V, hence to regular curves on $A_1 \times$const. Therefore: $V^* \mid A_1^*$ is tangent to A_1^* and moreover $\dim_y A_1^* = \dim_x A_1 \leq \dim T_y^i A_1^*$, $T\bar{q}_1(y) : T_y^i A_1^* \to T_x A_1 = T_x^i A_1$ is surjective (: $v \in T_x A_1 \setminus \{o\}$ was chosen arbitrarily). 10), ε) now gives: A_1^* is a manifold at y. Then $\dim T_y^i A_1^* = \dim T_y A_1^* = \dim T_x A_1$, hence $T\bar{q}_1(y)$ is bijective, and $\bar{q}_1$ is a bimorphism at y.

b) There exist holomorphic retractions $r_i : \mathbb{C}_o^n \to A_i^*$, respecting G-orbits. The A_i^* are therefore G-invariant complex manifolds, and the induced mappings $r_i \mid (T_o A_1^*)_o : (T_o A_i^*)_o \to A_i^*$ are biholomorphic and respect G-orbits.

Because: A point $x=(x_1,x_2) \in A_1 \times A_2$ is regular iff each $x_i \in A_i$ is regular. Using a) we see, that in regular points the projection $A_1 \times A_2 \to A_1$ can be lifted locally and can then be extended holomorphically to a map $\mathbb{C}_o^n \setminus S \to A_1^*$, then to a map $r_1 : \mathbb{C}_o \to A_1^*$ (: codim $S \geq 2$, Riemannscher Hebbarkeitssatz):

$$\begin{array}{ccc} \mathbb{C}_o^n & \longrightarrow & \mathbb{C}_o^n/G = A_1 \times A_2 \\ \Big\downarrow r_1 & & \Big\downarrow \text{projection} \\ A_1^* & \longrightarrow & A_1^*/G = A_1 \end{array}$$

The diagram commutes, therefore $r_1|A_1^* = g|A_1^*$ for some $g \in G$: first locally over a regular point x of A_1 (a)), then globally by the identity-theorem. Changing from r_1 to $g^{-1} \circ r_1$ if necessary we may assume $r_1|A_1^* = \mathrm{id}$. In a similar way we find r_2. b) now follows with [11].

c) $V_i/G \simeq A_i$ for $V_i := (T_o A_i^*)_o$

Because: $r_i|V_i : V_i \to A_i^*$ are biholomorphic; V_i, A_i^* are G-invariant, $r_i|V_i$ respect G-orbits. Therefore $V_i/G \simeq A_i^*/G \simeq A_i$.

d) Let $G_i := G|V_i$. Then $V_1 \oplus V_2 = \mathbb{C}_o^n$, $G = G_1 \times G_2$ (up to conjugation), and 15) follows.

Because: $A_1^* = r_1(\mathbb{C}_o^n)$, $A_2^* = r_1^{-1}(o)$. Then A_1^*, A_2^* are transversal at $\{o\} = A_1^* \cap A_2^*$. This gives $V_1 + V_2 = \mathbb{C}_o^n$, $V_1 \cap V_2 = \{o\}$, $V_1 \oplus V_2 = \mathbb{C}_o^n$. The quotientmappings $q : \mathbb{C}_o^n \to \mathbb{C}_o^n/G = A_1 \times A_2$ and $p : V_1 \times V_2 \to (V_1/G_1) \times (V_2/G_2) = A_1 \times A_2$ (: 11)) are not branched outside of sets S of codimension ≥ 2 (a)). Biholomorphisms on $\mathbb{C}_o^n$, that push down by q to id, are exactly those given by the $g \in G$ (this is obvious locally over regular points of $A_1 \times A_2$, then also globally). A similar result holds for $V_1 \times V_2$ and p (using a)). Because of $\operatorname{codim} S \geq 2$, the identity $A_1 \times A_2 \leftrightarrow A_1 \times A_2$ can be lifted to mappings $\varphi : \mathbb{C}_o^n \to V_1 \times V_2$, $\psi : V_1 \times V_2 \to \mathbb{C}_o^n$ with $p \circ \varphi = q$, $q \circ \psi = p$:

$$
\begin{array}{ccc}
\mathbb{C}_o^n & \overset{\varphi}{\underset{\psi}{\rightleftarrows}} & V_1 \times V_2 \\
q \downarrow & & \downarrow p \\
\mathbb{C}_o^n/G & & (V_1/G_1) \times (V_2/G_2) \\
\parallel & & \parallel \\
A_1 \times A_2 & \longleftrightarrow & A_1 \times A_2
\end{array}
$$

If follows: $\psi \circ \varphi \in G$ and $\varphi \circ \psi \in G_1 \times G_2$. So φ is biholomorphic. In the same way: $\varphi^{-1} \circ g \circ \varphi \in G_1 \times G_2$, hence $T\varphi^{-1}(o) \circ g \circ T\varphi(o) \in G_1 \times G_2$ for each $g \in G$; similarly $T\varphi(o) \circ g \circ T\varphi^{-1}(o) \in G$ for each $g \in G$. ✓

The theorems 7), 8) now follow for $K = \mathbb{C}_o$:

For "7), β) ⤳ α)" see 11) and the def. of ${}^\infty A$ in this case (bottom p. 2).

"α) ⤳ β)": $A = \mathbb{C}_o^n/G$ is an irreducible complex analytic germ. Let

${}^{\infty}A = {}^{\infty}B_1 \times \cdots \times {}^{\infty}B_r$ and ${}^{\omega^*}A = {}^{\omega^*}C_1 \times \cdots \times {}^{\omega^*}C_s$ be p-irreducible factorisations. By [14] or [2] we have $s = r$ and up to numbering ${}^{\infty}B_i \simeq {}^{\infty}C_i$. Using 15) by induction we obtain up to conjugation $G = G_1 \times \cdots \times G_r$ and ${}^{\omega^*}(\mathbb{C}^n_o)/G_i \simeq {}^{\omega^*}B_i$, hence ${}^{\infty}(\mathbb{C}^n_o)/G_i \simeq {}^{\infty}B_i \simeq {}^{\infty}C_i$. Thus "$\alpha) \curvearrowright \beta)$" is proved for p-irreducible factorisations. In the general situation use 1') and proceed as in the proof for $K = \mathbb{R}$ in § 2. To prove 8), α) note, that ${}^{\infty}A_i$ is nonsingular iff ${}^{\omega^*}A_i$ is nonsingular (for ex. [11]), iff $\mathbb{C}^{n_i}/id = \mathbb{C}^{n_i}/G_i$, iff $G_i = id$ (see the proof of 15) above). 8), β) follows similarly. ✓

Also theorem 9) follows for $K = \mathbb{C}$:

Start with p-irreducible factorisations according to 1):
${}^{\omega^*}A_1 \times \cdots \times {}^{\omega^*}A_r \simeq {}^{\omega^*}A$, ${}^{\omega^*}\tilde{A}_1 \times \cdots \times {}^{\omega^*}\tilde{A}_s \simeq {}^{\omega^*}\tilde{A}$, and without restriction again: the A_i, $\tilde{A}_j$ are complex analytic. Then $r = s$, ${}^{\infty}A_1 \times \cdots \times {}^{\infty}A_r \simeq {}^{\infty}A \simeq {}^{\infty}\tilde{A} \simeq$ $\simeq {}^{\infty}\tilde{A}_1 \times \cdots \times {}^{\infty}\tilde{A}_r$ are p-irreducible factorisations (3)), and we have up to numbering ${}^{\infty}A_i \simeq {}^{\infty}\tilde{A}_i$ (1)), even ${}^{\omega^*}A_i \simeq {}^{\omega^*}\tilde{A}_i$ or ${}^{\omega^*}\bar{A}_i \simeq {}^{\omega^*}\tilde{A}_i$ (3)). Here $\bar{A}_i \subset \mathbb{C}^n_o$ is the image of A_i under $z \to \bar{z}$. To the p-irreducible factorisasations of ${}^{\omega^*}A$ resp. ${}^{\omega^*}\tilde{A}$ correspond irreducible factorisations $G \simeq G_1 \times \cdots \times G_r$, $\tilde{G} \simeq \tilde{G}_1 \times \cdots \times \tilde{G}_r$ (8)). Moreover we have up to numbering ${}^{\omega^*}(\mathbb{C}^{n_i}_o)/G_i \simeq {}^{\omega^*}A_i$, ${}^{\omega^*}\tilde{A}_i \simeq {}^{\omega^*}(\mathbb{C}^{n_i}_o)/\tilde{G}_i$, therefore also ${}^{\omega^*}(\mathbb{C}^{n_i}_o)/\bar{G}_i \simeq {}^{\omega^*}\bar{A}_i$. Because: $f : \mathbb{C}^{n_i}_o \to \mathbb{C}^s_o$ is holomorphic iff $\bar{f} : \mathbb{C}^{n_i}_o \to \mathbb{C}^s_o$ is holomorphic with $\bar{f} : z \to \bar{z} \to f(\bar{z}) \to \overline{f(\bar{z})}$. One also obtains: f is G_i-invariant iff $\bar{f}$ is $\bar{G}_i$ invariant. This finally implies: $q_i : \mathbb{C}^{n_i}_o \to \mathbb{C}^{s_i}_o$ is a quotientmap with respect to G_i iff $\bar{q}_i$ is a quotientmap with respect to $\bar{G}_i$. And this was to be shown. As in the proof of 15), d) (or with [4] or [6]) we obtain now up to conjugation: $G_i = \tilde{G}_i$ or $\bar{G}_i = \tilde{G}_i$. With this, 9), β), γ) are established. The other direction of the statement is obvious. ✓

Finally 5) and 6) hold:

5): The identity $G = \max^{\omega^*} G$ follows as in the proof 15), d) above. Or realising first, that $\max^{\omega^*} G$ has no reflections (proof 15), d)) one also may use [4] or [6]. In case $\max^{\infty} G$ or $\max^{\omega} G$, G has to be considered as "real group"; 5) follows then from § 2. 6) follows from 4) and 5). ✓

§ 4 Examples

The assumptions made for 7) and 8) are necessary:

I) Case $K = \mathbb{R}$: Let $G := \{g \in O(2) \times O(2) \mid \det g = +1\} \subset O(4)$, where $\mathbb{R}^4 = \mathbb{R}^2 \times \mathbb{R}^2$. G is a compact Liegroup, which is not maximal, because the G-invariant functions are generated by x^2, y^2, where (x,y) are coordinatefunctions of $\mathbb{R}^2 \times \mathbb{R}^2$. Therefore $\max^\infty G = O(2) \times O(2) \neq G$. We have $\mathbb{R}^4/G = \mathbb{R}_+ \times \mathbb{R}_+$; but there is no splitting $\mathbb{R}^4 = V_1 \times V_2$ as in 15), d), because the only G-invariant subspaces of $\mathbb{R}^4$ are $\mathbb{R}^2 \times \{o\}$ and $\{o\} \times \mathbb{R}^2$. And on these spaces G induces already all of O(2). So G can not truely be of the form $G = G_1 \times G_2$. However, for any compact group $G \subset GL(n,\mathbb{R})$ with $\mathbb{R}^n/G \simeq A_1 \times A_2$, there always exist $\mathbb{R}^n = V_1 \oplus V_2$ with V_i G-invariant and $V_i/G \simeq A_i$ (but in general not $G = (G|V_1) \times (G|V_2)$): Use $\mathbb{R}^n/G = \mathbb{R}^n/\max G$ and apply 7) for max G.

II) Case $K = \mathbb{C}$: Let $G \subset GL(2,\mathbb{C})$ be a Diedergroup of atleast 6 elements, generated by two reflections. The quotient $\mathbb{C}^2_o/G$ is regular, so $\mathbb{C}^2_o/G = \mathbb{C}^2_o = \mathbb{C}^1_o \times \mathbb{C}^1_o$ ([6]). But this product does not lead to a decomposition of $\mathbb{C}^2_o$ into G-invariant subspaces, because there are only trivial such spaces.

This shows that, despite of their formal similarity, the factorisation-theorems 7), 8), 9) are fundamentally different in content for the case $K = \mathbb{R}$ or $K = \mathbb{C}$.

Literatur:

[1] Cartan, H.: Quotient d'un espace analytique par un groupe d' automorphismes. In: Algebraic Geometry and Topology. A symposion in honor of S. Lefschetz. (Princeton 1954); 90-102, Princeton University Press 1957.

[2] Ephraim, R.: C^∞ and analytic equivalence of singularities. Proc. of the Conference of Complex Analysis 1972, Rice Univ. Studies.

[3] Ephraim, R.: The cartesian product structure of singularities. Trans. Amer. Math. Soc. 224, 299-311 (1976).

[4] Gottschling, E.: Invarianten endlicher Gruppen und biholomorphe Abbildungen. Inv. Math. 6, 315-326 (1969).

[5] Luna, D.: Fonctions differentiables invariantes sous l'operation d'un groupe rêductif. Ann. Inst. Fourier 26.1, 33-49 (1976).

[6] Prill, D.: Local Classification of Quotients of Complex Manifolds. Duke Math. J. 34, 375-386 (1967).

[7] Reichard, K.: Quotienten analytischer und differenzierbarer Räume nach Transformationsgruppen. Preprint Bochum (1978).

[8] Reichard, K.: C^∞-Diffeomorphismen semianalytischer und subanalytischer Mengen. Compositio Math. 42; Fase 3, 401-416 (1981).

[9] Reichard, K.: Lokale Klassifikation von Quotientensingularitäten reeller Mannigfaltigkeiten nach diskreten Gruppen. Math. Z. 179, 287-292 (1982).

[10] Schwarz, G.W.: Smooth functions invariant under the action of a compact Liegroup. Top. 14, 63-68 (1975).

[11] Spallek, K.: Über Singularitäten analytischer Mengen. Math. Ann. 172, 249-268 (1967).

[12] Spallek, K.: Differenzierbare Räume. Math. Ann. 180, 269-296 (1969).

[13] Spallek, K.: Geometrische Bedingungen für die Integrabilität von Vektorfeldern auf Teilmengen im $\mathbb{R}^n$. Manuscripta Math. 25, 147-160 (1978).

[14] Spallek, K.: Produktzerlegung und Äquivalenz von Raumkeimen I. Der allgemeine Fall. Lecture Notes in Math., 1014, 78-100 (1983). und: Produktzerlegung und Äquivalenz von Raumkeimen II. Der komplexe Fall. Lecture Notes in Math., 1014, 101-111 (1983).

Dr. K. Reichard
Prof. Dr. K. Spallek
Fakultät für Mathematik
Ruhr-Universität Bochum
Universitätsstr. 150

D-4630 Bochum 1

LEAFSPACES AND INTEGRABILITY

by

Hans-Jörg Reiffen
University of Osnabrück, West Germany

This paper is based on my joint work with G. Bohnhorst ([B/R]). I use the definitions and results of [B/R].

In this publication let X be a paracompact connected complex manifold; $\dim X = n$. $\Omega = \Omega_X$ is the sheaf of holomorphic 1-forms on X and $\Theta = \Theta_X = \mathrm{Hom}_{\mathcal{O}}(\Omega, \mathcal{O})$ the sheaf of holomorphic vector fields on X. Θ is the sheaf of holomorphic sections in the tangent space T(X) too. Let F be a coherent analytic sheaf on X; then F is locally free almost everywhere. By $S(F)$ we denote the singular locus of F, i.e. the set of points, where F is not free. The maximalization $\tau(F)$ of the zero subsheaf of F is the torsion sheaf of F; the sections in $\tau(F)$ are the sections in F, which are zero almost everywhere. In the following let β be a coherent foliation on X; $\dim \beta = p$. We denote its sheaf of 1-forms by Ω_β, its sheaf of vectorfields by Θ_β, its cone by K_β and its linear space by $T\beta$. $T\beta$ is the linear space defined by Ω_β. Let $\mathcal{O}_\beta := d^{-1}\Omega_\beta = \{f \in \mathcal{O}_x : x \in X,\ df \in \Omega_{\beta,x}\}$. $S := S(\beta)$ is the singular locus of β; $X^* := X \setminus S$ is the set of points, where β is regular. Let $\beta^* := \beta | X^*$; β^* is the corresponding regular foliation.

We develop a general theory of leaves and apply this theory to get topological criterions for the integrability of foliations.

In § 1 we define integral varieties of β. There are two natural conceptions. Considering that our goal is to get integrals, it is reasonable to define integral varieties by the aid of $\mathcal{O}_\beta$. We deal with special types of integral varieties and derive some criterions for integral varieties.

In § 2 we define leaves by the aid of integral varieties. The definition is similar to the classical one. If β is locally open resp. F-integrable then β induces a decomposition of X into leaves. In this case we get a leaf space; this is a ringed space in a natural manner. We give examples to illustrate our definitions.

If there is a global open resp. F-integral of β then there is a leaf space of β with leaves, which are analytic subspaces of X. Foliations of this special type are called elementary. They were studied in § 3. We deal with special types of these foliations and derive topological criterions for elementary foliations.

In § 4 we derive topological criterions for the existence of integrals resp. H-integrals of β. For instance we get: If codim $S \geq 2$ codim β then the existence of an open local integral $f: (X,x^o) \to (Y,y^o)$ of β, which is simple in x^o is determined only by topological data. We get our results by technics of H.Holmann ([Ho,2]).

In § 5 we study topological isomorphismus of foliations. If dim $S < p$, then S is invariant under topological isomorphismus. To reach this result we use our results of § 4 and the generalized theory of J. Milnor ([Mi],[Ha]).

In [Re,2] I have published a more general and more explicit version of this paper; especially I study there foliations on complex spaces.

§ 1 Integral varieties

In this paragraph let Y be a reduced complex space and $\varphi: Y \to X$ a holomorphic mapping. φ is an immersion iff for every $y \in Y$ there are open neighborhoods W of y, U of $\varphi(y)$ and a (closed) analytic subspace V of U, such that $\varphi|W: W \to V$ is biholomorphic. The following is well known:

1.1 Let $\varphi: Y \to X$ be an injective immersion. Then the following are equivalent:

(a) $\varphi(Y)$ is a closed subset of X and $\varphi: Y \to \varphi(Y)$ is locally proper;

(b) φ is proper;

(c) $\varphi(Y)$ is an analytic subset of X and $\varphi: Y \to \varphi(Y)$ is biholomorphic, $\varphi(Y)$ supplied with the reduced structure. -

If φ is a embedding, i.e. proper injective immersion, we identify Y and $\varphi(Y)$.

<u>1.2 Definition.</u> Let Y be connected and let $\varphi: Y \to X$ be an injective immersion. φ is called an integral variety of β, iff the germ $f \circ \varphi$ is constant for every $y \in Y$ and for every $f \in \mathcal{O}_{\beta,\varphi(y)}$. φ is called a strong integral variety of β, iff $\varphi^*(\Omega_\beta) \subset \tau\Omega_Y$. -

<u>1.3</u> Every strong integral variety is an integral variety. -

<u>1.4</u> Let Y be connected and $\varphi: Y \to X$ an injective immersion. Then the following are equivalent:

(a) φ is an integral variety;

(b) there is a dense subset W of Y, such that the condition in 1.2 is fulfilled for every $y \in W$;

(c) for every irreducible component Y' of Y the mapping $\varphi|Y'$ is an integral variety of β.

Further the following are equivalent:

(a) φ is a strong integral variety;

(b) for every irreducible component Y' of Y there is a point $y \in Y' \setminus Y_s$, such that $\varphi^*(\Omega_{\beta,\varphi(y)}) = 0$;

(c) for every irreducible component Y' of Y the mapping $\varphi|Y'$ is a strong integral variety of β. -

<u>1.5</u> If $\varphi: Y \to X$ is an integral variety of β and $\varphi^{-1}(S)$ nowhere dense in Y then φ is a strong integral variety. -

Let Y be an analytic subspace of X, such that $Y \hookrightarrow X$ is an integral variety of β, then we call Y a proper integral variety of β in X. A local integral variety of β in a point $x \in X$ is the germ (Y,x) of a proper integral variety Y of β in an open neighborhood of x. The following is easy to prove.

<u>1.6 Proposition.</u> Let Y be an analytic subspace of pure codimension q, $x_o \in Y$. Let $\mathcal{I} \subset \mathcal{O}$ denote the ideal sheaf of Y. Then the following are equivalent:

(a) (Y,x_o) is a strong local integral variety of β in x_o;

(b) $\omega \wedge df_1 \wedge \ldots \wedge df_q \in I_{x_o} \cdot \overset{q+1}{\wedge} \Omega_{x_o} \quad \forall\, \omega \in \Omega_{\beta,x_o}, \forall\, f_1,\ldots,f_q \in I_{x_o}$. -

We may test (b) with a generating system instead of I_{x_o}.

1.6 is a generalization of a result in [C/M], Chap. II.

Let $v \in \Gamma(X,\Theta_X)$. Then we can interpret $v: X \to T(X)$ and $v \circ \varphi: Y \to T(X)$ as mappings. Let $u \in \Gamma(Y,\Theta_Y)$. Then we can interpret $\varphi' \circ u: Y \to T(X)$ as a mapping too. For an immersion $\varphi: Y \to X$ we may interpret (Y,y) as a subgerm of $(X,\varphi(y))$ via φ. In this case the equation $\varphi' \circ u = v \circ \varphi$ means, that the vectorfield v can be restricted to a vectorfield u on Y.
u is uniquely determinated by v.

<u>1.7 Proposition.</u> Let Y be connected and let $\varphi: Y \to X$ be an injective immersion. We suppose, that $\varphi^{-1}(S)$ is nowhere dense in Y and that $\dim Y' = p$ for all irreducible compontents Y' of Y. Then the following are equivalent:

(a) φ is an integral variety of β;

(b) for every $y \in Y \setminus (Y_S \cup \varphi^{-1}(S))$ the germ $\varphi(Y,y)$ is the germ of the leaf of β^* in $\varphi(y)$;

(c) for every $y \in Y$ and for every $v \in \Theta_{\beta,\varphi(y)}$ there is an element $u \in \Theta_{Y,y}$ with $\varphi' \circ u = v \circ \varphi$. -

In [C/S] it is shown:
Let X be a connected 2-dimensional manifold, $x_o \in X$, $\dim \beta = 1$. Then there is a local integral variety of β in x_o.

<u>1.8 Definition.</u> An integral variety $\varphi: Y \to X$ of β is called big, iff for every irreducible component Y' of Y we have $\dim Y' \geq p$. -

In [Jo] it is shown, that

$$\omega := (z_1^n z_3 - z_2^{n+1})dz_1 + (z_2^n z_1 - z_3^{n+1})dz_2 + (z_3^n z_2 - z_1^{n+1})dz_3,$$

$n \geq 2$, defines a 1-codimensional foliation β on $\mathbb{C}^3$ without a big local integral variety of β in 0.

We complete our considerations with some remarks about the relation between big integral varieties of β and leaves of β^*.

<u>1.9 Definition.</u> A big integral variety $\varphi: Y \to X$ is called a separatrix, iff $\varphi^{-1}(S)$ is nowhere dense in Y. -
Separatrices are always strong integral varieties.

<u>1.10</u> Let $\varphi: Y \to X$ be a separatrix, then:

(a) $Y_s \subset \varphi^{-1}(S)$;

(b) for every irreducible component Y' of Y, $\varphi|(Y' \setminus \varphi^{-1}(S))$ is a part of a leaf of β^*. -

In [Ho,2] it is shown:

<u>1.11 Proposition.</u> A leaf L of β^* is a proper integral variety of β in X^*, iff L is closed as a subset of X^*. -
Let $k \in \mathbb{R}$, $k \geq 0$, $M \subset X$, $x \in M$. We say, that M has in a neighborhood of x a finite Hausdorff k-measure resp. the Hausdorff k-measure zero, iff there is an open chart neighborhood U of x, such that $H_k(M \cap U) < \infty$ resp. $H_k(M \cap U) = 0$, where we identify U with its image in the number space and take there the Hausdorff k-measure H_k. It is clear to say that M has a locally finite Hausdorff k-measure resp. the Hausdorff k-measure zero.

By 1.11 and by theorems of Bishop (comp. [St]) we get:

<u>1.12 Proposition.</u> Let L be a leaf of β^*. Then the following are equivalent:

(a) $\bar{L}$ is a proper separatrix of β in X;

(b) L is closed as a subset of X^* an $\bar{L} \cap S$ has the Hausdorff 2p-measure zero;

(c) L has a locally finite Hausdorff 2p-measure as a subset of X. -

In 1.12 we denote by $\bar{L}$ the closure of L as a subset of X.

The theorems of Bishop are a generalization of the Remmert-Stein-Thullen theorem (comp. [Ab], § 37). If $\dim S < p$ then $\bar{L} \cap S$ has the Hausdorff 2p-measure zero automatically; if $\dim S = p$ then $\bar{L} \cap S$ has the Hausdorff 2p-measure zero or there is an irreducible component S' of S with $S' \subset \bar{L}$.

In [B/R] it is shown:

1.13 Proposition. codim $S \geq 2$. -
If in this case codim $\beta = 1$ then dim $S < p$.

1.14 Definition. Let $M \subset X$. M is β-small, iff there is no big integral variety φ: $Y \to X$ with $\varphi(Y) \subset M$. -
If dim $S < p$ then S is β-small; for instance if codim $\beta = 1$.
If S is β-small, then all big integral varieties are separatrices.

§ 2 Leaves

Similar as in the case of regular foliations we define leaves.

2.1 Definition. Let $\varphi: Y \to X$ be an integral variety of β. φ is called locally maximal, iff φ has the following properties:

(1) φ is a big integral variety;

(2) for every $y \in Y$ and for every germ of a big integral variety $\psi:(Z,z) \to (X,\varphi(y))$ we have $\psi(Z,z) \subset \varphi(Y,y)$.

2.2 Definition. Let $x \in X$, x is a leaf point of β, shortly β-point, iff there are an open neighborhood U of x and a proper locally maximal integral variety Y of β in U with $x \in Y$. Y is called a β-neighborhood of x. Let $\Lambda(\beta)$ denote the set of all β-points and $\lambda: \Lambda(\beta) \to X$ the natural injection. We set $\Sigma(\beta) := X \setminus \Lambda(\beta)$.-
Obviously: $X^* \subset \Lambda(\beta)$, $\Sigma(\beta) \subset S$.

2.3 Proposition and definition. The family of β-neighborhoods forms a basis of a topology on $\Lambda(\beta)$ and an atlas of a reduced complex structure. We call this topology on $\Lambda(\beta)$ the β-topology and the complex structure β-structure; by $\mathfrak{T}(\beta)$ resp. $\mathcal{O}(\beta)$ we denote the β-topology resp. the structure sheaf of the β-structure. The connected components of $\Lambda(\beta)$ are called leaves of β. λ is an injective holomorphic immersion and $\lambda|L$ is a big integral variety for every leaf L.-

Proof. If $Y_j \subset U_j$, j=1,2, are β-neighborhoods and $x \in Y_1 \cap Y_2$ then by maximality $(Y_1,x) = (Y_2,x)$. Therefore the family of β-neighborhoods forms a basis of a topology. $\lambda: \Lambda(\beta) \to X$ is continuous. Especially $\mathfrak{T}(\beta)$ is a Hausdorff topology. The further assertions are obvious.-
We always supply $\Lambda(\beta)$ and the leaves of β with $\mathfrak{T}(\beta)$ and $\mathcal{O}(\beta)$.

By the theorem of Poincaré-Volterra (comp. [Bou]) every leaf is paracompact.

In the following we omit the sign β, iff no confusion can be expected.

If an integral variety $\varphi:Y \to X$ of β is isomorphic to $\lambda|L : L \to X$, L leaf, we call $\varphi:Y \to X$ a leaf. If $Y \subset X$ is a proper integral variety of β and a leaf, we call Y a proper leaf of β.

<u>2.4</u> Let $\varphi:Y \to X$ be a big integral variety. Then:

(a) φ is locally maximal, iff φ induces an open mapping $\varphi:Y \to \Lambda$;

(b) φ is a leaf, iff φ is locally maximal and there is no locally maximal integral variety $\psi:Z \to X$ with $\varphi(Y) \subsetneqq \psi(Z)$.-

<u>Proof</u>. (a) If φ is locally maximal, then $\varphi(Y) \subset \Lambda$ and $\varphi:Y \to \Lambda$ is open. If $\varphi(Y) \subset \Lambda$, then φ is continuous automatically. If $\varphi:Y \to \Lambda$ is also open, we may identify Y with an open part of Λ.
(b) Obviously by (a).-

<u>2.5 Proposition</u>. Let $Y \subset X$. The following are equivalent:

(a) Y is a proper leaf of β;

(b) Y is a proper locally maximal integral variety of β in X.

(c) Y is an analytic subset of X and there is a leaf L of β with $Y=\lambda(L)$.-

<u>Proof</u>. By 2.4 we get the equivalence of (a) and (b). We derive (b) from (c).

Let be $y \in Y \setminus Y_s$. Then there is an open neighborhood U of y in X, such that $U \cap Y$ is a connected submanifold of U. L is paracompact. Therefore there is a point $x \in U \cap Y$ with a β-neighborhood $Z \subset U \cap Y$, such that $\dim Z = \dim U \cap Y$. Now let $f \in \Gamma(U, \mathcal{O}_\beta)$. Then $f|Z$ and $f|U \cap Y$ are constant. Therefore Y is a proper integral variety of β in X and by the same argumentation Y is big. Now let $y \in Y$. Then $(L,y) \subset (Y,y)$ by the continuity of λ and $(Y,y) \subset (L,y)$ by the local maximality of L.-

For the following compare [B/R]. By the definition of a F-integral we get:

2.6 Proposition. Let β be locally F-integrable; then $\Sigma = \phi$. If $f:X \to Z$ is a F-integral then the leaves of β are the level sets of f, i.e. the connected components of the fibers of f.-

Especially we get:

2.7 Proposition. Let β be locally H-integrable; then $\Sigma = \phi$. The leaves according to 2.3 are the leaves according to [Ho, 2].

In [Ho, 2] it is shown:

2.8 Proposition. Let β be locally H-integrable. Then a leaf L is proper, iff $\lambda(L)$ is a closed subset of X.-

2.8 is a more general version of 1.11.

2.9 Proposition. Let β be locally open integrable; then $\Sigma = \phi$. If $f:X \to Z$ is an open integral then the leaves of β are the level sets of f.-

Proof. Let $f:X \to Z$ be an open integral. Then by [B/R] every level set L of f is a proper integral variety of β in X. Obviously L is a locally maximal integral variety of β.-

We complete our considerations by some examples.

2.10 Let β be the foliation on $\mathbb{C}^3$ defined by the mapping $f:\mathbb{C}^3 \to \mathbb{C}^2$, $f(z) := (z_1, z_3^2 - z_1 z_2)$. Then:

(a) $S = \mathbb{C}_{z_2}$, $\Sigma = \phi$;

(b) the leaves are $\{(0, z_2, \tau) : z_2 \in \mathbb{C}\}$, $\tau \in \mathbb{C}$,

$$\{(\sigma, \frac{z_3^2 - \tau}{\sigma}, z_3) : z_3 \in \mathbb{C}\},\ \sigma \in \mathbb{C}^*,\ \tau \in \mathbb{C};$$

(c) all leaves are manifolds and proper strong integral varieties, the singular locus S is a leaf.-

The mappings $f(z) := (z_3^2 - z_2 z_4, z_4^2 - z_1 z_3, z_1 z_2 - z_3 z_4)$, $g(z) := (z_4^2 - z_1 z_5, z_5^2 - z_2 z_6, z_6^2 - z_3 z_4, z_1 z_2 z_3 - z_4 z_5 z_6)$ (comp. [B/R]) give examples for the following:

2.11 Let β be defined by a generically open but not open holomorphic mapping $f:X \to Y$, Y manifold, with codim $S(f) \geq 2$. Then:

(a) $S=S(f)$, $\Sigma = \phi$;

(b) the leaves are the level sets of f;

(c) all leaves are proper strong integral varieties of β in D, there are leaves with dimension > p.-

Proof. Because of codim $S(f) \geq 2$ we get: f is a maximal and a F-integral of β (comp. [Ma], Theorem 2.1.1), $S=S(f)$; $\Omega_{\beta,z}$ is generated by $\mathcal{O}_{\beta,z}$ for every $z \in X$.-

2.12 Let $X := \mathbb{C}^2 \times \mathbb{C}^*$ and let β be the foliation on X defined by $v = z_1 \frac{\partial}{\partial z_1} + z_3 z_2 \frac{\partial}{\partial z_2}$. Then $S = (0,0) \times \mathbb{C}^*$, $\Sigma = \{(0,0,z_3): z_3 \in \mathbb{C}, z_3 > 0\}$.-

Σ is not analytic.

2.13 Definition. Let $\Sigma = \phi$. Then we denote by $B = B(\beta)$ the set of all leaves of β and by $\pi = \pi_\beta$ the natural mapping $X \to B$, which maps every point $x \in X$ onto the unique leaf L_x passing through x. We supply B with the quotient topology. For an open subset V of B let $H(V) = H_\beta(V)$ be the ring of all continuous mappings $f: V \to \mathbb{C}$ with $f \circ \pi \in \Gamma(\pi^{-1}(V), \mathcal{O})$. We denote the sheaf given by the system $(H(V))_V$ by $H = H_\beta$. The ringed space (B,H) is called the leaf space of β.-

The following remarks are obvious: for every open subset V of B we have $\Gamma(V,H) = H(V) \simeq \Gamma(\pi^{-1}(V), \mathcal{O}_\beta)$ in a natural way; π is a morphism from $(X,\mathcal{O})$ into (B,H).

In general (B,H) is no complex space. If (B,H) is a complex space then all leaves are proper. In [Ho, 1] it is shown:

2.14 Theorem. Let β be locally H-integrable. Then the following are equivalent:

(a) (B,H) is a complex space;

(b) B is a Hausdorff space.-

§ 3 Elementary Foliations

We need some conditions concerning leaves L of β^*. By $\overline{L}$ we denote the closure of L as a subset of X.

3.1 Conditions.

(1) For every $x\in S$ there is a $L\in B(\beta^*)$ with $x\in\overline{L}$.

(2) Every $L\in B(\beta^*)$ is closed as a subset of X^*.

(3) Every $L\in B(\beta^*)$ is closed as a subset of X^* and $\overline{L}\cap S$ has the Hausdorff 2p-measure zero.

(3') Every $L\in B(\beta^*)$ has a locally finite Hausdorff 2p-measure as a subset of X.-

By 1.12 the conditions (3) und (3') are equivalent; they mean, that every leaf L of β^* defines a proper separatrix of β in X. If $\dim S < p$ then (2) and (3) are equivalent.

3.2 Definition. β is elementary, iff $\Sigma = \phi$ and all leaves are proper. β is simple, iff β is elementary, S is β-small, there are only finitely many reducible leaves and these leaves have only finitely many components. -

If $S = \phi$ then β is elementary, iff all leaves are closed as subsets of X.

A family $L \subset B(\beta^*)$ is called connected, iff for every pair $L, L' \in L$ there is a finite sequence $L = L_o, L_1, \dots, L_m = L'$ in L with $\overline{L}_j \cap \overline{L}_{j+1} \neq \phi$.

We set $L^* = L^*_\beta = \{L\in B(\beta^*) : \exists L'\in B(\beta^*) \text{ with } L \neq L',\ \overline{L}\cap\overline{L}' \neq \phi\}$.

3.3 Let β be elementary. Then the condition (3) in 3.1 is fulfilled and

(4) for every relatively compact open subset U of X and for every connected family $L \subset B(\beta^*)$ the set $\{L \in L : L \cap U \neq \phi\}$ is finite. -

3.4 Let S be β-small; suppose, that $B(\beta^*)$ fulfilles the conditions (1) and (3) in 3.1. Then:

(a) β is elementary, iff condition (4) in 3.3 is fulfilled;

(b) β is simple, iff L^* is finite. -

For technical reasons we need the following definition.

3.5 Definition. β is special, iff β is elementary and there is a nowhere dense analytic subset A of X, $S \subset A$, such that the natural mapping $B(\beta|X \setminus A) \to B(\beta)$ is bijective. -

3.6 Let β be simple. Then β is special.

Proof. Let $L_1, \ldots, L_r$ be the reducible leaves of β and let L_j be composed by $\overline{L}_{jl}$, $1 \leq l \leq r_j$, $L_{jl} \in B(\beta^*)$. Set

$$A := S \cup \bigcup_{\substack{1 \leq j \leq r \\ 2 \leq l \leq r_j}} \overline{L}_{jl} \ .-$$

3.7 Proposition. Let $\operatorname{codim} S \geq 2 \operatorname{codim} \beta$.
Suppose, that the conditions (1), (2) in 3.1 are fulfilled. Let $x^o \in S$ and suppose, that

(5) there are only finitely many leaves L of β^* with $x^o \in \overline{L}$.

Then β is simple on a neighborhood of x^o. -

Proof. Let $L, L' \in B(\beta^*)$, $L \neq L'$, $x \in \overline{L} \cap \overline{L'}$. Then (comp. for instance [Wh,2], theorem 12 C, p.70): $\dim_x \overline{L'} \cap \overline{L'} \geq \dim S$. Therefore every component of $\overline{L} \cap \overline{L'}$ is a component of S. By that and by 3.4, (b) we get 3.7. -

If $\operatorname{codim} \beta = 1$, then $\operatorname{codim} S \geq 2 \operatorname{codim} \beta$ by 1.13. In this case the condition (1) in 3.1 is also fulfilled by [M/M], Théorème 2,p.508, and [C/S].

3.8 Notation.

$\Lambda^e = \Lambda^e(\beta) := \{x \in X : \exists$ open neighborhood U of x, such that $\beta|U$ is elementary$\}$; $\Lambda^s = \Lambda^s(\beta) := \{x \in X : \exists$ open neighborhood U of x, such that $\beta|U$ is simple$\}$;

$\Sigma^e = \Sigma^e(\beta) := X \setminus \Lambda^e$, $\Sigma^s = \Sigma^s(\beta) := X \setminus \Lambda^s$. -

Obviously: $\Sigma \subset \Sigma^e \subset \Sigma^s \subset S$.

If $f: X \to Z$ is an open resp. F - integral of β then β is elementary.

Let β be the foliation on $\mathbb{C}^2$ defined by $v = z_1 \frac{\partial}{\partial z_1} + iz_2 \frac{\partial}{\partial z_2}$.

Then: $\phi = \Sigma, \Sigma^e = \Sigma^s = S = \{o\}$.

Let β be the foliation defined by $f: \mathbb{C}^3 \to \mathbb{C}^2$, $f(z) := (z_1 \cdot z_2, z_3)$.
Because $z_1 \cdot z_2$ is locally simple (comp. [B/R]) we get:

(a) f is an H-integral of β;
(b) the leaves of β are the sets
$\{z \in \mathbb{C}^3 : z_1 \cdot z_2 = \sigma,\ z_3 = \tau\}$, $\sigma, \tau \in \mathbb{C}$;
(c) $\phi = \Sigma = \Sigma^e$, $\Sigma^s = S = \mathbb{C}_{z_3}$;
(d) β is special.

§ 4 Leaf space and integrability

Let K,M,N be sets and $g: K \to M$, $f: K \to N$ mappings. f is g-invariant, iff f is constant on the fibers of g. In this case there is a unique mapping $f_g : g(K) \to N$ with $f = f_g \circ g$. f_g is injective, iff g is f-invariant too. In this case g_f is the inverse mapping for $f_g : g(K) \to f(K)$. Now let K,M,N be topological spaces and let f be g-invariant again; we suppose, that g is surjective and that M is supplied with the quotient topology. Then: If f is continuous then f_g is too; if f is open then f_g is too.

Let $\Sigma = \phi$, $K = X$, $M = B$, $g = \pi$.

If $f : X \to N$ is π-invariant, i.e. if f is constant on the leaves of β then $\tilde{f} := f_\pi : B \to N$ exists. It is continuous resp. open, if f has this property. Suppose, that N is a ringed space with a structure sheaf A of germs of continuous complex valued functions (comp. for instance [Ho,1] Def.1, p.332) and that $f : X \to N$ is a morphism. Then obviously $\tilde{f} : (B, H) \to (N, A)$ is a morphism.

Let U be an open subset of X. We denote all objects relating to $\beta|U$ by the index U. The mapping $\pi|U : U \to B$ is constant on the leaves of $\beta|U$. We set $\pi^U := \widetilde{\pi|U}$. $\pi^U : (B_U, H_U) \to (B, H)$ is a morphism.
If U is β-saturated, i.e. $\pi^{-1}\pi(U) = U$, then π^U is injective; if U is β-saturated and π open the $\pi^U : B_U \to \pi(U)$ is a bimorphism.
Let $f : X \to Y$ be an integral of β. Then $\tilde{f} : B \to Y$ is defined and a morphism.

In the following the definitions of simplicity, simplicity in a point and local simplicity are important (comp. [B/R]).

Let $f: X \to Y$ be an open or a F-integral of β. Then $\Sigma = \phi$ and the level sets of f are the leaves of β. If f is simple then $\tilde{f}: B \to Y$ is injective.

Let $f: X \to Y$ be an open integral. Then we may suppose, that f is surjective and that Y is normal. $\tilde{f}: B \to Y$ is a surjective open mapping. By [Ho,2] Lemma 1.6 we get:

<u>4.1</u> Let $f: X \to Y$ be an integral of β. Suppose, that f is surjective, open, simple and that Y is normal. Then $\tilde{f}: B \to Y$ is a bimorphism; especially (B, H) is a normal complex space. -

<u>4.2 Definition:</u> β is called s-elementary, iff

(1) β is elementary;

(2) B is a Hausdorff space;

(3) π is open. -

The definition of s-simplicity resp. s-speciality is obvious.

By 4.1 we get:

<u>4.3</u> Let an open simple integral $f: X \to Y$ of β exist. Then β is s-elementary. -

<u>4.4 Theorem:</u> Let β be s-special. Then there is an open simple integral $f: X \to Y$ of β.

<u>Proof.</u> Let A be as in 3.5. Set $X := \tilde{X} \setminus A$. Then $\pi^{\tilde{X}}: B_{\tilde{X}} \to B$ is a homeomorphism. Thus $B_{\tilde{X}}$ is a normal complex space by 2.14. It is easy to see, that $\pi^{\tilde{X}}$ is a bimorphism. -

Let $x^o \in X$. If there is an open integral $f: U \to Y$ of β on an open neighborhood U of x^o then there is an unique integral $g: (X, x^o) \to (Z, z^o)$ of β, which is factorial in x^o. We say, that β is open and simple integrable in x^o, iff there is an open integral $f: U \to Y$, which is simple in x^o. If we choose Y normal, then $f|(X, x^o) = g$ (up to an isomorphism).

<u>4.5 Proposition.</u> Let codim $S \geq 2$ codim β .
Let $x^o \in X$. Then the following are equivalent:

(a) β is open and simple integrable in x^o;
(b) there is an open neighborhood U of x^o, such that $\beta|U$ is s-elementary and $\pi_U : U \to B_U$ simple in x^o. -

<u>Proof.</u> By 4.1 we get (b) from (a). We derive (a) from (b). We may assume, that $U = X$. By 3.7 there is an open neighborhood U of x^o, such that $\beta|U$ is simple and $\pi|U : U \to B$ is simple. It follows, that $\pi^U : B_U \to \pi(U)$ is an homeomorphism. By 4.4 $\pi_U : U \to B_U$ is an integral of β. -

The open and simple integrability is a topological property for foliations with codim $S \geq 2$ codim β .

<u>4.6 Proposition.</u> Let codim $\beta = 1$. Let $x^o \in X$. Then the following are equivalent:

(a) β is integrable in x^o;
(b) β is open and simple integrable in x^o;
(c) there is an open neighborhood U of x^o, such that $\beta|U$ is s-simple . -

<u>Proof.</u> (b) $\Rightarrow$ (c) by 4.3 and 3.7 .

(c) $\Rightarrow$ (a) by 4.4 .
(a) $\Rightarrow$ (b) by the result, p.105 [Re,1]. -

By 4.6 the integrability is a topological property for foliations with codim $S = 1$. Theorem B in [M/M] is stronger than the conclusion (c) $\Rightarrow$ (a) in 4.6. It is possible to prove the following generalization of 2.14 and 4.6:

<u>4.7 Proposition:</u> Let codim $\beta = 1$.
We assume:
(1) All leaves of β^* are closed as subsets of X^*;
(2) L^* is finite (comp. § 3);
(3) the topological space, which we get from $B(\beta^*)$ by identifying the elements of maximal connected families, is a Hausdorff space.

Then there is a simple open integral $f : X \to Y$, Y Riemannian surface . -

The proof of 4.7 will be published elsewhere. It works with 2.14 and with extension theorems for holomorphic mappings with values in a Riemannian surface, which can be proven by technics of [Ko]. The equivalence of (a) and (b) in 4.6 is not longer true, if codim $\beta \geq 2$.

4.8 Let β be the foliation on $\mathbb{C}^3$ defined by the mapping $f : \mathbb{C}^3 \to \mathbb{C}^2$, $f(z) := (z_1 \cdot z_2, (z_1 + z_2) \cdot z_3)$. f is an open F-integral, but β is not open and simple integrable in O. (Comp. [Bo]). -

Now we will study the H-integrability of β. Analogously to 4.2 we define:

4.9 Definition: β is called H-elementary, iff

(1) β is elementary;

(2) B is an Hausdorff space;

(3) π is open and locally simple. -

By 4.1 get:

4.10 Let exist a simple H-integral $f : X \to Y$ of β.
Then β is H-elementary. -
In the following we need a result of [Bo] resp. [Sch]:

4.11 Theorem. Let $f : X \to Y$ be a holomorphic mapping, Y manifold. Then the following are equivalent:

(a) f is open and locally simple;

(b) f is open and reduced;

(c) f is generically open and no irreducible component of a fiber of f is contained in the singular locus S(f) of f.-

4.12 Definition. Let $f : X \to Y$ be an H-integral of β. If Y is a manifold we call f a good H-integral. -

4.13 Proposition. Let $f : X \to Y$ be a surjective H-integral of β. Then:

(a) If every fiber of f intersects X^* then f is a good H-integral;

(b) if f is a good H-integral then S is β-small, f maximal and β free. -

Proof. (a) Let $y_o \in Y$, $x_o \in f^{-1}(y_o) \cap X^*$. f is factorial in x^o. Comparing $f|(X,x_o)$ with a regular mapping defining β^* close to x_o we get $y_o \notin Y_s$.

(b) Because f is a F-integral we have $S(f) = S$.
Because of 4.11 S is β-small. We have $\operatorname{codim} S \geq 2$. Therefore f is maximal and β free. -

<u>4.14 Theorem:</u> Let β be H-special. Then there is a good H-integral $f : X \to Y$ of β.

<u>Proof.</u> We get 4.14 immediately by 4.4 und 4.13. We give another proof without using 2.14. We prove, that (B, H) is a complex manifold. Let $b_o \in B$ and $x_o \in b_o \setminus A$; A as in 3.5. Let $U_o \cong U_o' \times U_o''$ be a regular foliation chart of β^* close to x_o and let $\beta^* | U_o$ be defined by the natural projection $\pi_o : U \to U_o'$. There is an open neighborhood $V_o \subset U_o$ of x_o, such that $\pi | V_o$ is simple and an open neighborhood $W_o \subset V_o$ of x_o, $W_o \cong W_o' \times W_o''$. Let $x_o = (x_o', x_o'')$. It is easy to show, that the mapping $\sigma : W_o' \to \pi(W_o)$, $x' \to \pi(x_o', x_o'')$ is a topological morphism. Let $g \in \Gamma(W_o', \mathcal{O})$ and let $\hat{W}_o$ be the β-saturation of W_o in X. There is a unique continuous function $h : \pi(W_o) \to \mathbb{C}$ with $h \bullet \sigma = g$. Then $h \circ \pi : \hat{W}_o \to \mathbb{C}$ is continuous and $h | \hat{W}_o \setminus A$ is holomorphic. Therefore $h \in \Gamma(\pi(W_o), H)$. -

Analogously to 4.5 we get:

<u>4.15 Proposition.</u> Let $\operatorname{codim} S \geq 2 \operatorname{codim} \beta$. Then the following are equivalent:

(a) β is locally H-integrable;
(b) β is locally H-elementary. -

<u>4.16 Proposition.</u> The following are equivalent:

(a) S is β-small and β is H-elementary with leaves, which are normal complex subspaces of X;
(b) β is H-simple and $\dim S \cap \bar{L} \leq p-2$ for every $L \in B(\beta^*)$;
(c) there is a good H-integral $f : X \to Y$ of β with fibers, which are normal complex spaces;
(d) there is an integral $f : X \to Y$ of β, Y manifold, such that $\dim S(f) \cap L \leq p-2$ for every $y \in f(X)$ and for every irreducible component L of $f^{-1}(y)$. -

<u>Proof.</u> (a) $\Rightarrow$ (c) : $L^* = \phi$. By 4.14 we get (c).

(c) $\Rightarrow$ (a) : By 4.13, (b) S is β-small.
(b) $\Rightarrow$ (c) : By 4.14 there is a good H-integral $f : X \to Y$.

By 4.11 we get $f^{-1}(y)_s \leq p-2$ for every $y \in f(X)$. Because $f^{-1}(y)$ is a locally complete intersection we get, that $f^{-1}(y)$ is normal by a well known theorem (comp. for instance [Ab], p.435).

(c) ⇒ (b) by 4.11 .

(d) ⇒ (c) by 4.11. -

4.17 Corollary. Let $\dim S \leq p-2$. Then the following are equivalent:

(a) β is locally H-simple;

(b) β is locally H-integrable;

(c) β is free . -

Proof. (c) ⇒ (b) follows by the Frobenius - Malgrange theorem [Ma] . -

§ 5 Some topologically invariant properties of foliations

In this paragraph let $\tilde{X}$ be another paracompact connected complex manifold and $\tilde{\beta}$ a coherent foliation on $\tilde{X}$; $\dim \tilde{X} = \tilde{n}$, $\dim \tilde{\beta} = \tilde{p}$. We denote the objects relating to $\tilde{X}, \tilde{\beta}$ by a tilde. Let $\psi : X \to \tilde{X}$ be an homeomorphism. Then $n=\tilde{n}$. Wet set $T := S \cup \psi^{-1}(\tilde{S})$, $\tilde{T} := \tilde{S} \cup \psi(S)$; $\mathring{X} := X \setminus T$, $\mathring{\tilde{X}} := \tilde{X} \setminus \tilde{T}$.

5.1 Definition. ψ is called an isomorphism from (X,β) onto $(\tilde{X},\tilde{\beta})$, iff $\psi|\mathring{X} : \mathring{X} \to \mathring{\tilde{X}}$ is an homeomorphism relatively to the $(\beta|\mathring{X})$- resp. $(\tilde{\beta}|\mathring{\tilde{X}})$-topology.

If $\Sigma = \phi$, $\tilde{\Sigma} = \phi$ we call ψ a strong isomorphism from (X,β) onto $(\tilde{X},\tilde{\beta})$, iff $\psi : X \to \tilde{X}$ is a homeomorphism relatively to the β-resp. $\tilde{\beta}$-topology.-

5.2 For every leaf $L \in B(\beta|\mathring{X})$ let $\psi(L)$ be part of a leaf $\tilde{L} \in B(\tilde{\beta}|\mathring{\tilde{X}})$ of the same dimension. Then ψ is an isomorphism.

Proof. We may assume $S = \phi$, $\tilde{S} = \phi$. Let $x \in X$, $\tilde{x} := \psi(x)$. We consider neighborhoods $U, \tilde{U}$ of $x, \tilde{x}$, which are chart neighborhoods relatively to the foliations, i.e. $\beta, \tilde{\beta}$ are defined on $U, \tilde{U}$ by regular holomorphic mappings onto domains in the number space. Let $\tilde{U} = \psi(U)$. We may assume, that U is a product of discs and that β is defined on U by the projection on some of these discs. Let L be the product of the other discs.

Then the local leaves of β in U are of the form $a \times L$. We consider a local leaf $\tilde{L}$ of $\tilde{\beta}$ in $\tilde{U}$. Then $\tilde{L}$ is the disjoint union of sets $\psi(a \times L)$. By 1, p.46 in [H/W] this union is countable. Therefore there must exist one leaf $a \times L$ $\tilde{L}=\psi(a \times L)$.-

<u>5.3</u> Let ψ be a strong isomorphism. If L is a proper leaf of β then $\psi(L)$ is a proper leaf of $\tilde{\beta}$.-

<u>Proof</u>. By 1.1, b.-

By 5.4 and 3.4, b we get:

<u>5.4</u> Let ψ be a strong isomorphism.

(a) β is elementary, iff $\tilde{\beta}$ is elementary.

(b) Let $S, \tilde{S}$ be β - resp. $\tilde{\beta}$-small then β is simple, iff $\tilde{\beta}$ is simple.-

<u>5.5 Proposition</u>. Let ψ be an isomorphism. Suppose that $\dim S < p$, $\dim \tilde{S} < \tilde{p} = p$. Then: $\psi(\Sigma^e) = \tilde{\Sigma}^e$, and $\psi : \Lambda^e \to \tilde{\Lambda}^e$ is a strong isomorphism.-

<u>Proof</u>.

Special case: $S = \phi$.

We may assume, that X is a product of discs and that β is defined by the projection on some of these discs. Let L be the product of the other discs. We consider a leaf $a \times L$ of β and the image $\tilde{L} := \psi(a \times L)$. Because of Cor. 1, p. 48 in [H/W] $(a \times L) \setminus T$ is a connected open and dense subset of $a \times L$. Therefore $\tilde{L} \setminus \tilde{S}$ is a proper leaf of $\tilde{\beta}^*$ and $\overline{\tilde{L} \setminus \tilde{S}} = \tilde{L}$. $\tilde{\beta}$ fullfills the conditions (1) and (2) of 3.1 and $L^*_{\tilde{\beta}} = \phi$. Therefore $\Sigma^s = \phi$ and ψ is a strong isomorphism.

General case:

Let β be elementary. We show, that $\tilde{\beta}$ is elementary too and that ψ is a strong isomorphism. Because of the special case and 5.4 (a) $\tilde{\beta}^*$ is elementary; especially condition (2) in 3.1 is fullfilled. ψ induces a bijective mapping $B(\beta | \dot{X}) \to B(\tilde{\beta} | \dot{\tilde{X}})$. The forming of intersection gives a bijective mapping $B(\beta^*) \to B(\beta | \mathring{X})$ and analogously $B(\tilde{\beta}^*) \to B(\tilde{\beta} | \mathring{\tilde{X}})$. We get a bijective mapping $B(\beta^*) \to B(\tilde{\beta}^*)$. Therefore the conditions (1) of 3.1 and (4) of 3.3 are fullfilled and obviously ψ is a strong isomorphism.-

By a proof similar to 5.5 we get

5.6 Proposition. Let ψ be an isomorphism. Suppose, that $\beta, \widetilde{\beta}$ are locally integrable by good H-integrals. Then ψ is a strong isomorphism.-

5.7 Let ψ be a strong isomorphism. Then ψ induces an homeomorphism $B(\beta) \to B(\widetilde{\beta})$.-

5.8 Definition. Let $x \in X$. We call x a topologically regular point of β, iff there are an open neighborhood U of x, a connected complex manifold $\widetilde{U}$ with a regular foliation $\widetilde{\beta}$ and an isomorphism $\psi: U \to \widetilde{U}$ from $(U, \beta|U)$ onto $(\widetilde{U}, \widetilde{\beta}|\widetilde{U})$. We call x a topologically singular point of β, iff x is not topologically regular. By $S^o = S^o(\beta)$ we denote the set of all topologically singular points of β.-

5.9 Suppose dim S<p. If $S^o = \phi$ then X is locally integrable by good H-integrals.-

Proof. We may assume, that there is an isomorphism $\psi: X \to \widetilde{X}$ of (X, β) onto $(\widetilde{X}, \widetilde{\beta})$, where $\widetilde{X}$ is a product of discs and $\widetilde{\beta}$ is defined by the projection on some of these discs. By 5.5 $\Sigma^e = \phi$ and ψ is a strong isomorphism. Because of 5.4 (b) β is simple and because of 5.7 β is H-simple. By 4.14 there exists a good H-integral $f: X \to Y$.-

5.10 Lemma. Let Y be a reduced locally complete intersection, and let $f: Y \to \mathbb{C}$ be a reduced open holomorphic function, $y^o \in S(f)$, $f(y^o) = 0$, $Y_s \subset f^{-1}(0)$. Further let $\widetilde{Y} = \overset{n}{\underset{j=1}{X}} \widetilde{Y}_j$ be a product of discs $\widetilde{Y}_j$ in $\mathbb{C}$ and $g: \widetilde{Y} \to \widetilde{Y}_1$ the natural projection. Then there is no homeomorphism $\psi: Y \to \widetilde{Y}$ mapping the fibers of f onto the fibers of g.-

By S(f) we denote the singular locus of f and by Y_s the singular locus of Y.

Proof. Indirectly. We may assume, that Y is an analytic subspace of a domain U in $\mathbb{C}^m$, $y^o = 0$ and that all $\widetilde{Y}_j$ have center 0, $\psi(0) = 0$. We set $S := S(f)$. $Y_s \subset f^{-1}(0)_s = S \subset f^{-1}(0) \subset Y$.

We may assume, that there is an open neighborhood W of 0 in U, such that $S \cap W = (\mathbb{C}^k_{z_1,\ldots,z_k} \times 0) \cap W$ and that both Y and $f^{-1}(0)$ are Whitney-regular along S. We may choose W of the form $W = \dot{W} \times \ddot{W}$, where $\dot{W}$ is a real cubus in $\mathbb{C}^k$ with center 0 and $\ddot{W}$ an open ball in $\mathbb{C}^{m-k}$. We denote by $\dot{z}$ the coordinates relatively to $\mathbb{C}^k$ and by $\ddot{z}$ the coordinates relatively to $\mathbb{C}^{m-k}$. By the theory of Thom-Mather (comp. for instance [R/T]) we may assume that there is a homeomorphism $\Phi: W \to W$ with the following properties:

$\dot{\Phi}(z) = \dot{z}$, $|\ddot{\Phi}(z)| = |\ddot{z}|$, $z \in W$, $|\cdot|$ denoting the euclidean norm;

$\Phi(\dot{W} \times Y_0) = Y \cap W$, $\Phi | 0 \times \ddot{W} = \mathrm{id}$;

$\Phi(\dot{W} \times f^{-1}(0)_0) = f^{-1}(0) \cap W$,

with $A_{\dot{z}} := A \cap (\dot{z} \times \ddot{W})$ for $\dot{z} \in \ddot{W}$, $A \subset \mathbb{C}^m$;

$\Phi | W \setminus S$ is a C^∞-mapping.

Further we may assume, that $f|Y_0 \cap W$ has an isolated singularity in 0. Therefore by the theory of Milnor and Hamm (comp. for instance [Gr]) there are numbers $\delta, \varepsilon, 0 < \delta \ll \varepsilon$, such that we get:

$$f : E^*_{\delta\varepsilon} \to D^*_\delta \quad (* \text{ means removing } (f^{-1}(0))_0 \text{ resp. } 0)$$

is a locally trivial C^∞-fiber bundle with connected fiber F with $\pi_r(F) \neq 0$ for $r = \dim(f^{-1}(0))_0$ for

$$D_\delta := \{t \in \mathbb{C} : |t| < \delta\}$$

$$E_{\delta\varepsilon} := \{y \in X_0 : |y| < \varepsilon, |f(y)| < \delta\}.$$

We get a diagram

$$\begin{array}{ccccc} & & Y & \xrightarrow[\cong]{\psi} & \tilde{Y} \\ & & \cup & & \cup \\ \hat{M} & \xrightarrow[\cong]{\Phi} & M & \xrightarrow[\cong]{\psi} & \tilde{M} \end{array}$$

with $\hat{M} := \dot{W} \times E_{\delta\varepsilon}$. Now we repeat our construction and get the following diagram; the notations go without further explanation:

$$\begin{array}{ccccc}
 & & Y & \longrightarrow & \tilde{Y} \\
 & & \cup & & \cup \\
\hat{M} & \longrightarrow & M & \longrightarrow & \tilde{M} \\
\cup & & \cup & & \cup \\
\hat{Y}_* & \longrightarrow & Y_* & \longrightarrow & \tilde{Y}_* \\
\cup & & \cup & & \cup \\
\hat{M}_* & \longrightarrow & M_* & \longrightarrow & \tilde{M}_*.
\end{array}$$

(with inclusions $i: \hat{M}_* \to \hat{M}$ and $j: \tilde{Y}_* \to \tilde{Y}$)

We remove the fibers through zero everywhere and retain the notations. The mappings in the rows are homeomorphisms; the inclusions i,j induce isomorphisms of the homotopy groups. We get:

$$\pi_1(Y_*) \simeq \pi_1(M), \quad \pi_1(\tilde{Y}) \simeq \pi_1(E^*_{\delta\varepsilon});$$

$$\pi_l(\tilde{Y}) = \begin{cases} 0 & , \ l\geq 2 \\ \mathbb{Z} & , \ l=1; \end{cases}$$

$$\pi_r(E^*_{\delta\varepsilon}) \simeq \pi_r(F) \neq 0, \text{ if } r \geq 2,$$

$$\pi_1(E^*_{\delta\varepsilon}) \neq \mathbb{Z}, \text{ if } r=1.-$$

<u>5.11 Theorem</u>. Let β be locally integrable by good H-integrals. Then $S=S^o$.-

<u>Proof</u>. We assume $S \neq S^o$; let $x^o \in S\setminus S^o$. Then we may assume, that X is a domain in $\mathbb{C}^n$, $x^o=0$, that there is an open reduced mapping $f:X \to \mathbb{C}^q$ defining β and that there is an homeomorphism $\psi: X \to \tilde{X}$ onto a product $X = \overset{n}{\underset{j=1}{\times}} \tilde{X}_j$ of open discs $\tilde{X}_j$ in $\mathbb{C}$ with center 0, $\psi(0)=0$, such that ψ is a strong isomorphism, where $\tilde{\beta}_q$ is the foliation on $\tilde{X}$ given by the natural projection $g:\tilde{X} \to \overset{q}{\underset{j=1}{\times}} \tilde{X}_j$. We may assume, that $S=X \cap (\mathbb{C}^k_{z_1,\ldots,z_k} \times 0)$ and that $f|S$ is a composition of a projection with an embedding. Therefore we may assume, that f is of the form $f=(z_1,\ldots,z_l,f_{l+1},\ldots,f_q)$ with $f^{-1}f(S)=\{z\in X: f_{l+1}(z)= \ldots=f_q(z)=0\}$. Obviously we have $q-l \geq 1$. $\tilde{f}:=(z_1,\ldots,z_l,f_{l+1},\ldots,f_{q-1})$ is an open reduced mapping too. Consider the complete intersection $Y=\{z \in X : \tilde{f}(z) = 0\}$ and the function $f_q|Y$. We have an homeomorphism

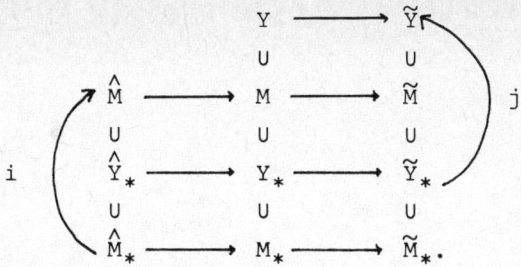

$$\Phi : f(X) \to \overset{q}{\underset{j=1}{X}} \widetilde{X}_j \text{ with: } \Phi \circ f = q \circ \psi.$$

Because f(Y) is a 1-dimensional manifold in a neighborhood of O, we may assume, that ψ maps the fibers of $f_q|Y$ onto the fibers of the projection of $\overset{n}{\underset{j=q}{X}} \widetilde{X}_j$ onto $\widetilde{X}_q$. By 5.10 we get a contradiction.

By 5.9 and 5.11 we get the following corollary:

5.12 Theorem. Let dim $S < p$. Then $S = S^{O}$.

Our result 5.11 is related to [G/L].

REFERENCES

[Ab] Abhyankar, S.S.; Local Analytic Geometry; Academic Press, New York, London (1964)

[Bo] Bohnhorst, G.; Einfache holomorphe Abbildungen; Math. Ann. 275, 513-520 (1986)

[B/R] Bohnhorst, G. und Reiffen, H.-J.; Holomorphe Blätterungen mit Singularitäten; Math. Gottingensis Heft Nr. 5 (1985)

[Bou] Bourbaki, N.; General Topology, Part 1; Addison-Wesley Reading (1966)

[C/S] Camacho, C. and Sad, P.; Invariant varieties through singularities of holomorphic vector fields; Ann. Math. 115 (1982), 579-595

[C/M] Cerveau, D. et Mattei, J.-F.; Formes intégrables holomorphes singulières; Astérisque 97 (1982)

[G/L] Gau, Y.-N. and Lipman, J.; Differential Invariance of Multiplicity on Analytic Varieties; Invent. math. 73, 165-186 (1983)

[Gr] Greuel, G.-M.; Der Gauß-Manin-Zusammenhang isolierter Singularitäten von vollständigen Durchschnitten; Math. Ann. 214, 235-266 (1975)

[Ha] Hamm, H.; Lokale topologische Eigenschaften komplexer Räume; Math. Ann. 191, 235-252 (1971)

[Ho,1] Holmann, H.; Komplexe Räume mit komplexen Transformationsgruppen; Math. Ann. 150, 327-360 (1963)

[Ho,2] Holmann, H.; Holomorphe Blätterungen komplexer Räume; Comm. Math. Helv. 47 (1972), 185-204

[H/W] Hurewicz, W. and Wallmann, H.; Dimension Theory; Princeton University Press (1969)

[Jo] Jouanoulou, J.P.; Equations de Pfaff algébriques; Springer Lect. Not. in Math. 708 (1979)

[Ko] Kobayashi, Sh.; Hyperbolic Manifolds and Holomorphic Mappings; Marcel Dekker, Inc., New York (1970)

[Ma] Malgrange, B.; Frobenius avec singularité II: le cas general; Inv. math. 39 (1977), 67-89

[M/M] Mattei, J.-F. et Moussu, R.; Holonomie et intégrales premières; Ann. scient. Ec. Norm. Sup. 4e ser., t.13 (1980), 469-523

[Mi] Milnor, J.; Singular points of complex hypersurfaces; Annals of Mathematics Studies 61, Princeton: Princeton University Press (1968)

[Re,1] Reiffen, H.-J.; Einfache holomorphe Funktionen; Math. Ann. 259 (1982), 99-106

[Re,2] Reiffen, H.-J.; Leafspaces and Integrability; Osnabrücker Schriften zur Mathematik, Preprints, Heft 99 (1987)

[R/T] Reiffen, H.-J. und Trapp, H.W.; Ein Beitrag zur Whitney-Regularität im unendlichdimensionalen Fall; Comment. Math. Helv. 54 (1979), 159-172

[Sch] Schumacher, G.; Ein topologisches Reduziertheitskriterium für holomorphe Abbildungen; Math. Ann. 220, 97-103 (1976)

[St] Stolzenberg, G.; Volumes, Limits and Extensions of Analytic Varieties; Springer Lect. Not. in Math. 19 (1966)

[Wh,1] Whitney, H.; Local Properties of Analytic Varieties; Diff. and Combin. Top., Princeton Univ. Press (1965), 205-244

[Wh,2] Whitney, H.; Complex Analytic Varieties; Addison-Wesley, Reading (1972)

STRUCTURAL STABILITY OF GERMS OF VECTOR FIELDS ON SURFACES WITH A SIMPLE SINGULARITY. *

Federico Sánchez-Bringas.

INTRODUCTION.

Let H be a finite subgroup of SU(2) acting freely in $\mathbb{C}^2-\{0\}$. Although the group H does not act freely in the origin, the quotient space of $\mathbb{C}^2$ with this action is an algebraic surface with an isolated singularity at the origin.

The germs of vector fields at the singularity are identified with the H-equivariant germs of holomorphic vector fields in the origin of $\mathbb{C}^2$.

We endow the space of germs of holomorphic vector fields with a singularity in the origin of $\mathbb{C}^2$, $\mathfrak{X}_0(\mathbb{C}^2)$, with the topology of power series [DU-RO] and regard $\mathfrak{X}_0(\mathbb{C}^2/H)$ as a subspace of $\mathfrak{X}_0(\mathbb{C}^2)$.

In this note we prove the following result:

Theorem 11. There exists an open and dense subset A of $\mathfrak{X}_0(\mathbb{C}^2/H)$ whose elements are structurally stable.

In section 1 we present the main tool: Poincaré's linearization theorem for singular germs of holomorphic vector fields in $\mathbb{C}^2$.

In section 2 we study the conditions for a germ of a vector field to be H-equivariant.

In sections 3 and 4 we prove the result for the two qualitatively different cases: whether H has or does not have a diagonal representation (Propositions 8 and 9).

Before ending I would like to thank Xavier Gomez-Mont and Jose Seade-Kuri for their great help in the elaboration of this note.

1. PRELIMINARIES.

Let X be an element of $\mathfrak{X}_0(\mathbb{C}^2)$. Let.

$$X(z_1,z_2)=(\Sigma a_{ij} z_1^i z_2^j, \Sigma b_{ij} z_1^i z_2^j) \tag{1}$$

be the power series expansion of X.

The linear part of X is $A_X(z_1,z_2)=\begin{pmatrix} a_{10} & a_{01} \\ b_{10} & b_{01}\end{pmatrix}\begin{pmatrix} z_1 \\ z_2\end{pmatrix}$

The linear vector field $A(z_1,z_2)=(a_{10}\ z_1+a_{01}\ z_2, b_{10}\ z_1+b_{01}\ z_2)$ is in the Poincaré domain if zero is not contained in the segment determined by the eigenvalues of the ---

* Supported by CONACYT.

matrix A, $\{\lambda_1,\lambda_2\}$. If besides, λ_1/λ_2 is not a real number we say that the vector field X is hyperbolic.

The field A is resonant if there exists an integral relation of the form:

$$\lambda_s = m_1\lambda_1 + m_2\lambda_2 \text{ where } s=1,2,\ m_k \in \mathbb{N}, \text{ and } m_1 + m_2 \geqslant 2$$

The subset of germs of holomorphic vector fields in the origin of $\mathbb{C}^2$ such that, their linear part is in the Poincaré domain and is non resonant, is open and dense - in $\mathfrak{X}_0(\mathbb{C}^2)$.

Let X and Y be two germs in $\mathfrak{X}_0(\mathbb{C}^2)$, X is topologically (differentially, biholomorphically) equivalent to Y if there exists a germ of homeomorphism (diffeomorphism, biholomorphism), $\phi: U \to V$, where U and V are open neighborhoods of the origin in $\mathbb{C}^2$ such that $\phi(0)=0$ and ϕ transforms the leaves of the foliation defined by X into the leaves of the foliation defined by Y. If besides ϕ sends the flow of the vector field X into the flow of the vector field Y, X is topologically (differentially, biholomorphically) conjugated to Y.

The following classic theorem is due to Poincaré [PO,ARN].

Poincaré's Theorem. If the linear part of a holomorphic vector field at a singular -- point belongs to the Poincaré domain and is non resonant, then the vector field is biholomorphically conjugate to its linear part.

2. H-EQUIVARIANT GERMS.

Let X be an element of $\mathfrak{X}_0(\mathbb{C}^2/H)$.

Developing X into power series (1) its linear part is non-degenerate if the determinant of A_X is non zero. In many cases the linear part determines the topological behavior of the germ X. The condition of being non-degenerate is not very restrictive:

1. Proposition. The subset G of germs X in $\mathfrak{X}_0(\mathbb{C}^2/H)$ such that its linear part is non-degenerate is open and dense.

Proof. Since the topology of $\mathfrak{X}_0(\mathbb{C}^2/H)$ is the power series topology, it is enough to - prove the statement for the germs of linear vector fields. If X is a degenerate germ, $X + {}^1/_n(z_1,z_2) = X_n$ is non-degenerate, X_n converges to X and is H-equivariant. Hence G is dense.

The openess of G is obvious. □

If X is H-equivariant then its linear part A_X is also H-equivariant.

If h is an element of H, $[A_X,h] = A_Xh - hA_X$. A_X is H-equivariant if and only if -- $[A_X,h]=0$ for every $h \in H$.

2. Lemma. Let X be an element of $\mathfrak{X}_0(\mathbb{C}^2/H)$ and $\phi:(\mathbb{C}^2,0)\to(\mathbb{C}^2,0)$ a germ of biholomorphism such that $D\phi(X)=Y$, then the germ Y is $\bar{H}$-equivariant, where $\bar{H}=\phi H\phi^{-1}$ i.e. $Y\in\mathfrak{X}_0(\mathbb{C}^2/\bar{H})$.

Proof. Let $\bar{h}$ be an element of $\bar{H}$, $\bar{h}=\phi h\phi^{-1}$ thus $D\bar{h}_z(Y(z))=D\phi h\phi_z^{-1}(Y(z))=D\phi h(X(\phi^{-1}(z)))=$ $D\phi(X(h\phi^{-1}(z)))=Y(\phi h\phi^{-1}(z))=Y(\bar{h}(z))$. □

If X is a diagonalizable linear vector field, due to the lemma, it is possible to find a linear conjugation of the group H, where the expression of X is diagonal.

The next proposition gives a characterization of linear germs according to the group representation. Its proof uses the following remark: If H is a finite subgroup of $GL(2,\mathbb{C})$ and $h=\begin{pmatrix} a & b\\ 0 & a\end{pmatrix}$ is an element of H, then $h^n=(h\circ\ldots\circ h)=\begin{pmatrix} a^n & na^{n-1}b\\ 0 & a^n\end{pmatrix}$. But h generates a cyclic group, so b=0 and a is a root of unity.

3. Proposition. i) If H is a finite, diagonal subgroup of SU(2) with an element $h=\begin{pmatrix}\lambda_1 & 0\\ 0 & \lambda_2\end{pmatrix}$, $\lambda_1\neq\lambda_2$, then a linear vector field A in $\mathfrak{X}_0(\mathbb{C}^2)$ is H-equivariant if and only if A is diagonal, i.e. $A(z_1,z_2)=(a_1z_1,a_2z_2)$ $a_i\in\mathbb{C}$.

ii) Let H be a finite subgroup of SU(2) and suppose that it is non diagonalizable, i.e. for each $\mu\in GL(2,\mathbb{C})$ the group $\mu H\mu^{-1}$ has a non diagonal element. Then a linear vector field A in $\mathfrak{X}_0(\mathbb{C}^2)$ is H-equivariant if and only if A is a multiple of the radial vector field, i.e. $A(z_1,z_2)=a(z_1,z_2)$ $a\in\mathbb{C}$.

Proof. i) If $A=\begin{pmatrix}a_{11} & a_{12}\\ a_{21} & a_{22}\end{pmatrix}$, then $[A,h]=0$ if and only if $a_{12}\lambda_1=a_{12}\lambda_2$ and $a_{21}\lambda_1=a_{21}\lambda_2$ - since $\lambda_1\neq\lambda_2$, we have $a_{12}=a_{21}=0$.

ii) Suppose that there is a vector field A which has two different eigenvalues λ_1, λ_2. Let $D=\begin{pmatrix}\lambda_1 & 0\\ 0 & \lambda_2\end{pmatrix}$ be its diagonalization. The same argument used in i), shows that the conjugated group $\bar{H}$ of H must be diagonal, since it satisfies the identity $[D,h]=0$, but this is a contradiction.

Hence the eigenvalues of A must be equal. If A has the Jordan canonical form $A=\begin{pmatrix}a & 1\\ 0 & a\end{pmatrix}$, then $[A,h]=0$ implies $h=\begin{pmatrix}\lambda & \beta\\ 0 & \lambda\end{pmatrix}$ and $\beta=0$; so H is diagonal, but this is another contradiction. Then the only possibility is $A=\begin{pmatrix}\lambda & 0\\ 0 & \lambda\end{pmatrix}$. □

Remark. If the group H is diagonalizable, its diagonal representation is the best coordinate chart to study the linear vector fields because of the knowledge of diagonal linear vector fields in the smooth case, namely in $\mathfrak{X}_0(\mathbb{C}^2)$.

It is interesting to point out that the H-equivariance does not only restrict vector fields but it can restrict also the group H itself.

Assertion: If there exists a linear vector field A, H-equivariant with two different eigenvalues then H is diagonalizable.

In the smooth case, the subset of germs which have their linear part in the Poincaré domain, is open and dense.

A similar result is true for H-equivariant germs.

4. Proposition. Suppose that H is a diagonal subgroup of SU(2), $P_H=\{X \in \mathfrak{X}_0(\mathbb{C}^2/H)$ X is in Poincaré domain$\}$ and $\mathcal{H}_H=\{X \in \mathfrak{X}_0(\mathbb{C}^2/H)$ X is hyperbolic$\}$.

Then P_H and $\mathcal{H}_H$ are open and dense subsets of the H-equivariant vector fields, -- $\mathfrak{X}_0(\mathbb{C}^2/H)$.

Proof. The openess is a direct consequence of the smooth case.

Since the topology of power series it is enough to prove the statement for the linear vector fields.

If A is a H-equivariant vector field, $A=\begin{pmatrix}\lambda_1 & 0\\ 0 & \lambda_2\end{pmatrix}$, proposition 3.

Suppose that A is not in the Poincaré domain, then $0 \in \mathcal{H}(\lambda_1, \lambda_2)$ the convex hull of λ_1, λ_2 in $\mathbb{C}$ but $0 \notin \mathcal{H}(\lambda_1 + {}^1/_n, \lambda_2)$ so $A_n=\begin{pmatrix}\lambda_1+{}^1/n & 0\\ 0 & \lambda_2\end{pmatrix}$ is in the Poincaré domain and - the sequence $\{A_n\}$ converges to A.

The proof for $\mathcal{H}_H$ is similar. □

3. DIAGONALIZABLE GROUPS.

The subset P_H is interesting because besides being open and dense, each of its - elements intersects transversally the sphere $S^3_r=\{(z_1, z_2) \in \mathbb{C}^2 ;\ \|z_1\|^2 + \|z_2\|^2 = r\}$ where r es a small positive real number.

More generally, if X in $\mathfrak{X}_0(\mathbb{C}^2)$ is a germ which intersects S^3_r transversally, then X induces a C^∞ foliation without critical points, in S^3_r.

5. Definition. If X in $\mathfrak{X}_0(\mathbb{C}^2)$ is a germ which intersects S^3_r transversally, then:

i) the precipitation vector field is defined by

$X_p(z_1, z_2)=\overline{\langle X(z_1, z_2)(z_1, z_2)\rangle}\ X(z_1, z_2)$

ii) the spherical vector field is defined by

$X_s(z_1, z_2)=i\overline{\langle X(z_1, z_2),(z_1, z_2)\rangle}\ X(z_1, z_2)$.

Where $\langle\ ,\rangle$ is the usual hermitian product of $\mathbb{C}^2$.

We note that this vector fields are C^∞ and the second one is tangent to S^3_r. The- flows (foliations) defined by them are the precipitation and spherical flows (folia - tions) respectively.

In case H is diagonalizable, Guckenheimer's proof [GU] of structural stability - for the smooth case, can be adapted to this case.

We have that in dimension 2 a hyperbolic linear vector field is in the Poincaré- domain.

6. Lemma. [GU] Let X be a hyperbolic vector field in $\mathbb{C}^2$. Then X induces non-vanishing Morse-Smale vector field X_S, in S^3_r, with only two closed orbits.

7. Corollary. Let H be a finite diagonal subgroup of SU(2). Let X be in $\mathfrak{X}_0(\mathbb{C}^2/H)$, such that its linear part is hyperbolic.

Then, the real analytic spherical vector field X_s defined by X over S^3_r, where- r is a small real number, is Morse-Smale without critical points and with just two -- closed orbits.

Proof. The spherical vector field $X_s(z_1,z_2)=i\overline{\langle X(z_1,z_2),(z_1,z_2)\rangle}X(z_1,z_2)$ is H-equiva-- riant in S^3_r. The fiber of the action is finite thus in S^3_r/H there are only two closed orbits, which are defined by the identification of C_1 and C_2 the closed orbits in S^3_r. □

8. Proposition. Let H be a finite diagonal subgroup of SU(2). Let X be in $\mathfrak{X}_0(\mathbb{C}^2/H)$ -- such that its linear part A_X is hyperbolic.

Then X is structurally stable.

Proof. Let X be as in the statement and let r be as in the corollary 7. Then there -- exists a neighborhood V_{r_0} of X in $\mathfrak{X}_0(\mathbb{C}^2/H)$ such that if Y is in V_{r_0}, Y is transversal- to S^3_r and the couple X_s, Y_s are topologically conjugate like vector fields in S^3_r -- [PA,SM].

Let $g:S^3_r/H\to S^3_r/H$ be the homeomophism which conjugates the foliations and ϕ_X and- ϕ_y the precipitation flows of X and Y.

Consider the function $F:(\mathbb{C}^2,0)\to(\mathbb{C}^2,0)$ defined in the following way: if (z_1,z_2) - is in the interior of the bounded ball defined by $S^3_{r_0}$ and $(z_1,z_2)\neq(0,0)$; there -- exists a unique number t in $\mathbb{R}^+$ and (w_1,w_2) in $S^3_{r_0}$ such that $(z_1,z_2)=\phi_X(t_0,w_1,w_2)$. --- Then $F(z_1,z_2)=\phi_y(t_0, g(w_1,w_2))$.

This way F turns up a homeomorphism which sends the foliation f_X to the folia -- tion f_y.

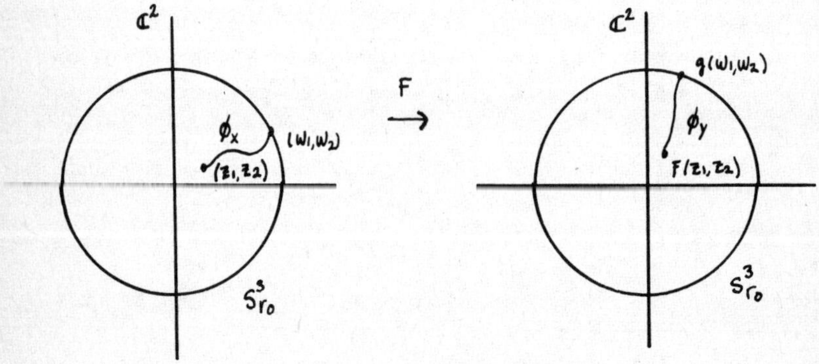

4. NON-DIAGONALIZABLE GROUPS.

In case of H is non diagonalizable, there exists an open and dense subset of -- $\mathfrak{X}_0(\mathbb{C}^2/H)$ such that any element X intersects transversally S^3_r/H , where r is a small - positive number. As we will see X defines a Seifert manifold structure on S^3_r/H [SE,MO] the proof of structural stability that we will give will use a uniqueness theorem for this Seifert Structure.

The Seifert invariants determined by X on S^3_r/H, are the following:

$$\{0,o,b,p,(\alpha_1,\beta_1)\ \ldots,\ (\alpha_n,\beta_n)\}$$

Where 0 means that S^3_r/H is orientable, o means that the quotient space is orientable, p is the genus of the quotient space and b is the Euler class. Finally the couples (α_i,β_i) determine in a unique way the exceptional fibres.

The Seifert Manifolds with finite fundamental group are classified [SE-TH] , a - direct corollary of this classification is the following.

9. Corollary. If H is a non diagonalizable, finite subgroup of SU(2), the Seifert manifold is determined by the following invariants:

i) If H is the dihedral group of order 4n
$M=\{0,o;-1,o;(2,1),(2,1),(n,1)\}$

ii) If H is the binary tethrahedral group of order 24
$M=\{0,o,-1,o;(2,1),(3,1),(3,1)\}$

iii) If H is the binary octahedral group or orden 48
$M=\{0,o,-1,o;(2,1),(3,1),(4,1)\}$

iv) If H is the binary icosahedral group of order 120
$M=\{0,o,-1,o;(2,1),(3,1),(5,1)\}$

10. Proposition. Let H be a finite, non diagonalizable subrgroup of SU(2).

Then any pair of H-equivariant germs X, Y with a non degenerated linear part are topologically equivalent.

Proof. In the first part we prove the following: If X is a germ in $\mathfrak{X}_0(\mathbb{C}^2/H)$ with a non degenerated linear part, X defines a foliation f_X in S^3_r/H which is differentially equivalent to the foliation defined by the radial vector field in the same manifold, f_R.

Suppose that X is a germ in $\mathfrak{X}_0(\mathbb{C}^2/H)$ with a non degenerated linear part, then -- $X(z_1,z_2)=\lambda R(z_1,z_2)+X^2(z_1,z_2)$ where $R(z_1,z_2)$ is the radial vector field, $X^2(z_1,z_2)$ is - the non-linear part of $X(z_1,z_2)$ and λ is in .

There exists a real number $r_0>0$ such that if $0\leq r\leq r_0$, X is transversal to S^3_r. Thus

X defines a C^∞ foliation without critical points in S^3_r .

Let $\psi:(\mathbb{C}^2,0)\to(\mathbb{C}^2,0)$ be the biholomorphism which linearize X, namely - $D\psi(X(z_1,z_2))=R(\psi(z_1,z_2))$.

The precipitation flow of the radial field defines a diffeomorphism between $S^3_{r_0}$ and $\psi(S^3_{r_0})$ which sends f_R to f_R, so X defines a foliation of circles in $S^3_{r_0}/H$.

Now there is a differentiable, effective action of S^1 on $S^3_{r_0}/H$ such that its orbits are the leaves of the foliation defined by X, [EP]. Now, because of corollary 1.8 $S^3_{r_0}/H$ only admits one structure of Seifert manifold, then the Seifert manifolds defined by X - and R in $S^3_{r_0}/H$ are the same.

So there exists a diffeomorphism f: $S^3_{r_0}/H\to S^3_{r_0}/H$ which sends the foliation defined- by X to the foliation defined by R.

Now consider the precipitation flows $\rho_X,\rho_y:\mathbb{R}\times B_r\to B_r$ of the fields X and R, where - $B_r=\{(z_1,z_2)\in\mathbb{C}^2; |z_1|^2+|z_2|^2\leq r\}$ and $r>0$ is a small real number.

Let $g:B_r/H\to B_r/H$ be the function defined by $g(z_1,z_2)=\rho_X(t_0,f(\rho_R(t_0,(\bar z_1,\bar z_2))))$, where $(\bar z_1,\bar z_2)$ is the unique point in $S^3_{r_0}$ which is sent by ρ_R to (z_1,z_2) at time t_0, and - $g(0,0)=(0,0)$.

g is a homeomorphism. □

Now the following theorem is a direct consequence of all previous results.

11. Theorem. Let H be a finite subgroup of SU(2). $\mathfrak{X}_0(\mathbb{C}^2/H)$ the space of H-equivariant -- germs of vector fields in the origin of $\mathbb{C}^2$.

Then there exists an open and dense subset A of $\mathfrak{X}_0(\mathbb{C}^2/H)$ such that its elements are structurally stable. If H is non diagonalizable, for any X in A, X is homeomorphically - equivalent to the radial vector field $R(z_1,z_2)=(z_1,z_2)$.

REFERENCES.

[ARN] ARNOLD, V.
Chapitres supplementaires de la theorie des equations differentielles ordinai-- res. Editons MIR 1980.

[DU-RO] DUMORTIER, F. et ROUSSARIE, R.
Etude locale des champs de vecteurs a parametres, Asterisque 59-60 1978.

[EP] EPSTEIN, D.
Periodic flows on three-manifolds, Ann. of Math. 95 1972, 66-82.

[GU] GUCKENHEIMER, J.
Hartman's theorem for complex flows in the Poincaré domain. Compositio Mathema tica vol. 24, Fasc. 1 1972, 75-82.

[MO] MONTESINOS, J.
Variedades de mosaicos.Sexta escuela latinoamericana de matemáticas, IPN 1982

[PA-SM] PALIS, J. and SMALE, S.
Structural Stability theorems. AMS Proceedings of Symposia of Pure Mathematics. Vol. XIV 1970, 223-234.

[PO] POINCARE, H.
Sur les propriétés des fonctions définies par les équations aux différences - partielles,These, Paris, 1879. Ouvres Completes, I.

[SE] SEIFERT, H.
Topologie dreidimensionaler gefaserter Raume, Acta Mathematica 60 1933, 147-238.

[SE-TH] SEIFERT, H. and THRELFALL, W.
Topologische Untersuchungen Diskontinuitatsbeiche endlicher Bewegungsgruppen - des dreimensionalen spharischen Raumes I, Math. Ann. 104, 1931, 1-70; II Math. Ann. 107, 1933, 543-596.

Federico Sánchez-Bringas
Instituto de Matemáticas, UNAM
c.p. 04510, México, D.F.

ATIYAH SEQUENCES AND COMPLETE CLOSED PSEUDOGROUPS PRESERVING A LOCAL PARALLELISM

Ana Maria F. Silva (*)
Université de Genève, Suisse

There are interesting relations between Atiyah Sequences and foliations (see [Al-Mo]). The aim of this paper is to establish a bijection between the equivalence classes of Atiyah Sequences and those of complete closed pseudogroups preserving a local parallelism (as in [Hae]).

The author wants to express her best acknowledgements to André Haefliger who suggested the main ideas of this work.

(*) The author thanks the INIC for its financial support

§1. Atiyah Sequences

1.1 Definition of the Atiyah Sequence associated to a Principal Bundle

Let $E \to W$ be a differentiable G-principal bundle over a manifold W, where G is a Lie group (not necessarily connected).

Let $L(E) = TE/G$ be the quotient of its tangent bundle by the right action of G

and $p^*: L(E) \to TW$ the natural projection.

If we denote by I(E) the kernel of p^*, we get an exact sequence of fibre bundles over W:

$$0 \to I(E) \to L(E) \to TW \to 0 .$$

We call $\underline{I}(E)$, $\underline{L}(E)$, $\underline{TW}$ the sheaves of germs of sections of these fibre bundles; over each open set $U \subset W$, the spaces of sections above U are modules over $A^o(U)$, the ring of differentiable functions on U.

The bracket of vector fields provides $\underline{L}(E)$ with a Lie algebra structure, and so

$$0 \to \underline{I(E)} \to \underline{L(E)} \to \underline{TW} \to 0$$

is an exact sequence of sheaves of Lie-algebras and $A^o(W)$-modules. It is the *Atiyah sequence associated to the G-principal bundle* $p: E \to W$.

Note that the dimension of each fibre of I(E) is equal to the dimension of G.

1.2 Abstract Atiyah Sequence

Based on the previous notion Pradines introduced the notion of Abstract Atiyah Sequence (in his terminology: "Transitive Lie Algebroid"). Consider an exact sequence of fibre bundles over W

$$0 \to I \to L \to TW \to 0$$

$\underline{L}$, $\underline{I}$, $\underline{TW}$ will again denote the sheaves of respective germs of sections and $\underline{p}$: $\underline{L} \to \underline{TW}$ is the homomorphism of sheaves associated to p.

We suppose that $\underline{L}$ has a Lie algebra structure such that

$$(1)\ \underline{p}([\xi,\eta]) = [\underline{p}(\xi),\ \underline{p}(\eta)]$$
$$\forall\ \xi,\eta \in \underline{L},\ f \in A^{o}(W)$$
$$(2)\ [\xi,\ f\eta] = f[\xi,\eta] + \underline{p}(\xi)\ (f)\eta$$

(Note that these conditions are satisfied by the Atiyah sequence associated to a principal bundle.)

Then we obtain again an exact sequence of sheaves of Lie algebras

$$0 \to \underline{I} \to \underline{L} \to \underline{TW} \to 0$$

which is an *abstract Atiyah sequence* with base space W.

Since the bracket of $\underline{L}$ induces over I another one which is $A^{o}(W)$-linear (applying condition 2), it provides each fibre of I with a Lie algebra structure in the following way:
Let ξ_o,η_o be two vectors of I projecting onto x by the fibration π_I: I → W; then take two local sections ξ and η of I such that $\xi(x) = \xi_o$ and $\eta(x) = \eta_o$. We define $[\xi_o,\eta_o]:=[\xi,\eta](x)$ which is well defined because of the $A^{o}(W)$-linearity of the bracket when restricted to I.

Up to an isomorphism (of $A^{o}(W)$-modules) the structure of Lie algebra in the fibres of I does not depend on the choice of the fibre ([Al-Mo]); then the typical fibre $\mathfrak{g}$ is called the *structural Lie algebra* of the Atiyah sequence.

Notation: We will use the notation $(A.S.)_{W,\mathfrak{g}}$ to abbreviate "Atiyah sequence with base space W and structural Lie algebra $\mathfrak{g}$; or simply A.S. for Atiyah sequence, when neither the base space nor the structural Lie algebra are specified.

1.3 Atiyah Sequence associated to a transversally complete foliation

1.3.1. Recall that a foliation (X,F) over a connected manifold X is said to be transversally complete if the foliated and complete vector fields (a vector field is foliated if its flow preserves the foliation), span the tangent space of X at any point x of X.

Molino described the geometry of these foliations: the closure of the leaves are the fibres of a locally trivial fibration $\pi: X \to W$ over a *basic* manifold W (π is called the basic fibration). Moreover the generic fibre of this fibration is a transversally complete $\mathfrak{g}$-Lie foliation, with dense leaves ($\mathfrak{g}$ is a Lie algebra and it is called the structural Lie algebra of F).

1.3.2. Let then (X,F) be a transversally complete foliation and $\pi: X \to W$ its basic fibration. For an open set U in W denote by $L(U,F)$ the vector space of the foliated vector fields on $\pi^{-1}(U)$.

Let $l(U,F) = L(U,F)/L_t(U,F)$ be the quotient of the foliated vector fields by those which are tangent to the foliation on $\pi^{-1}(U)$ (denoted by $L(U,F)$). Let $\underline{l}(F)$ be the sheaf of Lie algebras over W associated to the presheaf $U \to l(U,F)$; the projection π induces a surjective map $\underline{p}: \underline{l}(F) \to \underline{TW}$; $\underline{E}$ will denote its kernel - then we get an exact sequence

$$0 \to \underline{E} \to \underline{l}(F) \xrightarrow[\underline{p}]{} \underline{TW} \to 0$$

which is the Atiyah sequence associated to the transversally complete foliation (X,F). ([Al-Mo])

1.4 Isomorphism of Atiyah Sequences and principal realisation

An isomorphism between two Atiyah sequences over W is an isomorphism of exact sequences of sheaves which induces the identity on $\underline{TW}$. In such a case we say that the two Atiyah sequences are equivalent or isomorphic.

A principal realisation of a given A.S. over W is an isomorphism between that A.S. and the A.S. associated to a principal bundle with the same base space W.

In general it is not true that any abstract Atiyah sequence admits a principal realisation, as shown by Almeida and Molino ([Al-Mo]), but this is true locally.

1.5 Local Case

Lemma ([Al-Mo], [Ma]): (Existence of local realisations)
Let

$$0 \to \underline{I} \to \underline{L} \to \underline{TW} \to 0$$

be an $(A.S.)_W$ and U an open set in W diffeomorphic to R^n. Then

$$0 \to \underline{I}(U) \to \underline{L}(U) \to \underline{TU} \to 0$$

is isomorphic to the A.S. associated to the principal bundle $U \times G_o$ where G_o is a 1-connected Lie group with Lie algebra isomorphic to $\mathfrak{g}$.

1.6 Lemma on local isomorphisms

Let U be an open set diffeomorphic to R^n and G_o be a 1-connected Lie group.

a) If ϕ: $U \times G_o \to U \times G_o$ is a diffeomorphism of the form $(x,g) \longmapsto (x,h(x)\alpha(g))$ where $\alpha \in \mathrm{Aut}(G_o)$ and h: $U \to G_o$ is a differentiable map, then the differential $d\phi$ of ϕ induces an isomorphism of the A.S. associated to the principal bundle $U \times G_o$ (product bundle).

b) Conversely, any isomorphism of that Atiyah sequence is obtained in this way.

c) If ϕ_1, ϕ_2 are two such isomorphisms of the form $\phi_1(x,g) = (x,h_1(x)\alpha_1(g))$, $\phi_2(x,g) = (x,h_2(x)\alpha_2(g))$ where $\alpha_1,\alpha_2 \in \mathrm{Aut}(G_o)$ and h_1,h_2: $U \to G_o$ such that $d\phi_1 = d\phi_2$, then there exists $\gamma \in G_o$ satisfying $\begin{cases} h_2 = R_\gamma \cdot h_1 \\ \alpha_2 = \mathrm{Ad}_{\gamma^{-1}} \cdot \alpha_1 \end{cases}$.

Proof

a) α being an automorphism of G_o and h(x) acting by left translation (for x in U), it is clear that $d\phi$ gives an isomorphism on $\underline{L}(U \times G_o)$. Now, $d\phi$ projects onto the identity of $\underline{TU}$, therefore it induces, in a natural way, an isomorphism of the Atiyah sequence of $U \times G_o$.

b) Conversely, let Φ: $\underline{L}(U \times G_o) \to \underline{L}(U \times G_o)$ be an automorphism of the A.S. associated to the principal bundle $U \times G_o$.

Let $x_1,\dots x_n$ be the coordinates on $U \simeq \mathbb{R}^n$.

$$\underline{L}(U\times G_o) = \underline{H}(U) \oplus \underline{I}(U)$$

where $\underline{H}(U)$ is the submodule of $\underline{L}(U\times G_o)$ formed by the (right) invariant horizontal vector fields (they are of the form

$$\Sigma a_i\ \partial/\partial x_i \quad \text{where } a_i: U \to \mathbb{R} \text{ are differentiable maps)}$$

$\Phi(\underline{H})$ defines another distribution which is involutive since ϕ commutes with the bracket, and also right invariant. As U is simply connected there is a map $x \mapsto (x,h(x))$ of U in $U\times G_o$ which is a solution of this distribution.

For each $g \in G_o$, the map ψ_g: $x \mapsto (x,(h(x)\cdot g))$ is also a solution. Let Ψ be the differential of ψ: $(x,g) \mapsto \psi_x(g)$ of $U\times G_o \to U\times G_o$.

Then $\Psi(\partial/\partial x_i) = \Phi(\partial/\partial x_i)$ $(i = 1,\dots,n)$. Hence if

$$\Phi' = \Psi^{-1}\Phi, \text{ we have } \Phi'(\partial/\partial x_i) = \partial/\partial x_i \quad (i = 1,\dots,n).$$

So Φ' preserves $\underline{H}$; therefore it induces an isomorphism of the subalgebra V of $\underline{L}(U\times G_o)$ which commutes with $\underline{H}$ - this subalgebra is included in $\underline{I}(U)$ and is isomorphic to the structural Lie algebra $\mathfrak{g}$ of the Atiyah sequence of $U\times G_o$, i.e., isomorphic to the algebra formed by the right invariant vector fields on G_o.

G_o being 1-connected, Φ' induces an automorphism α of G_o such that the differential of $\phi' = 1_U\times\alpha$ induces Φ' on V. So the diffeomorphism $\phi = \psi\cdot\phi'$

$$\phi: U\times G_o \longrightarrow U\times G_g$$

$$(x,g) \mapsto (x,h(x)\alpha(g))$$

has differential d such that

$$d\phi/\underline{L}(U\times G_o) = \Psi\cdot\Phi' = \Phi.$$

c) Moreover a map ϕ: $U\times G_o \to U\times G_o$ of the form

$$(x,g) \longmapsto (x,h(x)\alpha(g))$$

induces the identity on $\underline{L}(U\times G_o)$ if h is a constant map of U with value

$\gamma \in G_o$ and α is equal to $Ad_{\gamma^{-1}}$; in other words if is the right translation by the element γ.

So it follows that if ϕ_1, ϕ_2 are diffeomorphisms of UxG_o of the form

$$\phi_1(x,g) = (x,h_1(x)\alpha_1(g)), \quad \phi_2(x,g) = (x,h_2(x)\alpha_2(g))$$

inducing the same Φ on $\underline{L}(UxG_o)$, then there is $\gamma \in G$ such that $\phi_2 = R_\gamma \cdot \phi_1$ which implies

$$\begin{cases} h_2 = R_\gamma \cdot \dot{\phi}_1 \\ \alpha_2 = Ad_{\gamma^{-1}} \cdot \alpha_1 \end{cases} .$$

This was essentially contained in [Al-Mo] (and probably in [Ma]).

§2. Complete Pseudogroups preserving a local parallelism

2.1 Definitions

2.1.1. A pseudogroup H of local transformations of a differentiable manifold S is a collection of local diffeomorphisms from open sets of S to open sets of S, such that

i) if h, h' $\in H$ then the composition $h \cdot h'$ (when it is defined) belongs to H, and also $h^{-1} \in H$. The identity of S belongs to H.

ii) if h: U → V is an element of H then the restriction of h to any open set of U belongs to H.

iii) if h: U → V is a diffeomorphism of open sets of S and for each $u \in U$ there is an open neighbourhood U_x of x such that the restriction of h to U_x is in H then h is in H.

Example: *Holonomy pseudogroups of a foliation*

A foliation F in a differentiable manifold is given by

a) an open covering $\{U_i\}_{i \in I}$ of S

b) $\{T_i\}_{i \in I}$ manifolds, all with the same dimension which is the codimension of F

c) submersions $f_i: U_i \to T_i$ with connected fibres such that there are local diffeomorphisms

$$h_{ij}: f_i(U_i \cap U_j) \to f_j(U_i \cap U_j) \quad \text{with } f_j = h_{ji} \cdot f_i .$$

The elements h_{ji} acting on $T = \coprod_{i \in I} T_i$ generate a pseudogroup of local diffeomorphisms which is by definition the holonomy pseudogroup of F.

2.1.2. Equivalence of Pseudogroups

Let H and H' be pseudogroups of local diffeomorphisms of differentiable manifolds S and S' respectively.

An equivalence between H and H' is a maximal collection Φ of diffeomorphisms of open sets of S on open sets of S' such that

i) $\Phi = H' \cdot \Phi H$, that is, is stable under the composition by elements of H and H'.

ii) H is generated by elements of the form $\phi_1^{-1} \cdot \phi_2$ with $\phi_1, \phi_2 \in \Phi$ and H' is generated by elements of the form $\phi_1 \cdot \phi_2^{-1}$ with $\phi_1, \phi_2 \in \Phi$.

Remarks:

a) An equivalence Φ between two pseudogroups H and H' induces a bijection between the respective space orbits.

b) The self-equivalence of a pseudogroup H form a group $E(H)$ which depends only on the equivalence class of H.

2.1.3. A pseudogroup H acting on a manifold S is *generated by the action of a group* G (acting by diffeomorphisms) if the elements of H are diffeomorphisms $f: U \to V$ of open sets of S such that for any $x \in U$ the restriction of f to a neighbourhood of x is the restriction of an element of G.

We say that G *acts quasi-analytically* on S if any two elements of G are equal whenever their germs at a point x of S are equal.

2.2 Lemma ([Hae])

An equivalence $\Phi: H \to H'$ between two pseudogroups H and H' generated respectively by groups G and G' acting quasi-analytically on 1-connected manifolds S and S' is generated by a diffeomorphism $\phi: S \to S'$ such that $\phi(x\gamma) = \phi(x)\alpha(\gamma)$ for any $x \in S$, $\gamma \in G$ where $\alpha: G \to G'$ is an isomorphism.

Two such pairs (ϕ, α) and (ϕ', α') define the same equivalence Φ, if and only if there is an element $\gamma \in G$, such that

$$\phi' = \phi \cdot \gamma \quad \text{and} \quad \alpha' = Ad_{\gamma^{-1}} \cdot \alpha \quad .$$

2.3 Complete closed pseudogroups preserving a local parallelism

2.3.1. Definition: A pseudogroup $\mathcal{H}$ of local diffeomorphisms of a differentiable manifold S is *complete* if for any two points $x,y \in S$ there are open neighbourhoods U of x and V of y such that any germ of an element of $\mathcal{H}$, with source in U and target in V is the germ of an element of $\mathcal{H}$ defined on the whole of U.

The motivation for this definition is due to the

Proposition ([Rei])
The holonomy pseudogroup of a riemannian foliation with bundle-like metric, in a complete manifold is complete.

The complete pseudogroup is said to be closed if it is closed in the C^1-topology ([Hae]).

2.3.2. Definition: A pseudogroup $\mathcal{H}$ acting on a differentiable manifold S of dimension n preserves a local parallelism if for each point x of S there are n independent vector fields $Y_1,\ldots,Y_n$ on a neighbourhood U of x which are invariant by the restriction of $\mathcal{H}$ to U.

2.3.3. Proposition ([Sa])
Let $\mathcal{H}$ be a complete closed pseudogroup of local diffeomorphisms of a manifold S, of dimension n, preserving a parallelism. Then there is a 1-connected Lie group G, a differentiable manifold W and a submersion $\P: S \to W$ such that for any $x \in W$, there is an open neighbourhood U of x such that $\mathcal{H}_{\P^{-1}(U)}$ is equivalent to the pseudogroup H(U) generated by G acting on UxG by right translations on G and projecting onto the identity of U, (that is $(x,g)\cdot h = (x,gh)$ for $x \in U$, and $g,h \in G$), and the equivalence projects onto the identity.

The next proposition describes the group of self-equivalence of H(U) which project onto the identity of U - which is an immediate consequence of lemma 2.2.

2.4 Proposition ([Hae])

The group $\mathcal{E}(U)$ of self-equivalences of H(U) projecting onto the identity of U is formed by pairs (h,α) where $h: U \to G$ is a differentiable map and $\alpha \in \mathrm{Aut}(G)$, which define a diffeomorphism $\psi: U\times G \to U\times G$ of the form

$(x,g) \mapsto (x,h(x)\alpha(g))$; by construction we have $\psi((x,g)\gamma) = \psi(x,g)\alpha(\gamma)$ for any $x \in U$, and $g,\gamma \in G$. Two such pairs (h,α) and (h',α') represent the same equivalence if there is an element $\gamma \in G$ such that

$$\begin{cases} h' = h\cdot\gamma \\ \alpha' = Ad_{\gamma^{-1}}\cdot\alpha \end{cases} .$$

§3. Comparison between Atiyah Sequences and complete closed Pseudogroups

3.1 Let $\mathfrak{g}$ be a Lie algebra and W a differentiable paracompact connected manifold.

For each open set $U \subset W$ let $H(U)$ be the pseudogroup generated by the action of a 1-connected Lie group G_o (whose Lie algebra is isomorphic to $\mathfrak{g}$) acting on UxG_o by $(x,g)\cdot\gamma = (x,g\gamma)$ for any $x \in U$ and $g,\gamma \in G_o$.

We consider now the set $AS(W,\mathfrak{g})$ of quivalence classes of Atiyah sequences with base space W and structural Lie algebra $\mathfrak{g}$, as well as the set $P(W,\mathfrak{g})$ of equivalence classes of pseudogroups H of the following type: an element of $P(W,\mathfrak{g})$ is represented by a pseudogroup H of local diffeomorphisms of a manifold S with a submersion $¶: S \to W$ and an open covering $\{U_i\}_{i \in I}$ of W, such that, for any $i \in I$, there is an equivalence of $H_{|¶^{-1}(U_i)}$ on $H(U_i)$ projecting on the identity of U_i (see 2.3.3.). This characterizes complete closed pseudogroups preserving a local parallelism.

Note that ¶ induces a bijection of the space of orbits of H onto W. Two such couples $(H,\ ¶: S \to W)$ and $(H',\ ¶': S' \to W)$ represent the same element of $P(W,\mathfrak{g})$ if there is an equivalence $\Phi: H \to H'$ compatible with ¶ and ¶'; namely $¶'\cdot\phi = ¶$.

To each couple $(H,\ ¶: S \to W)$ we may associate in a natural way an A.S. with base W and structural Lie algebra $\mathfrak{g}$ (where $\mathfrak{g}$ is isomorphic to the Lie algebra of G_o)

$$0 \to \underline{I} \to \underline{L} \to \underline{TW} \to 0$$

where for each open set $U \subset W$ $\underline{L(U)}$ is the module of H-invariant vector fields on $\pi^{-1}(U)$ and $\underline{I(U)}$ is the submodule of $\underline{L(U)}$ formed by the vertical vector fields, that is, those which are tangent to the orbits of H.

3.2 Theorem

The natural map

$$A: P(W,\mathfrak{g}) \to AS(W,\mathfrak{g})$$

associating to each equivalence class of a pseudogroup H the isomorphism class of the corresponding Atiyah sequence, is a bijection.

The proof of this theorem follows from the lemma:

3.3 Lemma

Let U be an open set, diffeomorphic to R^n and $H(U)$ as in 3.1. Then there is a one-to-one correspondence between the group $E(U)$ of self-equivalences of $H(U)$ projecting on the identity of U and the group of isomorphisms of Atiyah sequences associated to the principal bundle $U\times G_o$ (where G_o is the group appearing in the definition of $U\times G_o$.

Proof:
Let (h,α) represent an element of $E(U)$ and $\psi: U\times G_o \to U\times G_o$ be the diffeomorphism $\psi(x,g) = (x,h(x)\alpha(g))$ (2.3.3.). Then $d\psi$ induces an isomorphism of the Atiyah sequence associated to $U\times G_o$, which doesn't depend on the choice of the representant because of 1.6 c). Passing to equivalence classes, we get a bijection according to the same lemma. □

Proof of the theorem:
Let $0 \to \underline{I} \to \underline{L} \to \underline{TW} \to 0$ be an $(A.S.)_{W,\mathfrak{g}}$ and take an open covering $\{U_i\}_{i\in I}$ of W, where each U_i is contractible and for each $(i,j) \in I\times I$ $U_i \cap U_j$ is still contractible.

Then the A.S. over W induces over each U_i, $i \in I$, an A.S. and by lemma 1.5 there is a 1-connected Lie group G_o with Lie algebra isomorphic to the structural Lie algebra of the given A.S. such that the Atiyah sequence on U_i is isomorphic to the A.S. associated to the principal bundle $U_i\times G_o$.

For each $i \in I$, let $\Phi_i: L(U_i) \to L(U_i\times G_o)$ be such an isomorphism, and for

i,j such that $U_i \cap U_j \neq \emptyset$ define

$$\Phi_{ij} = \Phi_i \cdot \Phi_j^{-1} : \underline{L}(U_i \cap U_j \times G_o) \to \underline{L}(U_i \cap U_j \times G_o) \quad .$$

By lemma 1.6 there are isomorphisms $\phi_{ij}: U_i \cap U_j \times G_o \to U_i \cap U_j \times G_o$ such that $d\phi_{ij}|_{L(U_i \cap U_j \times G_o)} = \Phi_{ij}$ and such two elements differ by a right translation of an element of G_o.

Now let's construct a pseudogroup H acting on S as follows: We take S to be the product of G_o by the disjoint union of the U_i, $i \in I$, $S = \coprod_{i \in I} U_i x G_o$ with projections $\P_i: U_i x G_o \to U_i$ $(i \in I)$ and $\P = \coprod_{i \in I} \P_i$. Let H be the pseudogroup acting on S generated by G_o acting by right translations on each $U_i x G_o$ $(i \in I)$, and by the $(\phi_{ij})_{i,j \in I}$

$$\phi_{ij}: \P_j^{-1}(U_i \cap U_j) \simeq U_i \cap U_j \times G_o \to U_i \cap U_j \times G_o \simeq \P_j^{-1}(U_i \cap U_j) \quad .$$

It is clear that this pseudogroup is of the kind we consider in the theorem, and by the previous lemma its A.S. is isomorphic to the initial one. To an equivalence of this pseudogroup it corresponds an equivalence of the A.S. and vice-versa. □

Remark 1. Let H be a pseudogroup representing a class of $P(W,\mathfrak{g})$ and let $0 \to \underline{I} \to \underline{L} \to \underline{TW} \to 0$ be the Atiyah sequence associated to H. It follows from the proof of the theorem that there is a bijection between the self-equivalences of H projecting on the identity of W and the isomorphisms of the above Atiyah sequence.

Remark 2. Given a realizable element of A.S. $(W,\mathfrak{g})$ it is clear that the pseudogroup H associated to it is equivalent to a pseudogroup generated by the action of a group (here the Lie group of the principal realisation) and the converse is also true for pseudogroups in $P(W,\mathfrak{g})$.

3.4 Some invariants associated to an A.S.

3.4.1. Let $\underset{\sim}{E}$ be the sheaf of groups associated to the presheaf $E(U)$, where $E(U)$ (see 3.3) is described by the following exact sequence ([Hae])

(1) $$1 \to G_o \to G_o(U) \rtimes \mathrm{Aut}(\mathfrak{g}) \to E(U) \to 1$$

where G_o is identified to the normal subgroup of $G_o(U) \rtimes Aut(g)$ of elements of the form $(g, Ad_{g^{-1}})$. The fact that $P(W,\mathfrak{g})$ is in one-to-one correspondence with $H^1(W,\underset{\sim}{E})$ ([Hae]) combined with 3.2 gives the following diagram of bijections:

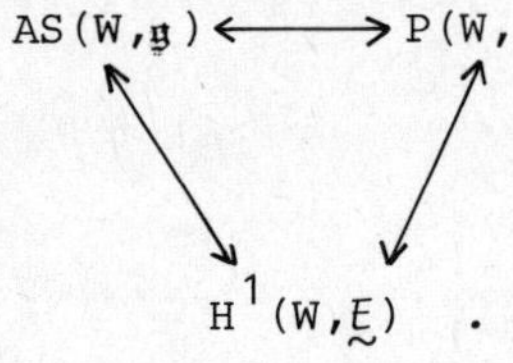

3.4.2. Moreover, from (1) one obtains the exact sequence of sheaves

$$1 \to \underset{\sim}{G_o}/C \to \underset{\sim}{E} \to Out(\mathfrak{g}) \to 1$$

where $\underset{\sim}{G_o}$ is the sheaf of germs of differentiable functions of W in G_o, C the sheaf of germs of constant functions of W in the center C of G_o and $Out(\mathfrak{g}) = Aut(\mathfrak{g})/Int(\mathfrak{g})$ is viewed as a constant sheaf.

Although they are not sheaves of abelian groups, one can construct the "exact sequence"

$$H^o(W,\underset{\sim}{G_o}/C) \to H^o(W,\underset{\sim}{E}) \to H^o(W,Out(\mathfrak{g})) \to H^1(W,\underset{\sim}{G_o}/C) \to H^1(W,\underset{\sim}{E}) \to H^1(W,Out(\mathfrak{g})).$$

And, since $Out(\mathfrak{g})$ is a constant sheaf, $H^1(W,Out(\mathfrak{g}))$ is isomorphic to $Hom(\pi_1(W),Out(\mathfrak{g}))$/conjugacy. Hence to each element s of $AS(W,\mathfrak{g})$ we associate a homomorphism $\alpha: \pi_1(W) \to Out(\mathfrak{g})$ up to conjugacy.

This invariant α, associated to each equivalence class of A.S. over a space W, will decide either the Poincaré duality holds or not. Applying proposition 2.9.1. of [Hae] we may conclude that if W is compact, oriented, then the cohomology of the invariant forms for the pseudogroup H corresponding to s verifies the Poincaré duality if and only if the elements of the subgroup of $Aut(\mathfrak{g})$ inverse image of Im α by the projection $Aut(\mathfrak{g}) \to Out(\mathfrak{g})$ preserve a volume form on $\mathfrak{g}$; this implies that $\mathfrak{g}$ must be unimodular.

3.4.3. A second invariant associated to an element s of A.S. (W,g) is a homomorphism

$$\delta: \pi_2(W,x_o) \to C$$

which is equivariant for the natural action of $\pi_1(W,x_o)$ on $\pi_2(W,x_o)$ and the action of $\pi_1(W,x_o)$ on C via the homomorphism α (note that Out($\mathfrak{g}$) operates on C). Almeida and Molino characterize the realizability of an A.S. in terms of this invariant.

This homomorphism appears in the exact sequence

$$0 \to \pi_2(BH) \to \pi_2(W) \overset{\delta}{\to} G_o \to \pi_1(BH) \to \pi_1(W) \to 1$$

where BH is the classifying space of the pseudogroup H corresponding to s.

Assuming that α is trivial, we know from 3.4.2. that s comes from an element s_1 of $H^1(W,\underset{\sim}{G}_o/C)$, well defined up to an action of Out($\mathfrak{g}$), and, from the ecaxt sequence

$$H^1(W,C) \to H^1(W,\underset{\sim}{G}_g) \to H^1(W,\underset{\sim}{G}_o/C) \overset{\partial}{\to} H^2(W,C)$$

we obtain an element $\beta = \partial(s_1) \in H^2(W,C)$ (which gives the homomorphism δ: $\pi_2(W,x_o) \to C$ of 3.4.3.).

Assume that there is a discrete subgroup Γ of C such that β is the image of an element $\gamma \in H^2(W,\Gamma)$ by the homomorphism induced by the inclusion $\Gamma \hookrightarrow C$. Then these two assumptions give a necessary and sufficient condition for an element s of AS(W,$\mathfrak{g}$) to be realisable as the A.S. of a G-principal bundle, where G is a connected Lie group with Lie algebra $\mathfrak{g}$ (in fact $G = G_o/\Gamma$).

References

[Al-Mo] - R. Almeida and P. Molino: "Suites d'Atiyah, feuilletages et quantification géométrique". Colloque de Géométrie différentielle 1984-1985 de Montpellier

[Alm-Mol] - R. Almeida and P. Molino: "Suites d'Atiyah et feuilletages transversalement complets". C.R. Acad. Sci., Paris, Sér. I 300, 13-15 (1985)

[At] - M. Atiyah: "Complex analytic connections on fibre bundles". Trans. Am. Math. Soc., 85 (1957) pp 181-207

[Hae] - A. Haefliger: "Pseudogroups of local isometries". Colloquio de Geometria Differencial de Santiago de Compostela, Sept. 1984. Research Nores 131, Pitman (1985) pp 174-197

[Ma] - K. Mackenzie: "Cohomology of local trivial groupoids and Lie algebroids". Bull. Austral. Math. Soc., vol 20 (1979)

[Mol] - P. Molino: "Feuilletages transversalement complets et applications". Ann. Ec. Norm. Sup. Paris, 10 (1977) pp 289-307

[Pra] - J. Pradines: "Troisième théorème de Lie pour les groupoids différentiables". C.R.Ac. Sc. Paris, 267 A (1968) pp 21-23

[Pra] - J. Pradines: "Théorie de Lie pour les groupoids différentiables". C.R.Ac. Sc. Paris, 264 A (1967) pp 245-248

[Rei] - B. Reinhart: "Foliated manifolds with bundle-like metrics". Ann. of Math. 69 (1959) pp 119-132

[Sa] - E. Salem: "Une généralisation du théorème de Myers-Steenrod aux pseudogroups d'isométries locales". Colloque de Géométrie Différentielle (1985-1986) de Montpellier.

Sur les bouts d'une feuille d'un feuilletage au voisinage d'un point singulier isolé

R. Thom

Institut des Hautes Etudes Scientifiques
Bures-sur-Yvette, France

On considère un germe de feuilletage $\mathcal{F}$, de codimension k , en un point 0, origine de $\mathbb{R}^n$, qui est une singularité isolée (*). On suppose le feuilletage lisse (C^r) sur $\mathbb{R}^n - 0$. On fera sur $\mathcal{F}$ des hypothèses extrêmement restrictives :

1) $\mathcal{F}$ est invariant par homothétie $x \to \lambda^+ x$, $x \in \mathbb{R}^n$, $\lambda^+ \in \mathbb{R}^+$.

2) Sur une tranche $a \leq |x| \leq b$, le feuilletage a un quotient Hausdorff : toute feuille est propre (compacte), et admet un voisinage saturé.

On se propose d'étudier comment varie la topologie d'une feuille dans la boule B_r , $|x| \leq r$, lorsque r tend vers zéro. Comme on veut évaluer la croissance de la topologie d'une feuille (L) lorsqu'on s'approche de l'origine, il est nécessaire de supposer que la feuille de départ, au voisinage d'une sphère $|x| = r_o$, est à homologie de type fini. Ce qui justifie l'hypothèse de compacité faite en 2°).

Selon l'hypothèse 2 le quotient $\phi = S_{ab/\mathcal{F}}, S_{ab} = a \leq |x| \leq b$ est un espace de Hausdorff; en fait, c'est "génériquement" un ensemble stratifié, et l'application $\phi : S_{ab} \to Q_{ab}$ est un morphisme stratifié sans éclatement (toutes les strates de la source sont soit régulières, au quel cas le noyau a la dimension de la feuille, soit singulières, au quel cas Φ est injective).

On retiendra de cette assertion le fait qu'il est possible de paramétriser les feuilles <u>globales</u> dans S_{ab} par une carte locale U de dimension k dans le quotient Q de telle manière que pour une plaque Π d'une feuille

L , l'espace produit $\Pi \times U$ s'envoie canoniquement par un difféomorphisme sur un voisinage de la plaque Π dans $\mathbb{R}^n$, la feuille globale L n'ayant qu'un représentant (central) dans la carte locale $U \times \Pi$, soit $(0 \times \Pi)$.

Fonctions réelles génériques sur un feuilletage

Une fonction lisse $F : (\mathbb{R}^n - 0) \to \mathbb{R}$ est dite générique sur (F) , si la restriction de F à toute feuille (L) de (F) ne présente que des singularités algébriquement isolées de codimension k au plus, et si l'application projection π au lieu singulier $\Sigma^k \xrightarrow{\pi} U$ est générique au sens de la théorie des singularités de fonctions. (Ceci implique, rappelons-le que (π) peut être stratifiée).

Si l'hypothèse (2) est satisfaite, la fonction distance $|x|$ n'est pas nécessairement générique, mais elle peut être C^k approchée par une fonction distance d générique sur $F|(\mathbb{R}^n - 0)$. Les variétés de niveau $d = cst$ sont alors des sphères strictement convexes, transverses aux rayons vecteurs issus de 0 .

Considérons alors une couronne sphérique $\Sigma_{a,\sigma a}$ de $\sigma < 1$ définie par $a \geq d(x) \geq \sigma a$. On à la

Proposition : Si (d) est générique sur le bord $d^{-1}(a)$ et $d^{-1}(\sigma a)$ alors : la fonction d/L restreinte à toute feuille $L \cap \Sigma_{a,\sigma a}$ admet un nombre fini de points critiques, majoré par un entier μ indépendant de L .

Preuve : Si on pouvait trouver une suite infinie de feuilles (L_i) dont le nombre de points critiques de d irait à l'infini, cette suite (L_i) aurait une feuille d'accumulation (L_∞) , et sur cette feuille, il existerait un point critique à nombre de Milnor infini (non algébriquement isolé). Mais ceci contredit l'hypothèse de généricité de d sur (L_∞) .

Graphe de Reeb d'une fonction sur une variété.

Soit f une fonction numérique sur une variété M , à valeur dans $\mathbb{R} = Ou$; on considère dans M la relation d'équivalence (ρ) $x = y$ si

1) $f(x) = f(y)$

2) x et y appartiennent à la même composante connexe de $f^{-1}(f(x))$.

Alors "génériquement" le quotient de M par la relation (ρ) est un graphe qui s'envoie surjectivement sur son image dans Ou . Il y a des points singuliers qui sont génériquement (pour f seule) la naissance  , la mort  , la scission dichotomique  , la confluence  . Si on introduit des singularités de codimension $\leq k$, il peut y avoir des singularités plus compliquées, comme des points triples  ou des  croisements

Considérons maintenant $\mathcal{F}|\Sigma_{a,\sigma a}$. $\mathcal{F}$ est supposé couper génériquement les bords $d = a$, $d = \sigma a$. Mais le feuilletage $\mathcal{F}$ n'en présente pas moins des singularités. Ceci est nécessaire, si l'on veut éviter le cas trivial d'un feuilletage transverse aux sphères $d = cst$, auquel cas toutes les feuilles seraient de la forme $\Lambda \times \mathbb{R}^+$, et le problème de leurs bouts serait trivialement résolu par la permanence topologique des sections.

Lemme. Pour toute tranche $\Sigma_{a,\sigma a}$, le graphe de Reeb de $d|L$ restreint à toute feuille $(L) \cap \Sigma_{a,\sigma a}$ rencontre chacun des bords $d = a$, $d = \sigma a$, en au plus q points, où chaque nombre q est majoré par un entier (Q) valable pour toute L .

Ceci résulte immédiatement de l'hypothèse (2). Si une feuille était feuille d'accumulation d'une suite de feuilles (L_i) pour laquelle le nombre q_i des composantes connexes d'intersection avec $d = a$ irait à l'infini, cette feuille ne pourrait être que singulière, mais, en vertu de l'hypothèse (2), cette singularité est de type Haefliger, et la feuille correspondante ne serait pas propre : en une telle singularité, de type local algébrique, une feuille n'a qu'un nombre fini de composantes connexes.

Définition.

Bout transverse d'une feuille (L).

On appellera bout transverse d'une feuille (L) de (F) la contre-image dans (L) d'un arc du graphe de Reeb qui se projette sur l'axe des a dans le sens des u décroissants (et tendant vers zéro).

On appellera "voisinage transverse" de (L) en 0 l'ensemble des bouts transverses de (L) issu d'un sommet bien défini du graphe (par exemple sur $d = a$) . On a alors les résultats suivants :

1°) Si l'on suit un bout transverse pour d décroissant, le type homologique de la composante connexe de la feuille considéré croît au plus polynomialement en $(\frac{1}{d})$.

2°) De même le nombre des bouts (composantes connexes) d'un voisinage transverse $d = \varepsilon$ croît au plus polynomialement en $(\frac{1}{\varepsilon})$.

La proposition 1° vient du fait que chaque composante d'un bout transverse restreinte à une tranche de la forme $\sigma^p a \geq d \geq \sigma^{p+1} a$ a une homologie (somme des nombres de Betti donnée) bornée par un nombre fixe B , B étant lui-même majoré par (μ) . La proposition 2° résulte du fait que la branche issue d'un sommet de $d = \sigma^{p-1} a$ donne naissance à au plus μ branches de bouts transverses sur $d = \sigma^p a$.

Remarque 1.

Dans l'exposé oral fait au Congrès de Chapala, j'avais énoncé ces résultats pour la topologie de toute feuille comprise entre $a \geq d \geq \sigma^k a)$, sans me restreindre aux "bouts transverses". Mais toutes les fois qu'un arc du graphe de Reeb rebrousse chemin (en traversant une section $\sigma^j a$ dans le sens des d croîssants) on introduit ultérieurement de nouveaux arcs, et je ne vois aucun moyen de majorer ce nombre de rebroussements permis sur une tranche $d = \sigma^i a$ par une constante indépendante de k. Il y a là une question ouverte. Bien entendu, les résultats seraient vrais de toute famille d'arcs ne comportant qu'un nombre fini de rebroussements.

Remarque 2.

Dans les exemples $\mathbb{C}$-analytiques, on sait d'après Malgrange [1], que si le feuilletage (F) est analytique complexe sur $\mathbb{C}^n-0$, et si ses singularités y sont toutes de type Haefliger (complexes) dans $\mathbb{C}^n-0$, $n \geq 2$, alors le prolongement analytique de (F) en 0 est possible et conduit à un feuilletage globalement analytique (et singulier en 0). Il en résulte que les feuilles passant par 0 sont des variétés analytiques (en 0), ce qui conduit à la trivialité topologique locale des bouts.

Il serait intéressant de savoir ce qui subsiste de ce résultat pour un feuilletage analytique réel. On pourrait également affaiblir l'hypothèse 1°) en supposant que pour un feuilletage analytique, la composante homogène de plus bas degré du développement de Taylor de F en 0 n'est pas trop dégénérée.

(*) Ceci suppose que le fibré en $(n-k)$ grassmanniennes sur $S^{n-1} \subset \mathbb{R}^n-0$ admet une section; on pourrait cependant généraliser le problème en admettant que le feuilletage (F) est de type Haefliger sur S^{n-1}, les singularités étant celles de morphismes locaux $\mathbb{R}^n \to \mathbb{R}^k$ génériques.

Références

[1] B. Malgrange, Frobenius avec singularités : le cas général, Inventiones Mathematicae, 39, 1, 1977, pp.67-89.

LECTURE NOTES IN MATHEMATICS

Edited by A. Dold and B. Eckmann

Some general remarks on the publication of proceedings of congresses and symposia

Lecture Notes aim to report new developments - quickly, informally and at a high level. The following describes criteria and procedures which apply to proceedings volumes. The editors of a volume are strongly advised to inform contributors about these points at an early stage.

§1. One (or more) expert participant(s) of the meeting should act as the responsible editor(s) of the proceedings. They select the papers which are suitable (cf. §§ 2, 3) for inclusion in the proceedings, and have them individually refereed (as for a journal). It should not be assumed that the published proceedings must reflect conference events faithfully and in their entirety. Contributions to the meeting which are not included in the proceedings can be listed by title. The series editors will normally not interfere with the editing of a particular proceedings volume - except in fairly obvious cases, or on technical matters, such as described in §§ 2, 3. The names of the responsible editors appear on the title page of the volume.

§2. The proceedings should be reasonably homogeneous (concerned with a limited area). For instance, the proceedings of a congress on "Analysis" or "Mathematics in Wonderland" would normally not be sufficiently homogeneous.

One or two longer survey articles on recent developments in the field are often very useful additions to such proceedings - even if they do not correspond to actual lectures at the congress. An extensive introduction on the subject of the congress would be desirable.

§3. The contributions should be of a high mathematical standard and of current interest. Research articles should present new material and not duplicate other papers already published or due to be published. They should contain sufficient information and motivation and they should present proofs, or at least outlines of such, in sufficient detail to enable an expert to complete them. Thus resumes and mere announcements of papers appearing elsewhere cannot be included, although more detailed versions of a contribution may well be published in other places later.

Surveys, if included, should cover a sufficiently broad topic, and should in general not simply review the author's own recent research. In the case of surveys, exceptionally, proofs of results may not be necessary.

"Mathematical Reviews" and "Zentralblatt für Mathematik" require that papers in proceedings volumes carry an explicit statement that they are in final form and that no similar paper has been or is being submitted elsewhere, if these papers are to be considered for a review. Normally, papers that satisfy the criteria of the Lecture Notes in Mathematics series also satisfy this

.../...

requirement, but we would strongly recommend that the contributing authors be asked to give this guarantee explicitly at the beginning or end of their paper. There will occasionally be cases where this does not apply but where, for special reasons, the paper is still acceptable for LNM.

§4. Proceedings should appear soon after the meeeting. The publisher should, therefore, receive the complete manuscript within nine months of the date of the meeting at the latest.

§5. Plans or proposals for proceedings volumes should be sent to one of the editors of the series or to Springer-Verlag Heidelberg. They should give sufficient information on the conference or symposium, and on the proposed proceedings. In particular, they should contain a list of the expected contributions with their prospective length. Abstracts or early versions (drafts) of some of the contributions are very helpful.

§6. Lecture Notes are printed by photo-offset from camera-ready typed copy provided by the editors. For this purpose Springer-Verlag provides editors with technical instructions for the preparation of manuscripts and these should be distributed to all contributing authors. Springer-Verlag can also, on request, supply stationery on which the prescribed typing area is outlined. Some homogeneity in the presentation of the contributions is desirable.

Careful preparation of manuscripts will help keep production time short and ensure a satisfactory appearance of the finished book. The actual production of a Lecture Notes volume normally takes 6 -8 weeks.

Manuscripts should be at least 100 pages long. The final version should include a table of contents and as far as applicable a subject index.

§7. Editors receive a total of 50 free copies of their volume for distribution to the contributing authors, but no royalties. (Unfortunately, no reprints of individual contributions can be supplied.) They are entitled to purchase further copies of their book for their personal use at a discount of 33.3 %, other Springer mathematics books at a discount of 20 % directly from Springer-Verlag. Contributing authors may purchase the volume in which their article appears at a discount of 33.3 %.

Commitment to publish is made by letter of intent rather than by signing a formal contract. Springer-Verlag secures the copyright for each volume.

LECTURE NOTES

ESSENTIALS FOR THE PREPARATION
OF CAMERA-READY MANUSCRIPTS

Springer

Springer-Verlag
Berlin Heidelberg New York
London Paris Tokyo Hong Kong

The preparation of manuscripts which are to be reproduced by photo-offset require special care. Manuscripts which are submitted in technically unsuitable form will be returned to the author for retyping. There is normally no possibility of carrying out further corrections after a manuscript is given to production. Hence it is crucial that the following instructions be adhered to closely. If in doubt, please send us 1 - 2 sample pages for examination.

General. The characters must be uniformly black both within a single character and down the page. Original manuscripts are required: photocopies are acceptable only if they are sharp and without smudges.

On request, Springer-Verlag will supply special paper with the text area outlined. The standard TEXT AREA (OUTPUT SIZE if you are using a 14 point font) is 18 x 26.5 cm (7.5 x 11 inches). This will be scale-reduced to 75% in the printing process. If you are using computer typesetting, please see also the following page.

Make sure the TEXT AREA IS COMPLETELY FILLED. Set the margins so that they precisely match the outline and type right from the top to the bottom line. (Note that the page number will lie outside this area). Lines of text should not end more than three spaces inside or outside the right margin (see example on page 4).

Type on one side of the paper only.

Spacing and Headings (Monographs). Use ONE-AND-A-HALF line spacing in the text. Please leave sufficient space for the title to stand out clearly and do NOT use a new page for the beginning of subdivisons of chapters. Leave THREE LINES blank above and TWO below headings of such subdivisions.

Spacing and Headings (Proceedings). Use ONE-AND-A-HALF line spacing in the text. Do not use a new page for the beginning of subdivisons of a single paper. Leave THREE LINES blank above and TWO below headings of such subdivisions. Make sure headings of equal importance are in the same form.

The first page of each contribution should be prepared in the same way. The title should stand out clearly. We therefore recommend that the editor prepare a sample page and pass it on to the authors together with these instructions. Please take the following as an example. Begin heading 2 cm below upper edge of text area.

MATHEMATICAL STRUCTURE IN QUANTUM FIELD THEORY

John E. Robert
Mathematisches Institut, Universität Heidelberg
Im Neuenheimer Feld 288, D-6900 Heidelberg

Please leave THREE LINES blank below heading and address of the author, then continue with the actual text on the same page.

Footnotes. These should preferable be avoided. If necessary, type them in SINGLE LINE SPACING to finish exactly on the outline, and separate them from the preceding main text by a line.

Symbols. Anything which cannot be typed may be entered by hand in BLACK AND ONLY BLACK ink. (A fine-tipped rapidograph is suitable for this purpose; a good black ball-point will do, but a pencil will not). Do not draw straight lines by hand without a ruler (not even in fractions).

Literature References. These should be placed at the end of each paper or chapter, or at the end of the work, as desired. Type them with single line spacing and start each reference on a new line. Follow "Zentralblatt für Mathematik"/"Mathematical Reviews" for abbreviated titles of mathematical journals and "Bibliographic Guide for Editors and Authors (BGEA)" for chemical, biological, and physics journals. Please ensure that all references are COMPLETE and ACCURATE.

IMPORTANT

Pagination. For typescript, number pages in the upper right-hand corner in LIGHT BLUE OR GREEN PENCIL ONLY. The printers will insert the final page numbers. For computer type, you may insert page numbers (1 cm above outer edge of text area).

It is safer to number pages AFTER the text has been typed and corrected. Page 1 (Arabic) should be THE FIRST PAGE OF THE ACTUAL TEXT. The Roman pagination (table of contents, preface, abstract, acknowledgements, brief introductions, etc.) will be done by Springer-Verlag.

If including running heads, these should be aligned with the inside edge of the text area while the page number is aligned with the outside edge noting that right-hand pages are odd-numbered. Running heads and page numbers appear on the same line. Normally, the running head on the left-hand page is the chapter heading and that on the right-hand page is the section heading. Running heads should not be included in proceedings contributions unless this is being done consistently by all authors.

Corrections. When corrections have to be made, cut the new text to fit and paste it over the old. White correction fluid may also be used.

Never make corrections or insertions in the text by hand.

If the typescript has to be marked for any reason, e.g. for provisional page numbers or to mark corrections for the typist, this can be done VERY FAINTLY with BLUE or GREEN PENCIL but NO OTHER COLOR: these colors do not appear after reproduction.

COMPUTER-TYPESETTING. Further, to the above instructions, please note with respect to your printout that
- the characters should be sharp and sufficiently black;
- it is not strictly necessary to use Springer's special typing paper. Any white paper of reasonable quality is acceptable.

If you are using a significantly different font size, you should modify the output size correspondingly, keeping length to breadth ratio 1 : 0.68, so that scaling down to 10 point font size, yields a text area of 13.5 x 20 cm (5 3/8 x 8 in), e.g.

Differential equations.: use output size 13.5 x 20 cm.

Differential equations.: use output size 16 x 23.5 cm.

Differential equations.: use output size 18 x 26.5 cm.

Interline spacing: 5.5 mm base-to-base for 14 point characters (standard format of 18 x 26.5 cm).
If in any doubt, please send us 1 - 2 sample pages for examination. We will be glad to give advice.

Vol. 1173: H. Delfs, M. Knebusch, Locally Semialgebraic Spaces. XVI, 329 pages. 1985.

Vol. 1174: Categories in Continuum Physics, Buffalo 1982. Seminar. Edited by F.W. Lawvere and S.H. Schanuel. V, 126 pages. 1986.

Vol. 1175: K. Mathiak, Valuations of Skew Fields and Projective Hjelmslev Spaces. VII, 116 pages. 1986.

Vol. 1176: R.R. Bruner, J.P. May, J.E. McClure, M. Steinberger, H_∞ Ring Spectra and their Applications. VII, 388 pages. 1986.

Vol. 1177: Representation Theory I. Finite Dimensional Algebras. Proceedings, 1984. Edited by V. Dlab, P. Gabriel and G. Michler. XV, 340 pages. 1986.

Vol. 1178: Representation Theory II. Groups and Orders. Proceedings, 1984. Edited by V. Dlab, P. Gabriel and G. Michler. XV, 370 pages. 1986.

Vol. 1179: Shi J.-Y. The Kazhdan-Lusztig Cells in Certain Affine Weyl Groups. X, 307 pages. 1986.

Vol. 1180: R. Carmona, H. Kesten, J.B. Walsh, École d'Été de Probabilités de Saint-Flour XIV – 1984. Édité par P.L. Hennequin. X, 438 pages. 1986.

Vol. 1181: Buildings and the Geometry of Diagrams, Como 1984. Seminar. Edited by L. Rosati. VII, 277 pages. 1986.

Vol. 1182: S. Shelah, Around Classification Theory of Models. VII, 279 pages. 1986.

Vol. 1183: Algebra, Algebraic Topology and their Interactions. Proceedings, 1983. Edited by J.-E. Roos. XI, 396 pages. 1986.

Vol. 1184: W. Arendt, A. Grabosch, G. Greiner, U. Groh, H.P. Lotz, U. Moustakas, R. Nagel, F. Neubrander, U. Schlotterbeck, One-parameter Semigroups of Positive Operators. Edited by R. Nagel. X, 460 pages. 1986.

Vol. 1185: Group Theory, Beijing 1984. Proceedings. Edited by Tuan H.F. V, 403 pages. 1986.

Vol. 1186: Lyapunov Exponents. Proceedings, 1984. Edited by L. Arnold and V. Wihstutz. VI, 374 pages. 1986.

Vol. 1187: Y. Diers, Categories of Boolean Sheaves of Simple Algebras. VI, 168 pages. 1986.

Vol. 1188: Fonctions de Plusieurs Variables Complexes V. Séminaire, 1979–85. Edité par François Norguet. VI, 306 pages. 1986.

Vol. 1189: J. Lukeš, J. Malý, L. Zajíček, Fine Topology Methods in Real Analysis and Potential Theory. X, 472 pages. 1986.

Vol. 1190: Optimization and Related Fields. Proceedings, 1984. Edited by R. Conti, E. De Giorgi and F. Giannessi. VIII, 419 pages. 1986.

Vol. 1191: A.R. Its, V.Yu. Novokshenov, The Isomonodromic Deformation Method in the Theory of Painlevé Equations. IV, 313 pages. 1986.

Vol. 1192: Equadiff 6. Proceedings, 1985. Edited by J. Vosmansky and M. Zlámal. XXIII, 404 pages. 1986.

Vol. 1193: Geometrical and Statistical Aspects of Probability in Banach Spaces. Proceedings, 1985. Edited by X. Femique, B. Heinkel, M.B. Marcus and P.A. Meyer. IV, 128 pages. 1986.

Vol. 1194: Complex Analysis and Algebraic Geometry. Proceedings, 1985. Edited by H. Grauert. VI, 235 pages. 1986.

Vol. 1195: J.M. Barbosa, A.G. Colares, Minimal Surfaces in $\mathbb{R}^3$. X, 124 pages. 1986.

Vol. 1196: E. Casas-Alvero, S. Xambó-Descamps, The Enumerative Theory of Conics after Halphen. IX, 130 pages. 1986.

Vol. 1197: Ring Theory. Proceedings, 1985. Edited by F.M.J. van Oystaeyen. V, 231 pages. 1986.

Vol. 1198: Séminaire d'Analyse, P. Lelong – P. Dolbeault – H. Skoda. Seminar 1983/84. X, 260 pages. 1986.

Vol. 1199: Analytic Theory of Continued Fractions II. Proceedings, 1985. Edited by W.J. Thron. VI, 299 pages. 1986.

Vol. 1200: V.D. Milman, G. Schechtman, Asymptotic Theory of Finite Dimensional Normed Spaces. With an Appendix by M. Gromov. VIII, 156 pages. 1986.

Vol. 1201: Curvature and Topology of Riemannian Manifolds. Proceedings, 1985. Edited by K. Shiohama, T. Sakai and T. Sunada. VII, 336 pages. 1986.

Vol. 1202: A. Dür, Möbius Functions, Incidence Algebras and Power Series Representations. XI, 134 pages. 1986.

Vol. 1203: Stochastic Processes and Their Applications. Proceedings, 1985. Edited by K. Itô and T. Hida. VI, 222 pages. 1986.

Vol. 1204: Séminaire de Probabilités XX, 1984/85. Proceedings. Edité par J. Azéma et M. Yor. V, 639 pages. 1986.

Vol. 1205: B.Z. Moroz, Analytic Arithmetic in Algebraic Number Fields. VII, 177 pages. 1986.

Vol. 1206: Probability and Analysis, Varenna (Como) 1985. Seminar. Edited by G. Letta and M. Pratelli. VIII, 280 pages. 1986.

Vol. 1207: P.H. Bérard, Spectral Geometry: Direct and Inverse Problems. With an Appendix by G. Besson. XIII, 272 pages. 1986.

Vol. 1208: S. Kaijser, J.W. Pelletier, Interpolation Functors and Duality. IV, 167 pages. 1986.

Vol. 1209: Differential Geometry, Peñíscola 1985. Proceedings. Edited by A.M. Naveira, A. Ferrández and F. Mascaró. VIII, 306 pages. 1986.

Vol. 1210: Probability Measures on Groups VIII. Proceedings, 1985. Edited by H. Heyer. X, 386 pages. 1986.

Vol. 1211: M.B. Sevryuk, Reversible Systems. V, 319 pages. 1986.

Vol. 1212: Stochastic Spatial Processes. Proceedings, 1984. Edited by P. Tautu. VIII, 311 pages. 1986.

Vol. 1213: L.G. Lewis, Jr., J.P. May, M. Steinberger, Equivariant Stable Homotopy Theory. IX, 538 pages. 1986.

Vol. 1214: Global Analysis – Studies and Applications II. Edited by Yu.G. Borisovich and Yu.E. Gliklikh. V, 275 pages. 1986.

Vol. 1215: Lectures in Probability and Statistics. Edited by G. del Pino and R. Rebolledo. V, 491 pages. 1986.

Vol. 1216: J. Kogan, Bifurcation of Extremals in Optimal Control. VIII, 106 pages. 1986.

Vol. 1217: Transformation Groups. Proceedings, 1985. Edited by S. Jackowski and K. Pawalowski. X, 396 pages. 1986.

Vol. 1218: Schrödinger Operators, Aarhus 1985. Seminar. Edited by E. Balslev. V, 222 pages. 1986.

Vol. 1219: R. Weissauer, Stabile Modulformen und Eisensteinreihen. III, 147 Seiten. 1986.

Vol. 1220: Séminaire d'Algèbre Paul Dubreil et Marie-Paule Malliavin. Proceedings, 1985. Edité par M.-P. Malliavin. IV, 200 pages. 1986.

Vol. 1221: Probability and Banach Spaces. Proceedings, 1985. Edited by J. Bastero and M. San Miguel. XI, 222 pages. 1986.

Vol. 1222: A. Katok, J.-M. Strelcyn, with the collaboration of F. Ledrappier and F. Przytycki, Invariant Manifolds, Entropy and Billiards; Smooth Maps with Singularities. VIII, 283 pages. 1986.

Vol. 1223: Differential Equations in Banach Spaces. Proceedings, 1985. Edited by A. Favini and E. Obrecht. VIII, 299 pages. 1986.

Vol. 1224: Nonlinear Diffusion Problems, Montecatini Terme 1985. Seminar. Edited by A. Fasano and M. Primicerio. VIII, 188 pages. 1986.

Vol. 1225: Inverse Problems, Montecatini Terme 1986. Seminar. Edited by G. Talenti. VIII, 204 pages. 1986.

Vol. 1226: A. Buium, Differential Function Fields and Moduli of Algebraic Varieties. IX, 146 pages. 1986.

Vol. 1227: H. Helson, The Spectral Theorem. VI, 104 pages. 1986.

Vol. 1228: Multigrid Methods II. Proceedings, 1985. Edited by W. Hackbusch and U. Trottenberg. VI, 336 pages. 1986.

Vol. 1229: O. Bratteli, Derivations, Dissipations and Group Actions on C*-algebras. IV, 277 pages. 1986.

Vol. 1230: Numerical Analysis. Proceedings, 1984. Edited by J.-P. Hennart. X, 234 pages. 1986.

Vol. 1231: E.-U. Gekeler, Drinfeld Modular Curves. XIV, 107 pages. 1986.

Vol. 1232: P.C. Schuur, Asymptotic Analysis of Soliton Problems. VIII, 180 pages. 1986.

Vol. 1233: Stability Problems for Stochastic Models. Proceedings, 1985. Edited by V.V. Kalashnikov, B. Penkov and V.M. Zolotarev. VI, 223 pages. 1986.

Vol. 1234: Combinatoire énumérative. Proceedings, 1985. Edité par G. Labelle et P. Leroux. XIV, 387 pages. 1986.

Vol. 1235: Séminaire de Théorie du Potentiel, Paris, No. 8. Directeurs: M. Brelot, G. Choquet et J. Deny. Rédacteurs: F. Hirsch et G. Mokobodzki. III, 209 pages. 1987.

Vol. 1236: Stochastic Partial Differential Equations and Applications. Proceedings, 1985. Edited by G. Da Prato and L. Tubaro. V, 257 pages. 1987.

Vol. 1237: Rational Approximation and its Applications in Mathematics and Physics. Proceedings, 1985. Edited by J. Gilewicz, M. Pindor and W. Siemaszko. XII, 350 pages. 1987.

Vol. 1238: M. Holz, K.-P. Podewski and K. Steffens, Injective Choice Functions. VI, 183 pages. 1987.

Vol. 1239: P. Vojta, Diophantine Approximations and Value Distribution Theory. X, 132 pages. 1987.

Vol. 1240: Number Theory, New York 1984–85. Seminar. Edited by D.V. Chudnovsky, G.V. Chudnovsky, H. Cohn and M.B. Nathanson. V, 324 pages. 1987.

Vol. 1241: L. Gårding, Singularities in Linear Wave Propagation. III, 125 pages. 1987.

Vol. 1242: Functional Analysis II, with Contributions by J. Hoffmann-Jørgensen et al. Edited by S. Kurepa, H. Kraljević and D. Butković. VII, 432 pages. 1987.

Vol. 1243: Non Commutative Harmonic Analysis and Lie Groups. Proceedings, 1985. Edited by J. Carmona, P. Delorme and M. Vergne. V, 309 pages. 1987.

Vol. 1244: W. Müller, Manifolds with Cusps of Rank One. XI, 158 pages. 1987.

Vol. 1245: S. Rallis, L-Functions and the Oscillator Representation. XVI, 239 pages. 1987.

Vol. 1246: Hodge Theory. Proceedings, 1985. Edited by E. Cattani, F. Guillén, A. Kaplan and F. Puerta. VII, 175 pages. 1987.

Vol. 1247: Séminaire de Probabilités XXI. Proceedings. Edité par J. Azéma, P.A. Meyer et M. Yor. IV, 579 pages. 1987.

Vol. 1248: Nonlinear Semigroups, Partial Differential Equations and Attractors. Proceedings, 1985. Edited by T.L. Gill and W.W. Zachary. IX, 185 pages. 1987.

Vol. 1249: I. van den Berg, Nonstandard Asymptotic Analysis. IX, 187 pages. 1987.

Vol. 1250: Stochastic Processes – Mathematics and Physics II. Proceedings 1985. Edited by S. Albeverio, Ph. Blanchard and L. Streit. VI, 359 pages. 1987.

Vol. 1251: Differential Geometric Methods in Mathematical Physics. Proceedings, 1985. Edited by P.L. García and A. Pérez-Rendón. VII, 300 pages. 1987.

Vol. 1252: T. Kaise, Représentations de Weil et GL_2 Algèbres de division et GL_n. VII, 203 pages. 1987.

Vol. 1253: J. Fischer, An Approach to the Selberg Trace Formula via the Selberg Zeta-Function. III, 184 pages. 1987.

Vol. 1254: S. Gelbart, I. Piatetski-Shapiro, S. Rallis. Explicit Constructions of Automorphic L-Functions. VI, 152 pages. 1987.

Vol. 1255: Differential Geometry and Differential Equations. Proceedings, 1985. Edited by C. Gu, M. Berger and R.L. Bryant. XII, 243 pages. 1987.

Vol. 1256: Pseudo-Differential Operators. Proceedings, 1986. Edited by H.O. Cordes, B. Gramsch and H. Widom. X, 479 pages. 1987.

Vol. 1257: X. Wang, On the C^*-Algebras of Foliations in the Plane. V, 165 pages. 1987.

Vol. 1258: J. Weidmann, Spectral Theory of Ordinary Differential Operators. VI, 303 pages. 1987.

Vol. 1259: F. Cano Torres, Desingularization Strategies for Three-Dimensional Vector Fields. IX, 189 pages. 1987.

Vol. 1260: N.H. Pavel, Nonlinear Evolution Operators and Semigroups. VI, 285 pages. 1987.

Vol. 1261: H. Abels, Finite Presentability of S-Arithmetic Groups. Compact Presentability of Solvable Groups. VI, 178 pages. 1987.

Vol. 1262: E. Hlawka (Hrsg.), Zahlentheoretische Analysis II. Seminar, 1984–86. V, 158 Seiten. 1987.

Vol. 1263: V.L. Hansen (Ed.), Differential Geometry. Proceedings, 1985. XI, 288 pages. 1987.

Vol. 1264: Wu Wen-tsün, Rational Homotopy Type. VIII, 219 pages. 1987.

Vol. 1265: W. Van Assche, Asymptotics for Orthogonal Polynomials. VI, 201 pages. 1987.

Vol. 1266: F. Ghione, C. Peskine, E. Sernesi (Eds.), Space Curves. Proceedings, 1985. VI, 272 pages. 1987.

Vol. 1267: J. Lindenstrauss, V.D. Milman (Eds.), Geometrical Aspects of Functional Analysis. Seminar. VII, 212 pages. 1987.

Vol. 1268: S.G. Krantz (Ed.), Complex Analysis. Seminar, 1986. VII, 195 pages. 1987.

Vol. 1269: M. Shiota, Nash Manifolds. VI, 223 pages. 1987.

Vol. 1270: C. Carasso, P.-A. Raviart, D. Serre (Eds.), Nonlinear Hyperbolic Problems. Proceedings, 1986. XV, 341 pages. 1987.

Vol. 1271: A.M. Cohen, W.H. Hesselink, W.L.J. van der Kallen, J.R. Strooker (Eds.), Algebraic Groups Utrecht 1986. Proceedings. XII, 284 pages. 1987.

Vol. 1272: M.S. Livšic, L.L. Waksman, Commuting Nonselfadjoint Operators in Hilbert Space. III, 115 pages. 1987.

Vol. 1273: G.-M. Greuel, G. Trautmann (Eds.), Singularities, Representation of Algebras, and Vector Bundles. Proceedings, 1985. XIV, 383 pages. 1987.

Vol. 1274: N. C. Phillips, Equivariant K-Theory and Freeness of Group Actions on C^*-Algebras. VIII, 371 pages. 1987.

Vol. 1275: C.A. Berenstein (Ed.), Complex Analysis I. Proceedings, 1985–86. XV, 331 pages. 1987.

Vol. 1276: C.A. Berenstein (Ed.), Complex Analysis II. Proceedings, 1985–86. IX, 320 pages. 1987.

Vol. 1277: C.A. Berenstein (Ed.), Complex Analysis III. Proceedings, 1985–86. X, 350 pages. 1987.

Vol. 1278: S.S. Koh (Ed.), Invariant Theory. Proceedings, 1985. V, 102 pages. 1987.

Vol. 1279: D. Ieşan, Saint-Venant's Problem. VIII, 162 Seiten. 1987.

Vol. 1280: E. Neher, Jordan Triple Systems by the Grid Approach. XII, 193 pages. 1987.

Vol. 1281: O.H. Kegel, F. Menegazzo, G. Zacher (Eds.), Group Theory. Proceedings, 1986. VII, 179 pages. 1987.

Vol. 1282: D.E. Handelman, Positive Polynomials, Convex Integral Polytopes, and a Random Walk Problem. XI, 136 pages. 1987.

Vol. 1283: S. Mardešić, J. Segal (Eds.), Geometric Topology and Shape Theory. Proceedings, 1986. V, 261 pages. 1987.

Vol. 1284: B.H. Matzat, Konstruktive Galoistheorie. X, 286 pages. 1987.

Vol. 1285: I.W. Knowles, Y. Saitō (Eds.), Differential Equations and Mathematical Physics. Proceedings, 1986. XVI, 499 pages. 1987.

Vol. 1286: H.R. Miller, D.C. Ravenel (Eds.), Algebraic Topology. Proceedings, 1986. VII, 341 pages. 1987.

Vol. 1287: E.B. Saff (Ed.), Approximation Theory, Tampa. Proceedings, 1985–1986. V, 228 pages. 1987.

Vol. 1288: Yu. L. Rodin, Generalized Analytic Functions on Riemann Surfaces. V, 128 pages, 1987.

Vol. 1289: Yu. I. Manin (Ed.), K-Theory, Arithmetic and Geometry. Seminar, 1984–1986. V, 399 pages. 1987.